GALÁPAGOS: WORLD'S END
by William Beebe

With Sections by Henry Fairfield Osborn,
Ruth Rose and Robert G. McKay

24 Drawings in Full Color by Isabel Cooper
75 Photographs by John Tee-Van and Others
15 Black-and-White Drawings

DOVER PUBLICATIONS, INC.
New York

To

HARRISON WILLIAMS

Published in Canada by General Publishing Company, Ltd., 30 Lesmill Road, Don Mills, Toronto, Ontario.
Published in the United Kingdom by Constable and Company, Ltd.

This Dover edition, first published in 1988, is an unabridged republication of the work originally published ("under the auspices of the New York Zoological Society") by G. P. Putnam's Sons (The Knickerbocker Press), New York and London, February 1924. In this Dover edition eight of the color plates, originally scattered throughout the text, have been reproduced in a color insert. A ninth appears on the covers. In addition, the location of many of the black-and-white illustrations has been altered to reduce the number of blank pages.

Manufactured in the United States of America
Dover Publications, Inc., 31 East 2nd Street, Mineola, N.Y. 11501

Library of Congress Cataloging-in-Publication Data

Beebe, William, 1877–1962.
   Galápagos, world's end / by William Beebe ; 24 drawings in full color by Isabel Cooper ; 75 photographs by John Tee-Van and others ; with sections by Henry Fairfield Osborn, Ruth Rose, and Robert G. McKay.
      p.   cm.
   Reprint. Originally published: New York : Putnam, 1924.
   Bibliography: p.
   Includes index.
   ISBN 0-486-25642-1
   1. Natural history—Galápagos Islands.   2. Galápagos Islands—Description and travel.   I. Title.
QH198.G3B44   1988
508.866'5—dc19   87-37526
                                                                    CIP

IN THE WAKE OF DARWIN

IN the wake of Charles Darwin! This expresses the prevailing spirit of the 1923 Harrison Williams Galápagos Expedition in contrast to the spirit of Charles Darwin's visit to the same islands in 1835. During the eighty-eight years which have intervened between the arrival of the *Beagle* and that of the steam yacht *Noma*, the mind and spirit of man have remained exactly the same, while a wonderful development has taken place in man's intellectual and mechanical environment.

According to the account of the immortal naturalist's voyage around the world, Darwin arrived at the Galápagos on September seventeenth, visited four islands, made his wonderful observations and departed October twentieth. Only five weeks, but five weeks of Darwin's eyes and Darwin's powers of observation and reasoning were equivalent to a whole previous cycle of human thought. At the time he was twenty-six years of age and was practically the only scientist in his party.

The intervening period between 1923 and 1835, the date of Darwin's visit, witnessed two revolutions; first, progress of steampower and of engineering by which in 1923 the party was carried from the Port of New York to the islands swiftly, independently of winds and currents, and enabled, en route, even to bisect two continents. Second, the Darwinian period of thought, in its infancy in 1835 even in Darwin's mind, is now at full tide in the mind of every naturalist in the world. Thus today, there is no time lost in seeking the direction to observe; the splendid group of young naturalists under the leadership of William Beebe knew what facts to look for and where each fact belongs in the still unfilled archives of present and past Evolution. Consequently it was possible in 1923 to accomplish as much well directed observation in an hour as was possible in a working day of ten hours in 1835.

Thus among the scientific wonders of our century, we have to record: first, that in less than one hundred actual hours on land, the *Noma* party accomplished results—artistic, photographic, observational—which are entirely without rival. William Beebe's own highly trained mind and powers of observation tend to work not only accurately but at top speed, surmounting obstacles and dangers of every kind. Second, that in a correspondingly brief time there were brought together an unparalleled accumulation of facts, which fill the pages of this book and have stimulated many scientific papers on which the leading naturalists of America are now busily engaged. It is like rubbing the Aladdin's lamp of science!

Many of Darwin's own views, like those on animal and plant dispersal and adaptive colouring, possess this power of inspiration, this inherent quickening force which no passing of years can diminish. We must never cease observing and observing, recording and recording, for we know not how near we may be to the ultimate solution of the mystery of adaptation.

As sponsor and patron of this expedition of the New York Zoological Society, Harrison Williams may well be proud of the results accomplished during this brief visit to the wonderland of the Galápagos Archipelago. This volume opens up rich vistas of further inquiry on other seas and among other islands, and cannot fail to stimulate other benefactors of science to further investment in the real things of our planet.

HENRY FAIRFIELD OSBORN,
President of the New York Zoological Society.
December 24th, 1923.

PREFACE

THIS volume might well be called *Six Thousand Minutes on the Galápagos*, for of the two and a half months' duration of the trip, less than one hundred hours were spent actually on the islands themselves. Although this brief expedition, to one of the least visited corners of the earth, was conceived and achieved in such record time, yet every hope was consummated, every expectation realized. First and last, the credit belongs to Harrison Williams, who initiated and financed the whole trip, and then to the twelve members of my party who made possible all that we accomplished during the limited time at our disposal.

We left New York on the steam yacht *Noma* on March 1st, 1923, and returned on May 16th. During the trip we steamed a total distance of nine thousand miles, and crossed the equator eight times. Besides the Galápagos we touched at Charleston, Key West, Havana, Colon, and Panama.

The Galápagos Archipelago is a tiny group of about sixty islands and islets, directly on the equator, in the Pacific, five hundred miles off the coast of Ecuador, to which country they belong. When I realized that our stay must be measured by hours, I knew that little could be gained by forced marches into the interior of the larger islands. Even expeditions which had spent a year or more here, had failed to penetrate to the central craters. The smaller islets—mere specks on the largest charts—revealed themselves as of infinite variety and charm as we steamed close past them, and it was on these that we spent most of our precious time. Hence it is Eden, Guy Fawkes, Daphne, Seymour, and Tower which dominate these pages, rather than the vast wastes of Indefatigable, or the lava-strewn uplands of James and Albemarle.

In the course of our very short expedition we were able to collect a number of living creatures which were brought north and added to the collections of the Zoological Park. Some of these species had never been in captivity before. From Panama we brought monkeys, opossums, jays, and parrakeets, and from the Galápagos, penguins, flightless cormorants, fork-tailed gulls, hawks, doves, and mockingbirds.

For the American Museum there was collected material for two giant iguana groups, including vegetation, rocks, shells, sand, and many photographs and paintings, together with a giant tortoise, forty-two iguanas, and a family of sea-lions.

For study by the department of Tropical Research of the Zoological Society, there was collected:

   90 water colour paintings by Isabel Cooper.
   40 oil paintings by Harry Hoffman.
   46 pen and ink drawings by Gilbert Broking.
  400 photographs and 11,000 feet of motion picture film by John Tee-Van.
  160 bird skins.
      many nests and eggs.
  150 reptiles.
  200 fish.
3,000 insects.
   40 jars of specimens.
   60 vials and jars of plankton.
  200 microscope slides of plankton.
  100 specimens of plants.
  300 pages of narrative, records, notes, and catalogues by Ruth Rose.

This material is remarkable both for its rarity, excellent preservation, and for the fact that it was collected within such a short period of time.

PREFACE

Within six months after our return, there are actually published, or nearing completion, the following twenty-two scientific papers by specialists of the various groups enumerated:

Heterocera, by Schaus
Biological Notes on Heterocera, by Beebe
Pisces, by Nichols
Apterygota, by Folsom
Homoptera, by Osborn
Mallophaga, by Ewing
Diptera, by Johnson
Arachnida and Neuroptera, by Banks
Formicidæ, by Wheeler
Triungulids, by Brues
Chilopoda, by Chamberlin
Termites, by Emerson
Coccidæ, by Morrison
Brachyuran Crabs, by Rathbun
Macrura and Anomura, by Schmitt
Heteroptera, by Barber
Aves, by Beebe
Coleoptera, by Mutchler
Hymenoptera, by Rohwer
Copepoda, by Wilson
Entomostraca, by Pearse
Isopoda, by Van Name

These are published in *Zoologica*, the scientific publication of the New York Zoological Society.

It is impossible to comment individually on the work of the various members of the party, for there was only one level of effort and achievement, which reached its maximum with the first day in the field, and was sustained without cessation until the expedition ended. This volume is a mosaic of the results of this splendid enthusiasm and co-operation.

The most evident contributions are the coloured and black and white paintings of Isabel Cooper, the photographs of John Tee-Van, the Game Fishing chapter by Robert G. McKay and

PREFACE

the historical and two other chapters by Ruth Rose. Figs. 9, 28, 41, 54, 76 and 77 were taken by Rollo H. Beck; Fig. 40 by Elwin R. Sanborn; and Figs. 34, 35, 42, 50 and 63 by James Mitchell. All others were taken by John Tee-Van.

Our personnel, as ultimately constituted, was as follows:

HARRISON WILLIAMS, Patron and Curator of Oceanography
WILLIAM BEEBE, Director of Scientific Work
WILLIAM MORTON WHEELER, Entomologist
JAMES F. MITCHELL, Surgeon
ROBERT G. McKAY, Executive Officer, Game Fish
JAMES F. CURTIS, Curator of Dredging and Diving
WILLIAM H. MERRIAM, Chief Hunter
HARRY HOFFMAN, Marine Artist
JOHN TEE-VAN, Photographer
ISABEL COOPER, Scientific Artist
RUTH ROSE, Historian, Curator of Catalogues and Live Animals
GILBERT BROKING, Artist and General Preparateur
H. FROBISHER, Physician
WALTER ESCHERICH, Taxidermist

This book is in no way intended as a technical contribution to our present knowledge of the fauna and flora of the Galápagos Archipelago. It is a record of spontaneous observation, a crystallization of the more obvious and characteristic impressions of the land and the sea, of birds, reptiles, fish and insects, whose forms and colours and personalities made us wish that every one of the six thousand minutes of our stay could be lengthened to hours or days.

WILLIAM BEEBE.

CONTENTS

ILLUSTRATIONS

IN BLACK AND WHITE

MOSTLY FROM PHOTOGRAPHS BY JOHN TEE-VAN

ILLUSTRATIONS

xi

ILLUSTRATIONS

ILLUSTRATIONS

COLOURED ILLUSTRATIONS
(See Note below.)

[**Note:** All of the color plates, originally appearing throughout the book, have been renumbered where necessary and placed together in the Dover edition, in an insert following page 16 (except for Plate VI, here reproduced on the covers). They have been reproduced here at 71 percent of their size in the original edition. This should be taken into account with captions that compare sizes of illustrations and subjects (all captions have been reprinted without alteration).]

GALÁPAGOS:
WORLD'S END

GALÁPAGOS

CHAPTER I

HALF-BACKS swathed in bath robes, butterflies in chrysalids, spads in their hangars,—these and other similes came to mind as I shivered on a dirty Brooklyn wharf one day in late February, and walked down between dense lines of unlovely snowy sheds,—winter camouflage of the aristocracy of yachts. A magic word had been spoken and had awakened one of these from a two years' hibernation, and when I saw it first it was stripped of its housing, and slowly rising from the icy waters, exposing the sheer heights of its crimson keel, encrusted with sea-moss and barnacles acquired in alien seas on voyages of other years.

Our first boarding was memorable, by means of a ladder let down into the water, with the top hauled back and forth by means of ropes and over-enthusiastic longshoremen, while the passengers clung tenaciously to this inverted pendulum, swinging through a sixty degree arc of bitter air.

She was the *Noma*, with a two hundred and fifty foot water line, luxuriously fitted, and a bowsprit with a real figurehead. For weeks we planned and made lists and crossed the items out; we weighed and appraised, rejected and chose among friends for a suitable staff and party. Our plan was to spend

3

two months and a half; to steam to Panama, on through the canal to the Pacific, and straight to the Galápagos Islands, six hundred miles off Ecuador, for notes and photographs and specimens alive or dead of all interesting creatures which might come to eye or ear or aquarium, to lens, net or gun, throughout the trip.

Of the preparations I shall tell nothing; the joys of these are too holy for cold phrases; nor of the imagination which can look through a camera lens in a New York store and see future bird or insect or atoll mirrored therein, or along an unpurchased gun barrel and see adumbrated shapes of trophies to come, or can fill the dry meshes of a net with living fish passing all jewels for beauty, and human affairs for interest,—before these, the written word pales.

At 1.30 on March first, 1923, the *Noma* was towed from her berth, out through a narrow lane of other yachts. Just across the pier the *Corsair*, boarded up in tenement garb for the winter, watched us dully through unshiplike windows, envious perhaps of the trim, cleared sister setting out for unknown seas. All winter the big yachts had slumbered here, bound by hemp and wire thongs, mere lifeless extensions of the mundane piers. Now as we passed, we realized that time as well as space had special meanings for these beings of the sea,—wholly different from that which passed over or separated the sordid buildings of the city's front. We tinkled three bells, and all the hooded, housed, cribbed and confined yachts sent back muffled echoes—three, three, always three. And yet all the clocks and chimes and churches of the great city were silent. These alert homes of the open waters marked even the half hours, while land folk were content to notice only hours.

As we cleared the pier, an English sparrow flew down upon our bowsprit, a smudged and sooty bird, humped with cold, and garrulous with the obscene gossip of longshore life. And only when he saw open water between our bow and the shore did he give a contemptuous flirt of his wings, and leave. He was of one mind with the three members of the crew who had signed on, drawn two days' pay, and deserted when they learned we were going to wild lands. We were well rid of them all, feathered and otherwise. If I could know the inmost feelings, each particular urge of going, of all of us who were left, this chapter would be an epic. Even at this moment I could tabulate idle curiosity, and love of travel without the curiosity, restlessness, the sea, science, besides which there were incipient explorers, whalers, gentlemen adventurers and buccaneers. The least conscious was one of these inhibited pirates, of whom I hoped great things in the way of mental atavisms.

Slowly we backed out into mid-stream and turned, all at the will of the fussy tugs. Then they relaxed their hold, drew off, waited, and like a queen ant on her one day of days, the great yacht awoke and started under her own power. Solicitously one big tug followed and in the lower bay stood by while we described slow, dignified circles, adjusting our compasses. As the last turn was made, some little insignificant thing happened deep, deep down in the vitals of the ship, and she drifted uselessly, helplessly—changed in an instant from a living, vital creature to a floating mass of impotent metal. Whatever it was, it required very painstaking adjustment and for two days we remained anchored within sight of our city's towers. In the daytime the Woolworth building rose like a dim, grey, smoke wraith, and Staten Island's water front brought no joy to our eyes. But at night the horizontal constellations of lights might have been the sampan lanterns of the Yangtze or the flares of a village of Sea Dyaks. And the full moon was the same which had in the past, and would again sift down to me through alien jungles, or be reflected from the glistening sands of trackless deserts. A few over-familiar herring gulls flew past on their way to some common roost on Long Island sands, and there seemed no thing of novelty in this humiliating anchorage—rocking in the wash of ferry-boats and of tandem garbage scows.

Idly I watched the harbor flotsam drift past on the tide, patches of straw, barrel hoops, crates and bottles, heart-warming relics of the attempts of worthy bootleggers. The gnomes in the heart of the vessel had ceased their pounding and anvilling for a while, and in the silence I heard sweet music from the polluted water below me. Looking closely I saw a host of little icebergs drifting past, none more than a yard across, each sending forth crystalline tinklings, some on a higher tone, some lower, until the air was filled with the sound of their tiny crashing cymbals. They had come into being far up the Hudson and were now rapidly disintegrating in the higher temperature of the salt water. Their eight under water parts were thick and pale grey, but to the eye they appeared flat and disklike, the ice in the air being thin as paper, scalloped and pocketed with circular cutouts. The ripples lapped against these tissue sheets of hardened water, and rang and tinkled and clinked the strange swan song of the berglets as they passed, mingling with the silent, drab jetsam of the city.

During the first few days on the *Noma* I realized the possibilities of a yacht when rushed into commission after a two years' rest. I also came to appreciate the fact that sailor's lore is achieved by conscientious trial and error, and not by instinct. Only one among the fifty-eight members of the officers and crew had ever sailed on the *Noma* before, and the multiplicity of levers, faucets, gadgets, handles, trapdoors, wheels and general unlabelled mechanical temptations to pull, twist or push, had all to be tried and learned by pragmatic methods. The engines were wonderful in their silence and lack of vibration, in fact they were the only part of the yacht's anatomy to which we never gave thought. Otherwise the *Noma* was for a time an habitation of ten thousand sounds. Dinner bells rang which had no connection with either stewards or meals, tappings came at doors from empty cabins, and squeaks from mouseless holes. So one morning when I distinctly heard a low sweet warble outside the sun parlor, I mentally added it to the list of Things heard and went on with my work. A few minutes later I found that this time the *Noma* was innocent; my ears had registered truly and a Maryland yellowthroat blown on board during the night had sung a little hymn of thanksgiving at finding sanctuary.

The water supply for the first week would alone have kept ennui at a distance. Four kinds of baths were possible in my tub,—hot dirty fresh, cold dirty fresh, hot dirty salt and cold dirty salt, and by comparing notes I found that I had vanilla in color while my neighbor had chocolate. I came to look upon my basin supply as a thing to respect, mistrust, even to fear. The faucet marked "cold salt" would usually yield a gentle flow of hot fresh, when suddenly without warning, a blast of mingled air and water would burst forth, with such force that I would be transfixed against the wall. Occasionally in the dead of night a faucet would begin to trickle all by itself, and so deeply was the crime of wasting fresh water implanted in our Noman moral code, that upon such an occasion several ghostly figures more or less awake would appear headed faucetwards to quench the inexplicable nocturnal flow.

The first morning my fresh water ran freely, but refused to be shut off without the aid of the ship's carpenter; in other cabins this fluid obstinately refused to leave its receptacle deep in the yacht, so the inmates shaved with Poland water, feeling like old Roman Exquisites with their baths of cream or other unusuality. The day we left Key West we took on forty tons of fresh water and promptly lost it in the depths of the ship.

It was there, everyone had seen it come, and the boat did not leak, but it hid until half the crew had searched diligently and at last located the outlet, gave it air, and made it available. The electric supply was constant but permitted the water supply no handicap in eccentricity, as when the lighting of the pseudo open coal fire caused the player piano to cease functioning.

All this was but the harmless hazing which school-boys practice upon a newcomer, or natives upon a tenderfoot, later to become closest of friends; and when the *Noma* had had her practical jokes, had spilled water upon our heads, frightened us with restless ghosts and fooled us with wrongly labelled directions, she gathered us close to her, set her face resolutely to the sea, unflinchingly braving the dangers of the open waters—the same unchanged ocean which once dashed its spray over the great carven bows of Viking ships, and just one hundred and fifty-seven thousand and ninety days ago floated the high poop of the *Santa Maria* within sight of its unnamed goal.

I had thought it would be an easy matter to write of the feel of the soul of a yacht, but I had believed the same of an active volcano, and both reduced me to stark, trite statements of fact, statistical mouthings, futile superficial gossip; while the deep, real emotions came and went, vivid, full of gasping realization, which settled and passed into experience, only to reawaken to new enthusiasms. There (I once said) was smoke and fire of actual Mother Earth, the same primeval conflagration raging in the volcano as in the sun and the comets, —by grace of the partial cooling of which we contrive to live and breathe and creep about; here was an ocean steamer in miniature, not owned by an impersonal, unknown company of strangers, upon which for a fixed sum of money I was vouchsafed a cubical and three meals a day.

NEW YORK TO PANAMA

The only imaginative satisfaction I ever gleaned from such reciprocity was the elaborate fine print absolution of the steamship company from responsibility for loss of or injury to the passenger arising from perils of the sea, rivers or navigation, mutiny, pirates or other public enemies, barratry of master and crew, negligence of the company's servants, defects in the machinery, gear or fittings, or Act of God. I have always loved the word "barratry" and its fascination was increased by my absolute ignorance of its meaning. Recently I decided to sacrifice one of its charms and I find it defined as what the master of a ship does when he injures or slays the owner or passengers by running away with the vessel, sinking or deserting her, deviating from the fixed course or embezzling the cargo. I have hopefully suggested barratry to several fat, jolly captains but with absolutely no success. As to the final threat of interest and excitement on the trip, an "Act of God" seems to mean anything thrilling that a reasonable man cannot think of. Only mayhem is omitted as far as I can see, and in spite of my newly aroused hopes at the beginning of each voyage, the list of possibilities proves as unreal and unrealized as the navy recruit posters on sidewalk easels, or the what-men-will-wear in the theatre programs.

Here, however, on the *Noma*, was a real home, for what period of time did not matter, for when the past and present of any given experience are perfect, the future, even before it filters into the present, potentially reflects its superlative. Here, with a dozen congenial friends, I shared an nth power raft, which made of the earth a mere revolving cyclorama. I had travelled around the world, I had gone to strange and distant places; now these were to come to me,—Mohammed's peripatetic mountain, Macbeth's ambulatory woods, Aladdin's magic carpet, these were become real, and not in any anæmic Chautauquan sense; with Monsieur Perrichon I felt like exclaiming with superlative conceit, "Moi et Mont Blanc!" Such magic had hitherto existed for me only in Jules Verne, or been laboriously achieved by dint of hired board and lodging in floating hotels, with hosts of "thrust-upon" companions. In my many exploring trips I had had infinitely greater responsibilities, more actual power of decision, but never such concentrated, unegotistical appreciation of the power of man, as now, having chosen and spoken aloud to my friend the word "Galápagos" and watching my circular vignette of the waters off Atlantic City, to know that one day in the near future, after filtering between two continents, I would look out of this same porthole and see framed some of the "two thousand craters" of which Darwin wrote so surpassingly. And all this magic was being wrought by a friendship which saw in me a poor instrument, with at least the will and desire to unlock a few new facts of science. The *Noma* was a perfectly appointed palace, the host ideal in thoughtfulness, patience and optimism; there remained as a never-failing drab background, the conviction of my brain's inadequacy, the untold opportunities which I would let slip, the trip of trips which I might achieve, had I only that inspiration which, to my poor mentality, in future, present and past is respectively always obscure, impotent and obvious.

It was close to the darkest hour when men die most easily that the *Noma* finally awoke to perfect life and carried us out to sea, and in that moment all her eccentricities passed from mind. Without vibration or tremor she slipped swiftly through the water and as even in this calm weather I lay in my berth and watched the water swirl green and white past and over the porthole glass, I first knew that real intimacy with the sea which only a submarine or a yacht such as this can give. The musical tinkle of the miniature ice floes was gone, but in its place throughout the night there came to my ears the soft

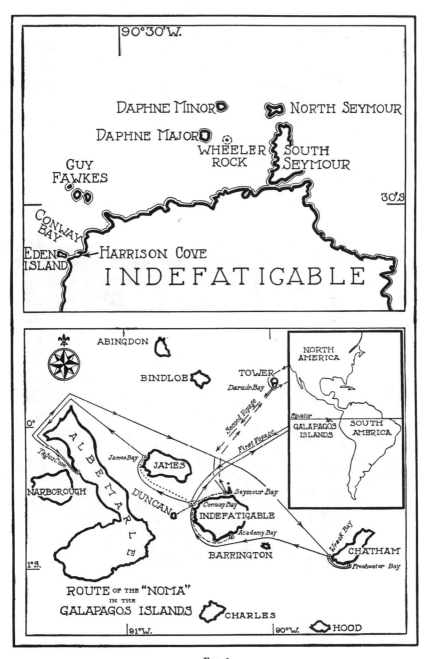

FIG. 1

ISLETS ON THE NORTH SHORE OF INDEFATIGABLE
Galápagos Archipelago showing route of the *Noma*

FIG. 2

THE STEAM YACHT "NOMA" IN TAGUS COVE, ALBEMARLE

swish and gurgle of troubled waters, mysterious and fascinating. It gave a thrill of adventure to look over the rail and see the surface so close, and to realize that at last we were in open ocean and headed for wild equatorial zones. To watch her bow and outreaching figurehead cut through the moonlight, compensated for such incongruities as brass beds and imitation open fireplaces. We already loved her, for she was taking us *through*, not *over* the sea, and no better phrase could be found for any experience in life.

On a trip of exploration such as ours, there are three phases of interest and excitement,—anticipation, realization and retrospection,—and if we had nothing else to do, and no more trips to look forward to, we might even subdivide these and pad our account with the varying emotions of each. But there is one single moment which is never quite duplicated, although higher fever-heat levels of enthusiasm may subsequently persist for weeks. This is the moment when the *first* specimen is secured. It may be a very common, ordinary creature, its accession may be by accident or intention, it may be carefully labelled and preserved, or it may die and be lost an hour after capture; but in its supreme effect, it is unique.

On our third day out, the first scoop of the long-handled surface net brought a small patch of sargassum weed. For a time it seemed to be barren of life, except for the tiny white whorls of worm shells and the waving fringe of hydroids. We were all gathered in the sun parlor when a bit of the half dead weed let go and swam off by itself, and with a rush we were all around the aquarium which held the *first specimen*. Seven pairs of eyes, artist, banker, scientist, lawyer, photographer, physician, microscopist, all focussed eagerly on this little waif which had come to our expedition. And out through the glass, wholly undisturbed, there gazed through wide, lidless eyes, a baby filefish. He even swam up to the sides of the glass

11

nearest the huge creatures outside, and raised aloft his little trigger guard, with a menacing motion. At the attack of a pencil point he gave water slowly, with dignity, always facing this inconceivable danger. When he had looked at us as long as he wished, he gently turned and made his way beneath the small bit of weed we had left him. To us, who only the night before had come through great storm waves, one of which had buried our decks under three feet of water, the little fish seemed of superlative courage, and when one of us named him "Joey" it was wholly without reason, except for that worldwide human tendency to give diminutives to beloved pets. With surprisingly little success the recently unpacked library was ransacked for news of Joey and Joey's relatives. The two facts that he and his kind do not readily take a hook, and that his flesh is bitter, if not indeed actually poisonous, seemed sufficient to blot him from human interest and his private affairs have suffered little in consequence, either from prying eyes or lenses. Painfully fresh note-books were produced and on our various catalogues of this trip Joey heads the list.

When some unpropitious star in his horoscope directed the path of his fate in the route of the *Noma*, he was cozily swimming close beneath a tiny ball of sargassum weed, which like the conventional approaching storm cloud, was "no larger than a man's hand." He was about a foot under water at three o'clock in the afternoon of March 5, 1923, headed slowly north on the edge of the Gulf Stream. The water was a comfortable 63° of temperature, the bottom was twenty fathoms deep below him, and he was ninety miles off Charleston, South Carolina.

Stimulated by enthusiastic aphasia, one of our zoologists rushed to a bathroom for a supply of water, forgetting that the faucet "cold fresh" yielded only hot salt, and in two minutes Joey showed signs of distress, but in a pailful from overside he recovered at once and faced his fate resolutely.

"What is he?" was as usual the first question and one of us was able to answer filefish, from a glance at the curious first ray of the dorsal fin, tall, isolated, movable and toothed, the lowering and raising of which registered all his visible emotions. Although fairly thin and normal in shape, yet in structure and motions he hinted at his affinities with the bizarre trunk- and cow-fish, and puffers. His scientific name *Monocanthus* is even more apt.

The first glance showed that Joey was a true member of the gulfweed fraternity, and an old enough inhabitant to have taken upon himself the colors and patterns of that seaweed. No matter how much sargassum I see, the marvel of it never lessens. Imagine patches of material cloud floating across the sky, some grey, some brown, others olive green, and beneath and in and through them, numberless strange creatures, which, dislodged by some sudden wind, at the coming of calm all rush back to their precarious protection; such is a fish's eye view of the floating weed. With greedy gulls and gannets above, and voracious fish beneath, the weed folk have need of all the protection which color, pattern, form, movement, and instinct can give.

Our filefish Joey was of especial interest, for to him the weed was only a boarding school of sorts, where he spent his youth. During this period he was presumably an acceptable mouthful to any passing fish, or at least, like a durian fruit, they would not know how bad he was until they had given a tentative nibble—leaving them with only a bad taste in their mouths, but making Joey a cripple for life or bringing immediate death. As freshmen for a time affect identical caps, so for the period of his youth the filefish is sargassum colored,—brown and yellow and olive-green, mottled with darker and with occasional streaks and patches of white like the calcareous worm shells scattered over the weed. Then when its file, or better, lance, is big and strong enough to be used as the horn of the unicorn,[1] with the poisonous mucus of its skin added to its powers of defence, the filefish can leave the shelter of its Alma Alga and shift for itself. How doubly tragic for the other weed folk, when the last strands of their shelter die and sink down into the chill northern waters and they are left, failures in the school of life,—with hopeless conditions in every subject through no fault of their own, while the filefish, armed by inheritance, leaves the derelict home *cum summum laude;* his barbed, poisoned lance forever ready, his body covered with spiny scales, his flesh sparse, bone-filled and bitter. And withal he has not even to risk his precious person in fights for food, for unlike many of his near relatives, he is herbivorous, content to mumble seaweed for sustenance.

Like whales and seacows among mammals, Joey, in the course of past evolution, had lost his nether limbs, the paired ventral fins having wholly disappeared, being replaced by a single median spine, half sheathed in skin. As he moved about we noticed that his pectoral fins and tail were almost motionless, unlike the general rule with fishes, and all his movements were controlled by the long median fins along his back and belly. These rippled and waved and sculled him along, raised or lowered him, while by a quick change, an emphasis on the opposite curves, he shifted into reverse and backed away. His skin was leathery but not bony, and while stiff, his body was far from immovable, and when we saw his mode of progress, which was that of the solid bony trunkfish, we wondered whether it was not in part an inheritance from a more specialized condition, with lack of realization of the possibilities of his

---

[1] Twice I have seen filefish, one adult, one half grown, erect their spines and charge, head-on, other fish, whose precipitous flight prevented any idea of the exact method of infighting.

more mobile form. Thus Joey moved through his element, a quaint little rigid form, with only the rippling fin fringes and his wide, round eyes rolling comically about to show that he was a thing of separate life and not a wandering bit of gulf-weed.

With his perfection of defence against all conceivable dangers, nothing in the past history of his life or race had warned against the sudden dip of a net from the deck of a yacht, in the hands of an enthusiastic scientist; and even when again set free in water, his new cosmos was bound by invisibility, and his little snout bumped and bumped helplessly against the glass which could never be understood. Yet even Joey could learn and the second day of his life with us, warned perhaps by some faint reflection, he seldom pushed against the glass, preferring to scull along near it, and neatly bank around the corners. The bottom of the aquarium seemed to produce a perfect reflection, and more than anything else he loved to skate along, leaning well over to one side, apparently watching the counterfeit which accompanied every movement. While doing this, his file remained stiffy upright, which may have indicated vanity, otherwise we have no clue to his feelings, for he has no way to express them other than by the raising, lowering, or vibrating of this single, roughened little spine.

Joey's neighbors in the sargassum kept us interested for days. Each had its little niche in which it had found sanctuary in this weedy world. Some, such as tiny sea worms, were encased in hard lime shells and behind their protection could laugh at danger. No camouflage being necessary, they were blatantly white, and so common and so integrally a part of the weed that they have been seized upon by the refined taste of Natural Selection and trade-marked into the physical insignia of edible, unprotected snails, crabs and fish.

Protective coloring was the most popular defence of the

gulfweed fraternity, and the practitioners so closely connected position with safety that they were to the weed like filings to a magnet. Pull off an olive yellow and white crab and he would make all speed possible back to the shelter of a strand; detach a naked mollusk and it struggled until death to regain the leaf which seemed so much a part of itself. And these creatures not only looked like and clung to the weed, but they swayed like it, and a crab would oscillate like a walking stick insect in the air, and a nudibranch let its tentacles blow with every watery current, or if the water was still, would wave them in imitation of gulf stream influence, as if they were muscleless fronds.

Other forms of life had chosen water instead of weed as a model, such as tiny shrimps, transparent as water or glass. Sometimes the sunlight struck a glint of purple from the carapace or telson, merging with mauve stains on the weed, but aside from these momentary prismatic gleams, the only opaque parts of the shrimp were its food, mere bits of formless detritus, and the mass of eggs deep within its crystal body, the yellow green ova being optically non-existent among the sargassum berries.

As we got farther south, more and more living creatures crowded the heads of sargassum which we fished up. The most interesting were plant-like growths, some real, such as the red alga *Polysiphonia*, with its thick, stiff, branched stems giving off a mass of fine, cobweb-like filaments at the ends. Beyond all count were the graceful, waving hydroids—the mothers and daughters of jelly-fish, but never themselves free to float off at will. A third common sargassum tenant was hosts of little green vases, sprouting from long, slender threads which wound in and out among the fronds. These seemed the very essence of moss or lowly vine, instead of which they represented one of the greatest tragedies of our own class of

PLATE I

GIANT LAND IGUANA
Conolophus subcristatus (Gray)

These brightly coloured lizards feed on blossoms and spiny cactus; they are very vicious, but easy to run down and capture. They live well in captivity.

(Natural size)

PLATE II

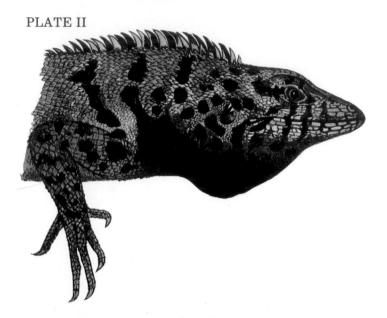

GALÁPAGOS LAVA LIZARDS
Tropidurus albemarlensis Baur

Common everywhere, running about the lava and sand, feeding on insects and berries. The male has a high dorsal crest, and the female a slash of scarlet on the sides of the face and neck.

(Enlarged three times)

PLATE III

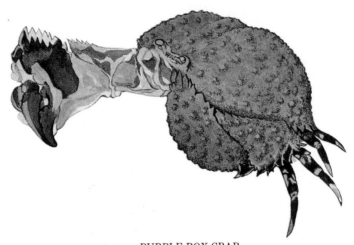

PURPLE BOX CRAB

Calappa flammea (Herbst)

With the claws drawn in the animal forms a symmetrical oval. This is an Atlantic species; a closely related one inhabits the Galápagos Islands.

SCARLET ROCK CRAB

Grapsus grapsus (Linné)

Abundant everywhere along the Galápagos shores, showing no fear of sea-lions or iguanas, but fed upon by herons.

(Natural size)

PLATE IV

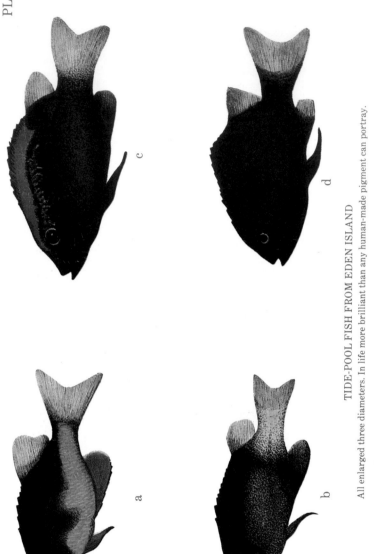

TIDE-POOL FISH FROM EDEN ISLAND

All enlarged three diameters. In life more brilliant than any human-made pigment can portray.

a—Orange-bellied Fishlet,
    Microspathodon bairdii (Gill)

b—Yellow-tailed Black Fishlet,
    Eupomacentrus flavilatus (Gill)

c—Beebe's Scarlet-backed Fishlet,
    Eupomacentrus beebei Nichols

d—Azure Fishlet,
    Microspathodon dorsalis (Gill)

PLATE V

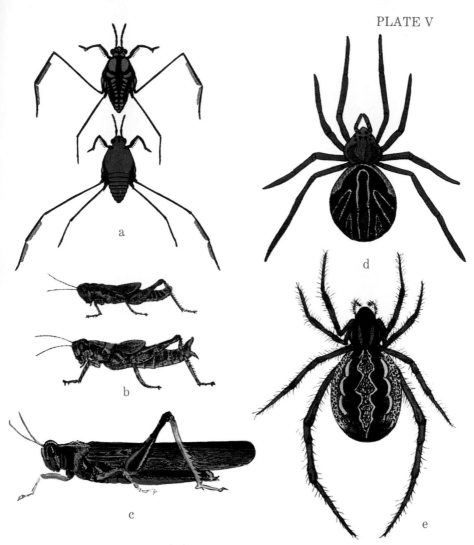

GALÁPAGOS INVERTEBRATES

a—GALÁPAGOS OCEAN-STRIDER, *Halobates* (new species).
    This group of insects is the only one which lives and breeds far out at sea.

b—FLIGHTLESS GRASSHOPPERS, *Halmenus robustus robustus* Scudder.
    The male is smaller than the female. These were found on Tower Island, and spent the day
    under lava to avoid the lizards and owls.

c—GIANT PAINTED GRASSHOPPER, *Schistocerca melanocera* Stal.
    Abundant, forming the food of many birds; occasionally flying two miles out to the yacht.

d—UNION-JACK SPIDER, *Lathrodectes apicalis* Butler.
    Lives under stones, and is related to an American spider which is supposed to be poisonous.

e—GALÁPAGOS ORB-WEAVER, *Epeira oaxensis* Keyserling.
    Thousands of the webs of these spiders were stretched between bushes and lava boulders,
    entangling grasshoppers, moths and even small birds.

(All enlarged except b and c)

PLATE VI

a

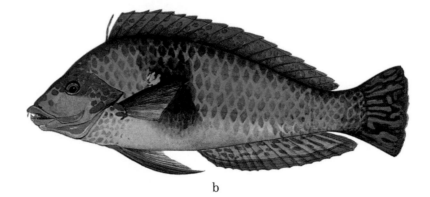

b

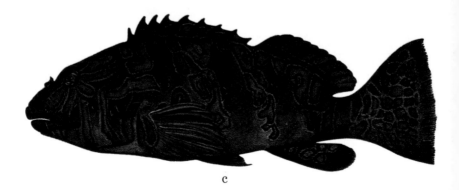

c

COMMON GALÁPAGOS FISH
Caught on hand lines from the deck of the *Noma*.
(Reduced one-half in size)

a—Blue-lined Golden Snapper, *Evoplites viridis* (Valen.)
b—Spectrum Fish, unnamed.
c—Hieroglyphic Fish, *Cirrhitus rivulatus* Valen.

PLATE VII

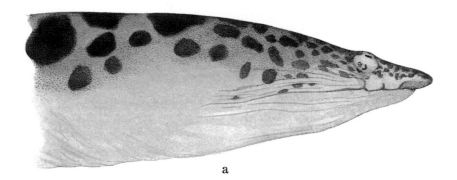

a

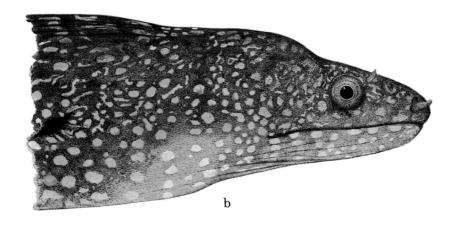

b

GALÁPAGOS EELS

a—Short-tailed Viper Fish, *Scytalichthys miurus* (J. & G.).
Living in holes in the sand, with eyes rotated for upward vision.

b—Brown Mottled Moray, *Gymnothorax dovii* (Günther).
A vicious eel, often attacking without provocation.

PLATE VIII

GALÁPAGOS SNAKE
Dromicus dorsalis (Steind.)

A dark phase of the species living on Indefatigable Island. These are harmless snakes, feeding on grasshoppers and other insects.

(Lower figure natural size)

creatures—for they actually belong to the same group, the backboned animals, as ourselves.

The microscope showed a rhythmical contraction, about ten every fifteen seconds, of the whole organism—tunicate or ascidian as it is called. At the upper end of the vase were two openings, into one of which water was drawn, to be expelled at the other. Thus the little being breathes and eats, and from time to time broadcasts its eggs. These hatch into tadpole-like creatures with sense organs, nerves, gills, a foreshadowing of a backbone—all the promise of becoming fish, frog, man himself. But the maelstrom of degeneration seizes upon it, thrusts it upon a sargassum frond, destroys its senses, obliterates its notochord—all badges of superiority—and wills it a safe, simple existence, in appearance calling the cabbage sister, and with its mighty ancestry, its unused opportunity, apparent only in the evanescent development of its own youth and of its offspring, whose destiny spells mere quiescent existence. It is one of the very few surviving failures of the early vertebrates, related to an enormous extinct company which was not in the single favorable line of ascent.

The greatest surprise to me on this part of the trip was to find a number of pelagic anemones, *Anemonia sargassensis*. They have, I found later, been known for many years, but I had never heard nor read of them, and when one of the little animal blossoms unfolded before my eyes, I had all the thrill of discovery of a familiar friend far out of its usual haunts. They were tiny, only an eighth of an inch across, but bravely colored in russet, tawny and translucent grey, with a dainty thread of bluish-violet drawn about the mouth. Their tentacles waved vigorously and they showed an amazing ability to travel, creeping up and down the sargassum fronds as freely as their nudibranch mollusk neighbors, and belying the static characteristic usual to their radial fellows.

17

GALAPAGOS

We and our friends had crept into this complex contraption of wood and steel, pushed off from our home town, wound up the machine and a whirling bit of metal pushed us through the water. From time to time we squinted at sun and stars, juggled with angles and figures, and rather astonishingly knew how far we had progressed over the surface of the waters. With speed and direction under control, we commanded climate, temperature, geography, the world. We left not the slightest trace, when once the churned-up emerald whey of the wake had quieted again into ultra-marine; we drove a momentary wedge through sunshine, wind, and water, but a few yards in any direction and our little contraption might never have existed or passed. If some little gadget went wrong, or the waves lapped a bit too high, we would not even cast a shadow upon the bottom towards which we slowly sank.

But although the elements were so utterly oblivious of our yacht's existence, we found that the *Noma* had a peculiar fascination for certain living creatures. This power of attraction was for several reasons,—by way of hunger, an apparent love of play, a wholly unreasonable attraction, both voluntary and involuntary, acceptance of the keel as home, and finally as sanctuary in the last extremity of a fight for life. The creatures who perceived us and more or less willingly approached or came aboard, included mammals, birds, fishes, and many of the lower forms of marine life. We had not left the harbor before we had collected our convoy of following gulls, who kept close behind, crossing over from time to time, and at sunset dropped away one by one. One immature herring gull I marked down individually,—it had lost the fifth right primary, and two tail-feathers were shorter than the rest. This bird remained until the afternoon of the third day, ·disappearing at nightfall and coming back each morning, twice at about seven o'clock and once at nine. I saw the bird first off the New Jersey coast, and last at the northern line of Florida, trailing us for seven hundred miles, sleeping whether on sea or land, I cannot tell. Probably ever since the first ships put to sea, gulls have followed in patient and never-failing hope of scraps from the galley, just as, far inland, they follow the plough in flocks to snatch the up-turned worms and grubs. Psychologically they are the crows of the sea, and no matter how apparently tame and familiar they appear, they are never off their guard. I trailed a line and baited hook one day, hoping to catch one in the tough skin of the throat, examine and liberate him. Down went the half dozen birds, their eyes having instantly caught the tiny bit of meat. One bird picked it up, but the weight of the dragging line pulled it out of its beak; another tried and failed, and then all the birds quietly drifted off, and did not return for two or three hours. They sensed something wrong, the feeling of suspicion won over that of hunger, and they scattered. We seldom realize how much individuality is hidden from our eyes by superficial identity of plumage. On the fourth day two adult gulls were following, when a third approached. The most careful scrutiny showed not a particle of difference among the three, and yet the third bird came on with hoarse cries and open, threatening mandibles, and pursued and harried the other two, until they were compelled to flee, and he had the field to himself for the rest of the afternoon.

The second companionship between the *Noma* and creatures of the sea was as common a phenomenon as the gulls,— common but none the less remarkable. Lying flat upon the bowsprit on the lookout for any living thing which might appear, I suddenly saw three dolphins beneath me, close to the yacht's bow, keeping steadily in place and showing no more movement or exertion than if they were torpedo tubes attached to the keel. This was exciting enough, but still more so was to see the newcomers arrive. A hundred yards away five or six animals rolling slowly along, suddenly turned and bounded in our direction. At the same moment, six, ten, fifteen might be seen all converging toward our bow, from port, starboard, or dead ahead, swift as catapults, strong as turbines and at the last moment turning as abruptly as a reined-in Arab, and joining the crowded mass of fellow cetaceans in their inexplicable play. I doubt very much if this is anything but a new-formed habit, as they can hardly do anything comparable in the case of surface barging whales. Like vultures or sheep they follow one another, only they detect and are drawn by the throbbing of the propellers, and from a considerable distance focus to their play.

One day I saw a particularly enthusiastic crowd of twenty-four packed as closely as possible in front of the *Noma*. Those beneath, rising to breathe, would push up through two or three layers, rolling their companions clear over, showing now their dark or mottled backs, now their paler undersides. Through all the splash and rush of water at the bow, there sounded now and then the soft, penetrating sigh of the dolphins' breath. On a pitch dark night, without phosphorescence, few stranger sounds can be imagined than the sighs of these mammals of the sea, now here, now there, gentle and sibilant with no hint of muscle or force in the author.

Since boats were first launched, dolphins, like the gulls, have evolved this habit, and for miles allow themselves to be half pushed along by the keel. Occasionally they actually hang back and let the bow rub against their backs—submarine blessings of the Duke of Argyle. They average one to two minutes beneath the surface, each dive being divided by one to two seconds of intake of breath. Just before they emerge, a rush of bubbles indicates the escaping old air, but there is no steamy breath or spout of spray, merely a quick opening of the single nostril, an immediate closing, and submergence. The crescent-shaped flippers are quiescent, held well out from the body, and only now and then move slightly as if for balancing. The closest scrutiny is necessary to detect any movement of the tail, and yet only by its vertical sweeps is the tremendous power of this strong creature attained. Like scores of other observers I have no alternative than to call the whole performance an exhibition of play or sheer excess of spirits. This habit is one thing which links them to ourselves as fellow mammals. No matter how unlike us and how similar to fish they appeared as I looked down at them, no fish would have just this seeming love of fun, this waste of energy having to do with something less or more than food or enemies. We know that young dolphins have on their bodies four real hairs which are another bond with us unscaled, hairy beings and we clearly detect fingers beneath the mittens of their flippers, but we have no exact clue as yet to their ancestral connections. As far back as we can go, fossil dolphins were dolphins and not part dolphin and part bear, seal or cat. So we must take them as they are today and rejoice in their jolly companionship. Their virility is infectious, and I have seen passengers thrilled by dolphins,—passengers who otherwise would consider a six day ocean trip as one hundred and forty-four hours of boredom. Dolphins find sailing ships too slow to remain for long in the bows, so they dive under, swim around and around, or leap clear of the water, sometimes landing broadside on with a slap which is audible half a mile away. The enviable place which dolphins have won for themselves in the sea, easily swimming down food and avoiding their enemies, is hinted at in the variety of their coloring, purplish grey, pied, mottled, and pale individuals being discernible in any good-sized school.

After the first rush of my crowd of four and twenty dolphins was over, and one by one they had dropped away, a mother shot up with her half-grown young, remained for a few seconds and then both disappeared *straight ahead*, although we were going thirteen knots at the time. A minute later a baby only two feet in length appeared, with its seven foot parent. They stayed for two minutes during which time the mother twice made complete revolutions about her offspring, while the infant had apparently no more trouble in keeping in place than the old one.

Louis, our old sailing ship helmsman, helped me with the harpoon I was preparing, and told me that dolphins were the souls of dead sailors, and often when a sailor had been drowned, the dolphin into which his soul entered would support the sailor's body, with the aid of some of its companions, and swim with it to shore. Hence the love of dolphins for ships. I asked him if he objected to having them harpooned, and he said no, for there were good and there were bad sailors, and most of the modern steam sailors were bad ones.

There is little excuse for the confusion which has arisen around the names dolphin and porpoise. No honest sea captain with whom I have ever talked knows the difference, yet any land-lubber can easily distinguish the two at a glance. Both are actually diminutive whales, and although there are a number of species, yet the two general groups are quite unlike one another. The dolphin (genus *Delphinus*) has a long pointed beak, easily seen when the animals execute their usual breathing roll or when they leap clear of the water. Porpoises (genus *Phocœnus*) have rounded heads in profile, more like typical whales, and lack the spirit of play which so often impels their relatives to spring high into the air. The fish known as dolphin, celebrated from the time of the Greeks for its marvellous ante-mortem color changes, was so called only from a fancied resemblance in shape and finage.

In the daytime we could easily detect the little organic swirl in the pelagic life caused by the *Noma*—gulls and dolphins going out of their way, even gannets, which, moved wholly by curiosity, came, circled about once or twice, then left. But at night we were unable, except very rarely, to know the effects of the tiny ship passing over the great expanse. When at anchor, and the searchlight was turned down into the water, fish assembled in schools, hosts of small ones which were instantly chased away by some great fellow, all returning in another moment to the lure of the bright circle.

That this strange attraction is a powerful one is proven by the flying fish which come aboard, as migrating birds are drawn to lighthouses. I do not remember a voyage of any length, when sooner or later a flying fish did not land on the ship at night, skimming high over the rails, held by the irresistible fascination and falling with a plop upon the deck. Six or eight boarded the *Noma* in the Atlantic and two in the Pacific. I once caught one on a deck of the *Lusitania* forty-five feet above the water line, and in the Pacific, many years ago, a flying fish shot clear into my cabin through the open port-hole. This was sheer luck or accident, as these creatures have no such exact power of direction.

Another class of creatures hailed the *Noma* as home—using hail in a very figurative sense indeed. This was not evident from the point of vantage of the deck, but though they worked quietly, and out of sight, yet day and night they were active and eventually exerted more permanent effect on the ship than all of the other organisms combined.

Clusters of barnacles on rocks along the shore seem well-nigh as firmly rooted and as inorganic as the boulders themselves. But if watched in an aquarium in the spring, hosts of new-born barnacles may be seen pouring forth from their parent's shell, actively swimming about in the water, resembling their parent about as much as a mosquito does a tree. At first they steer directly for the light and in those which drift far from land this ineradicable trait keeps them at the surface of the ocean, but soon a new desire obsesses them—they must seek a solid foot-hold, and of the countless myriads, while hosts are unable to find their respective Ararats, one in every few millions, by the nth chance, touched the keel of our yacht, and stuck. He turned on his little glue factory and his lime precipitator, lost his head, and lo, a barnacle. More and more ones-out-of-millions came, and in time the ship would be held back with the load of tiny barnacles, kicking their food into their mouths as we slipped swiftly through the water.

By far the most interesting of all the visitors to vessels at sea are the land birds which, driven out of their course by wind and fog, in the last extremity seek sanctuary on any ship which good fortune may reveal. Few of us realize what a terribly dramatic thing is the migration of birds. We see the bobolinks in our meadows, the bluebirds in our orchards; we watch the young reared, and then in the autumn, they and the swallows and the sandpipers and the geese, each in their different way and time, pass from view southward; and in the spring they return.

But do they return?

How many pairs of bobolinks nested in the meadow last season? Six. How many young were reared? Twenty. How many pairs are nesting this year? Probably six. Twelve from thirty-two leaves twenty. Two score birds are missing, most of them the young of last year, with a scattering of old birds. It is not likely these have gone elsewhere to nest—birds are fanatics in their love of birthplace. Almost two-thirds of the bobolinks which went south have not returned and they never will. Some of their little bodies are drying on the sands of far distant deserts, or slowly dissolving in grass or jungle; some were snatched in midair by hawks, or swept from sleep to death by yellow-eyed owls. But the majority were probably overcome in the midst of seething waters—after being blown and buffeted, choked by salty spray, and terrified by failing muscles over a landless horizon.

An abstract appeal to our imagination is through facts; such as the astounding annual journey of the American golden plover. In June we find these birds on the frozen tundras far north of the Arctic Circle and the limit of tree growth, making their nests near the ice-covered lakes. Here they incubate and rear their quartet of young, and by August have flown hundreds of miles to the southward—to Labrador. Crowberries in abundance attract and hold them, and here they rest from their labors of nesting, allow their new plumage to reach full perfection and go into training for their long, long voyage. This latter achievement consists in stuffing themselves day after day with the black, nutritious berries and thus acquiring astonishing layers of fatty tissue.

One night, when the spirit of migration moves them, they rise into the air and start over the cold, turbulent waters, on past the St. Lawrence and Nova Scotia straight out to sea. After nearly two thousand miles they sight the Bermudas, or if it is fair weather they may fly over the sea far to the east, passing the islands unseen. Unless forced ashore by storms, the plovers do not rest and recuperate until the Antilles are reached, thence continuing their journey over the South American jungles to the pampas of southern Argentine. After six months of idling and feeding through the southern summer, among the native birds all busy with their nests and young, the golden plovers again take wing along in April, and hasten back to northern Greenland. This return route is via the Andes, Central America and the heart of the United States and Canada, terminating in another six to eight weeks of home life. Each year eighteen thousands of miles are traversed by a bird whose weight is less than a half dozen ounces, and with a spread of wing of only two and twenty inches.

Still more realistic did migration become when I once spent a night in the head of the Statue of Liberty, and watched the tiny wrens and warblers shoot into the light, like golden bees, either to sheer outward and disappear into the safety of darkness, or to crash head-on against the glass and drift down, down—pitiful, disordered, lifeless fluffs of feathers.

Even more vivid became the marvel of migration when the first wind-blown waif came hurtling toward the *Noma* in the van of a purple-black doom, alighting just in time to cheat the cataclysm of whipping wind and churning sea. Thus came a purple martin and a robin to us off Hatteras. Mid-way down the coast of Florida, a moderately strong off-shore breeze blew throughout the night of March sixth. The next morning, when we were fifty miles off-shore, five birds sought refuge on the ship, a pair of Maryland yellowthroats, a parula warbler and two chuck-will's-widows. I have already spoken of the male yellowthroat which in the early morning sang his heart out, though even then, hunger was pinching him sorely. I secured one of the chuck-will's-widows and found that, on the contrary, it had dined bountifully on small beetles, apparently just before being driven off-shore. After the other birds had rested sufficiently they took to wing, and their different modes of orientation were most interesting. The warblers flew up and up in as narrow a spiral as the steep climbing would permit, and finally, when they had become mere eye-straining motes, each described a mighty circle and headed straight for the shore, invisible below the horizon, even to their lofty outlook. The goatsucker lifted itself on soft, owl-like wings, and zig-zagged back and forth aimlessly a few times, then, so close to the waves that it seemed to brush them now and then, it too set a compass-straight course for the Florida woods. The off-shore breeze had hardly slackened, yet what had confused the birds in the blackness of night was now overcome, while yet they had no direct guide of eye or ear. Unafraid and unfaltering they started to force their way into the stiff wind, over half a hundred miles of tossing sea. I have had in my possession a living European turtle dove rescued at sea seven hundred and fifty miles off the coast of Ireland, and have known of herons being blown two and three hundred miles over a stormy sea. The ability of these birds to survive such a terrible experience is no more wonderful than their methods of regaining land, knowing instinctively the right direction when daylight comes. As I watched the chuck-will's-widow disappear I wondered for a moment at the unreasonableness of my seeking such a distant goal in the Pacific, with its many unsolved problems, when mysteries such as these faced me on every hand.

In the course of our trip to the Galápagos and back, other unwilling voyagers met us and partook of our hospitality, but of all the creatures which we observed at sea I think the phalaropes made the most lasting impression on my imagination, perhaps for the same reason that human beings are so often moved and deeply attracted by a certain aloofness or self-sufficiency in their fellows. These were birds to which our yacht meant nothing—they neither asked nor gave. Far ahead, on the slowly heaving slopes of the great rollers, there appeared at times hundreds of tiny dots—a stippling of living creatures against the restless texture of the open sea. When our figurehead looked down upon the nearest of these marine sandpipers, all arose as one bird, snapped into that strange aerial bondage of the flock, and whirled far away from our path, to keep on out of sight, or to merge with other uncounted thousands of their kind.

GALAPAGOS

Sea-ducks haunt the shallower banks of the coast, gulls trust the air of the widest sea, gannets dive deep into the waves for their prey, while skimmers plough their tiny furrows through the ocean, but phalaropes actually live on the surface of the open sea. With their long legs for wading, with narrow lobes instead of broad webbing between their toes, without the slender, narrow wings upon which albatrosses hang through days of storm and stress of weather, phalaropes seem little adapted for pelagic life, and yet here and on the Pacific, hundreds of miles from land, I have seen them in enormous flocks—daring wind and water, spending the whole winter out of sight of land, trusting to the floating bounty of the sea for food, and to the buoyancy of their dense plumage and air-filled bodies for safety. I have never seen them in a full storm, but in a half gale, with spray blowing, and every watery hilltop fountaining into ugly lashing foam, I have watched their marvellous seamanship, paddling steadily up wind, able, by some perfected knowledge, to keep in the sliding, shifting valleys and free of the choking spume-drift.

Their winter life focusses our wonder, and their habits on the frozen tundras which are their home in summer are astonishing. The female phalaropes are dominant in size, color and character. It is they who do the courting, even fighting fiercely over a popular male bird, and then, having laid a quartet of gayly colored eggs for the favored one, they promptly desert, leaving all the duties of incubation and care of the downy phalaropelets to the male. They may not meet again until the great flocks gather far out on the southern seas. We know the facts of these strange little lives: the *why* ever eludes us.

On Thursday the eighth of March we put in at Key West to retrieve our surgeon guest who had come on from Washington by train, and we also hoped to get our preserving alcohol and thus save an extra day at Havana. For prohibition, like anti-vaccination and anti-vivisection and other anti-sanities, works blindly against scientific progress, and a scientist in New York in search of alcohol for his work must turn bootlegger and even then finds the fluid far beyond the limits of his slender purse.

Key West draws sustenance and life through a single slender thread of an artery—the Key West Railroad—and were this cut, one feels that it would all dissolve into coral and heat and water. It wished to be tropical, it strove to leave behind the temperate zone, but the sand was against it, and the bougainvillia and hibiscus were only faintly reminiscent of the glories of Dominica or Barbados. The houses were too assertive, not tumbled-down enough to be convincing, although the natives fulfilled all expectations. They sat or lay down and allowed time to pass over their heads as rapidly as might be. We over-lunched at the one vast hostelry, and watched the tourists, recently returned from Havana, who were waiting for their train north. Why, in heaven's name, should we be ashamed of enthusiasm? After we had sloshed gleefully through straggly but very real grass and flowers, exciting after New York's snow, and shouted aloud at our first man-o'-war bird, we entered the great hotel and became immobile masks in the presence of our fellow men. If anything of joy had accrued to these people in the great island of Cuba—if blossoms, and new birds, and a strange tongue and an alien race had aroused aught of enthusiasm, it had since been completely crushed out. They looked askance through their lorgnettes at our negligee yachting costumes, and I slid my field glasses into my pocket and quieted the tinkling of my collecting vials. When I came out again into the clear ocean air, I looked back fearfully before I raised my glasses to make sure of a vesper sparrow. After this I believe I went into a rather pleasant gambling house, visited the club and saw the tennis courts, but these conventional things passed from memory when I achieved the real joy of lying on my stomach on the smelly deck of a negro's fishing smack, and revelling in the gorgeous fishes swimming about in the well. The old fisherman's account of his catch, his piscatory gossip, was more absorbing than any table-talk at the hotel, and the dyes and patterns of the tourists' dresses paled before the wonderful pigments of these fish. When the fisherman learned we were interested, we were taken to see the catch of the day, a six-foot sailfish, *Istiophorus nigricans,* a sword-snouted, steel-blue torpedo, whose fins, of astounding size and blackness, could all be furled and stowed away, completely out of sight, in slits deep within the very flesh of back and belly.

The next day we were still held up by lack of stokers, a break in our radio outfit and the uncertainty of getting grain alcohol at Havana. Wireless settled the two latter problems and finally we succeeded in getting coal-heavers who did not require six hours of rest to one of work. Sharks' fins kept cutting the water astern, so we dropped a baited hook on a heavy cod-line. A shark took it and snapped it at the first rush. A second hook on a half-inch rope held, and we had one of the biggest, surging and splashing about, until quieted by several Lugar bullets. With block and tackle we hoisted her up and inboard. She was of that wonderful battleship grey, shading into white below, which is so characteristic of these Lamia—using the word in the old sense. This species, the requiem shark, *Carcharhinus lamia,* is known by its long and falcate pectoral fins, and is common both in the Caribbean and the Mediterranean, being described as a man-eater, notorious in warm regions as a greedy scavenger about wharves. Our specimen was exactly eight feet and half an inch in length, with teeth about an inch in height, and weighed at least two hundred and fifty pounds. It was a female with four full-sized embryos. One of these had been dead for some time, but the other three, about two feet in length, were very active and would have been born in a few days. We put one in an aquarium where it swam readily, breathed normally, thrashed about when disturbed and attempted to bite, although its teeth were the merest pin-points. In another which we dissected after apparent death, the heart beat long after removal from the body. No wonder these fish are dominant forms and successful in their realm of activity; with such a grip upon life they could not well be failures.

My passion for the study of a limited field, a sharply demarcated zone, such as a square yard of jungle or an isolated desert island, extends also to individual organisms isolated by size, special interest or temporary environment, or all three as in the case of this first shark. I was interested to find her the center of a little world of creatures in addition to her own offspring. The strange little crustacean parasites on her skin would alone repay prolonged study.

While she was still in the water several small shark-suckers, *Echeneis naucrates*, clung tenaciously to her skin, slid about over her body, or swam close alongside, and only when the long lobe of the shark's tail fin was drawn clear of the water, did the last of these little companions desert her. They reminded me for all the world of feather flies sliding and scurrying over and among the feathers of a bird. Like those flies they were exquisitely adapted to their strange life, the anterior dorsal fin of the little fish being metamorphosed into a great, corrugated, sucking disk which gripped with all the force of a vacuum. These fish had remarkable powers of changing color, and when excited would produce longitudinal bands of brilliant white along the entire body, or in a few seconds could again bury these so completely beneath pigment that they became of a monotonous neutral grey throughout. This latter color phase was of extreme interest in that it was equal in tone on all parts of the body, being thus wholly unlike the usual coloring of fish and terrestrial animals, darker above and paler below. The shark-sucker, spending its life slipping about the body of its host, was sometimes right side up, sometimes inverted, so the uniform coloring above and below rendered it quite inconspicuous from any viewpoint.

They do no harm to the shark's skin, only delaying its progress slightly, like barnacles on a ship's keel, and they feed not only on the bits that fall from their host's mouth, but also take small live fish on their own account, as we found when we hooked two on live bait over the ship's side.

From before the time of Aristotle, sucking fishes have been known, surrounded with mystery and endowed with miraculous powers. Dr. Gudger has recently assembled the legends relating to the shark sucker. Whether called *Echeneis* by the Greeks or *Remora* by the Romans, the significance of the title of these two closely related fishes lay in their supposed power to hold back a ship in spite of favorable winds, oars or tides.

Pliny the Elder in his "History of Animals," Book XXXII, Chapter I, gives the following account of this fish: "And yet all these forces . . . a single fish, and that of a very diminutive size . . . the fish known as the Echeneis . . . possesses the power of counteracting. A fish bridles the impetuous violence of the deep, and subdues the frantic rage of the universe—and all this by no effort of its own, no act of resistance on its part, no act at all, in fact, but that of adhering to the ship.

"At the battle of Actium, it is said, a fish of this kind stopped the prætorian ship of Antonius in its course, at the moment he was hastening from ship to ship to encourage and exhort his men, and so compelled him to leave it and go aboard another. Hence it was, that the fleet of Cæsar gained the advantage in the onset, and charged with redoubled impetuosity. In our own time, too, one of these fish arrested the ship of the Emperor Caius (Caligula) in its course when he was returning from Astura to Antium: and thus, as the result proved, did an insignificant fish give presage of great events; for no sooner had the Emperor returned to Rome than he was pierced by the weapons of his own soldiers. Nor did this sudden stoppage of the ship long remain a mystery; the cause being perceived upon finding that, out of the whole fleet, the emperor's five-banked galley was the only one that was making no way. The moment this was discovered some of the sailors plunged into the sea, and on making a search about the ship's sides, they found an Echeneis adhering to the rudder. Upon its being shown to the emperor, he strongly expressed his indignation that such an obstacle as this should have impeded his progress, and have rendered powerless the hearty endeavours of some four hundred men. One thing, too, it is well known, more particularly surprised him, how it was possible that the fish, while adhering to the ship, should arrest its progress, and yet have no such power when brought on board." It added great interest to the little shark-suckers which we caught today to know that they aroused such excitement nineteen hundred years ago. Natives of various countries are said to tether the sucking fish and by paying it out to catch fish and turtles, and small specimens are reported to take refuge in time of danger within the gills and even the mouth of the shark host. One of the strangest things is that the shark appears never to attack or injure its followers. Very little is known about the intimate life of these fish, and nothing at all about the development of the disk, either from eggs or embryos.

Havana was a wild jumble of mi-câreme parades and confetti, and a long argument with one of the principal merchants, in too jovial mood, with my hastily resurrected Spanish. I tried to obtain from him an order for ninety gallons of 98% alcohol, while he lamented with tears that for me to take with him only a single puncha crema was an international insult. I succeeded finally in obtaining his signature, and armed with an enormous wooden spigot in lieu of signet ring, I extracted from his warehouse slave the precious fluid.

Almost exhausted, I strolled into a bookstore which was open only because it was also a wine and chemist shop, and on the topmost shelf, balancing on a broken stepladder, half choked with dust, I made a wonderful find—a two volume set of Poey's "Historia Natural de la Isla de Cuba" with plates colored and perfect. The joy of this discovery almost overshadowed a splendid early edition of "Don Quixote." I returned on board jubilant—with my books and my keg of alcohol, all achieved at the climax of mi-câreme; ¡es la fin de un dia perfecto!

That night we left for Panama; running into the after waves of a mighty storm which kept us wallowing steadily for four days and nights, decks under, with only the old Latin meaning of the ocean-to-come to cheer us—Pacific.

Fig. 3

Catching Sea Life from the Boat Boom

Many rare creatures were thus caught while the yacht went at full speed

Fig. 4

Shark Caught from the Yacht

Sharks were abundant everywhere and took any bait which was offered

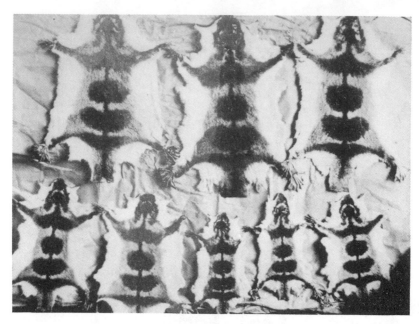

FIG. 5

SKINS OF PANAMA WATER OPOSSUM

FIG. 5a

JUNGLE HOME OF THE WATER OPOSSUM

CHAPTER II

SEVEN miles over a good motor road from Colon led first through low, settled country, then rolling hills with burned up vegetation, for this was mid-March and the height of the dry season, and finally we wound past the great fort and the officers' quarters at Gatun Dam. The great lines of electric light pillars gave the locks a semi-Egyptian look, rather worthily when we remember the achievements of the pyramids and desert irrigation. Far out in Gatun Lake there came into view a ripple which might have been made by the neck of a snakebird, then a low foamy wave like that in the bow of a swimming tapir, then, to continue our zoological similes, a whale surged up, its back lifting into the sunlight—and a submarine gave forth a dozen sailors and rose to full height, followed by another and another, the whole school headed for the locks and their final berth in Colon Bay. Overhead two planes droned, one well up, spiralling slowly in alternate whorls, and high above it, as if in mockery, a turkey vulture also spiralled, wings motionless, a noiseless, propeller-less parody of man's efforts.

Every picture post card, every tourist's kodak, every folder shows the canal and the locks, but few people take the trouble to walk a mile beyond to the spillway,—the waste spigot, the overflow of the canal, where all sorts of strange

35

things happen. Along this neglected channel, this jungle communication trench, the wild things approach and find rich largesse; their bounty begins where man's economic values cease. The enormous curved slopes of the spillway were no more than damp, except for a thin, silver sheet which tapestried down the centre. At the base, where the huge amphitheatre began, were two series of gigantic seats, which Dunsany could have filled with sonorously named gods, ready to witness some unthinkable play of Pegana. I went a little way down stream, found a small, humble seat befitting a mere scientist mortal, and here, together with the gods invisible on their seats behind, I waited. With the evidences of men all around us, yet no man was in sight for a long time, and birds and other wild things of Panama came and went freely. Fish-hawks in pairs swooped over the spillway rim, hung motionless a full second at the top of their swing, and dropped with loose wings like stricken birds into the shallow water. Before the splash, an instantaneous shift, a double forward reach of eight talons, and straining wings beat a thousand drops from the water before they whipped spray into the air and finally swept clear. Head on, the big fish wriggled in vain, sending brilliant heliographs of flashing silver and so delaying the bird's flight that an attacking pair of grey-breasted kingbirds seldom missed their mark, and, from the big bird's neck, feathers eddied slowly down. Five ospreys dropped, and five fish were carried off to perch or eerie. Here were tragedy, victory, bravery and comedy, from the various points of view of fish, osprey, kingbird and audience. Then unexpectedly came spectacular beauty,—an egret in full plumage with a feathery mist of aigrettes streaming like an aureole, alighted on the skim of green which lined the spillways, resting on the all but vertical slope with crest raised, wings spread. Slowly she climbed upward, daintily picking here and there as she went, a thistle-down of a heron,

36

an ivory cameo set in darkest emerald: at the sight the wind sent a breath across the gorge, or else the gods were moved to a murmur of applause, a faint *coup de sifflet* hovering on the verge of human inaudibility!

Close to my seat, barely beyond arm's reach, spotted sandpipers teetered, spotless as yet in their winter plumage, while in an outer circle, a zone of less familiarity, were Louisiana and little blue herons and egrets. Vultures swooped low overhead, envying the hawks their grasping talons, and hoping my lowly attitude presaged demise; hummingbirds came from nowhere, materialized on the rim of human sight and hearing, and vanished again into their particular fourth dimension; black ani cuckoos peered over the brink, whaleeped and blundered away.

Men now came upon the scene, five bare-legged specks wading across the low dam downstream, armed with the best tackle that money and Walton lore could achieve, and walked back and forth, whipping the troubled waters with cobweb lines. Now and then a great tarpon would jump clear of the rapids and shake its head like a bulldog, flicking out the hook and bringing despair to the tense nerves of an excited fisherman. For hours the enthusiasts paced untiringly, while fish after fish struck and escaped. And the gods and I watched and were not too sorry, and even the fishermen admired the cleverness which could baffle them and their utmost efforts with subtle tackle.

Beneath the water at my feet, golden-eyed shrimps danced upon carpets of marbled algæ, and watched me, as I watched the fish-hawks and they the gods. Across the rush of water a stranded log lay for a long time motionless, as logs should, then suddenly it moved upstream against the current, an action wholly foreign to logs, and with a twist of my glasses I saw two crocodile eyes of gold, watching everything through vertical slits; he saw me and the fishes with perhaps hungry desire, he saw the shrimps regardlessly, the ospreys with envy, and out of the corners of his eyes he must have seen the gods sitting dimly but expectantly, and his thoughts of them were, I am sure, those which crocodiles think always about gods.

The sight of sights was a mighty water tortoise, the sudden vision of which robbed me of a heartbeat, as did once the glimpse of a giant cat-bear, and as a migratory pterodactyl might. I had been thinking giant tortoises ever since I had whispered Galápagos to Roon, our god of going, but on one side of the spillway I now saw a tortoise as amazing as a twelve-inch ant,—a huge fresh-water chelonian nearly a yard long with feet as large as my hand. It saw me and dived, and although later pursued by enthusiastic negroes, at the urge of unheard-of remuneration, it avoided all efforts at capture. Never had I imagined, outside of marine turtles and snappers, any as gigantic as this. The sight of it made me feel that the canal and spillway were really worth while, more than statistics of shipping or patriotic emphasis of ownership had done. The sun sank and the swift tropic twilight passed into crescent light, and the gods became more distinct in the dusk, as all invisible things do, and I grew less substantial, until the gods and I were almost of one material, and the birds of the day left, and night things came forth, and I went slowly back to noise and talk, to electricity and efficiency.

The yapock is an aquatic opossum, with normal hands and huge webbed feet. It is the most beautiful of all opossums, as large as a very large rat, with soft silky fur and a double dumb-bell pattern of delicate gull grey on a background of purest white. It is also one of the rarest of mammals, both in actual specimens and in our knowledge of its life and habits. I learned of a tributary of the Chagres where twenty-five of these beautiful little creatures had been collected and others seen, and so under a blazing two o'clock sun we set out—my hunter friend and I—to spend the night in the tropical swamps of Panama. It was my last night before leaving and there was no one in the world with whom I would have changed places.

We landed on a beach of coral and shells and walked through a grove of cocoanut palms, with plumes as graceful and delicate as smoke, and on through the back yards of mighty gun emplacements. Here were unlovely clothes-lines of gunners' wash, their vestments being cleaned in order that they might continue in good health, with strength to oil and polish the sleeping tons of steel, to keep in readiness conical containers of thunders and lightnings, which, years from now, might be launched at a certain second of time, toward some distant square yard of space, and so save the wonder ditch for the good of the world. But now we thought only of dodging gunners' wet jeans, and of scrambling through thorny thickets, the natural wire entanglements which spring up over cut jungle.

Topping a ridge and down into steamy greenness took us from sight of man and his works, except that for a long time we crossed and recrossed narrow ditches draining the land of mosquito water. Whenever there stretched a bit of tide-washed mud, or a wind-fallen tree had left an open space, huge crabs abounded—land crabs which had clipped the bond which bound them to the sea of their ancestors, and found life pleasant and livable high and dry. The largest were of a strange greenish-blue, with great pincers of chalcedony tipped with maroon, and with them were smaller, swifter land-crabs, whose carapaces were jet black, bordered with garnet, with legs of carnelian. The earth was baked and dull, the leaves dry and dusty;—crabs and birds provided the brilliant spots of colour in this landscape, crabs and tanagers, hummingbirds and parrakeets.

Finally we came to low jungle and for a mile or more followed a dim trail. It may have been begun by wild animals, it was assuredly widened by Indians, and we know for certain that old Henry Morgan and his horde of buccaneers used it, and today it is passable for a horse. But yapock lived far even from this trail, and we soon launched out at left angles straight into the jungle. Sometimes we were helped by a long stretch of shallow water; again we had to creep ant-wise through a solid mass of agave, moving each thorn-lined leaf to one side, and paying for each misstep with shreds of clothing and drops of blood. In the centre of each plant rested a great scarlet bloom, a foot across, its reflection turning the heart leaves to a rosy pink, but its beauty was hardly compensation for the cruel gauntlet of spines. In the intervals between the zones of spines, more active, zoological thorns sought out unexpected places, and Azteca ants drove home their stings with enthusiasm. Soon we moved in an aura of formic acid emanating from those we killed. Two miles and a half of this going brought us to a ridge where we could look into clear sky, and hear the deep distant voice of a Royal Mailer calling for a canal pilot.

Down into a new stream we went, with its current towards the Chagres, listening to evening flocks of parrots and hearing the swish of vultures' wings as they took a last look at us, before despairing of our early death. Doves boomed at their water holes, motmots mumbled converse deep in side gullies, *uh-huh! uh-huh!* and tree-creepers' beaks dripped silvery cadenzas.

The kr-ump of a big gun came from the distance, but in place of an echo, there followed a still deeper rumble and we knew that the eight months of rain had begun. The jungle as a whole was green—the heavy nightly dews saw to that, but hosts of flowers and insects were waiting for the first downpour to stir and develop and fulfil their destiny.

PANAMA AND YAPOCKS

After about six miles of gruelling pushing through jungle which was more like story-book jungle than is usual in the tropics, we reached yapock country and sat down on a great flat expanse of stone, smoothed by the torrents of hundreds of rainy seasons. Here we ate our army rations and talked of past hunts with Indians and Dyaks, of silky anteaters and flying lemurs, of cat-bears and kinkajous, and always we came around to yapocks. As we sat there by the little river, the light softened and greyed with mist, and a complete circle of songs closed in upon us. They were antbirds, but wholly invisible in the dark undergrowth. The sweet dropping notes trickled continually, as if a score of tiny rills were tinkling down rocky ways toward us.

Upstream a wide slope of mossy rock stretched from bank to bank, with the dry season brook confined to a deep, even-sided trough down the centre. Everywhere were geometrically rounded potholes bored straight down into the rock, many dew or spray-filled, to the rims of which came occasional green midget kingfishers, who clicked at us, then dived, caught something too small for us to see, and finally flew on around the bend.

As I looked, a long-tailed lizard rushed out from shore and stopped, body flat, head and neck high arched. It made its way with quick sudden rushes across the open, snatched some titbit now and then, and, to my delight, became a biped at each rush. The first step took it up on finger tips, and then a short distance farther its forelegs rose clear, folded close to its breast, the tail upraised, and I saw before me a tiny dinosaur speeding toward the opposite bank.

With a warning grunt my companion reached slowly for his gun, looking behind me and downstream. Like a well-trained setter, I slowly flattened, rolled over and froze, and saw a good-sized, stoutly-built otter nosing over the open rock where we had passed. She was suspicious, and with dim, water-flattened eyes looked in our direction, striving to make us out. We secured her, *Lutia repanda*, and found she measured thirty-eight inches in length, full-grown but not breeding, with long flattened tail, and most excellent fur for a tropical otter. I traced her back and found that while she had slid gently down a steep bank, yet further on she had been in the river bed. In fact her food showed that she must have been feeding for a considerable period of daylight before we saw her at six o'clock. She had caught many small fish, mostly the dark mancholas, so abundant here, together with many shrimp and catfish.

The subdued report of the small collecting gun had no effect on the antbirds' chorus, but as the dusk settled down, they thinned out, not weakening with distance but silencing one by one. Before the last ceased, a new sound arose a few feet away, an ascending, liquid *whoooeep! whoooeep!* and the burden of song passed from the sleepy antbirds to the awakening world of small frogs.

Even the relative proximity of the canal and the batteries and human beings was forgotten when some jungle cat—ocelot or jaguar—snarled twice upstream, and this was reinforced by a muttering roll of thunder and a stray, momentary breeze which shook every leaf. In the last ten minutes of daylight I saw two nests overhanging the water. I reached the first and found it a hummingbird's old home, fashioned of cottony seeds, which were fraying out, and drifting a second time to earth. The other was a yellow flycatcher's pinch of tide-drift and it was shaking when I first discovered it. Before I reached it, a lizard shot forth and along the branch. It too was empty of eggs and old.

Night came with a rush and we hastily made all our preparations for darkness. My jack was electric, my companion's an acetylene bull's eye which buckled around his forehead.

After this our world consisted of the tiny circle which our shaft of light picked out on water, rock or foliage. All else was impenetrable, absolute darkness—apparent to us only through its feel, its sounds and smells. We had no lack of variety as regards the feel, for we worked up and down short stretches of two streams and connected each likely yapock zone with longer or shorter trips through dense jungle. The potholes, so distinctly outlined in the daytime, now merged with the slippery water-covered stone and each step must be felt, or chances taken of a plunge. When once wet to the skin we cut small holes in our pockets and haversacks to drain out the water and sought only to keep jack light and gun above the surface. Later when it began to rain we ceased to worry even about that. To all intents and purposes we became yapocks ourselves, and however little I know about them, I at least have shared many of their feelings. The air and water were of equally pleasant temperature, every moment was filled with driving interest, and every coming second with potential discovery. During one spell of watchful waiting I tried to think of some place in the world more preferable,—and I failed.

The pothole contingency was objectionable only because of the uncertainty of depth and diameter. If one plunged deep into a narrow hole and fell sideways a badly sprained ankle would result, and the inability to arrange correlation of muscles for a drop which might be six inches or six feet was trying. Early in the night I stepped out of the shallows on to a short, half submerged log, when the log turned into a steel spring, which flicked me backward into a seven-foot pothole. When I climbed out, my companion said that to the best of his knowledge the youthful crocodile which had exploded beneath me was still barging full speed downstream, even more startled than I. As for me, in spite of many past experiences,

I had still been trying to keep my solar plexus dry, using much valuable effort to this useless end, and now I was equally and pleasantly soaked all over, and throughout the rest of the night wasted no more thoughts on adventitious dry spots.

Just as the source of all our visual knowledge came from the narrow circle of direct lighting, so the exact whereabouts of the living beings of the jungle was indicated by reflection, —the pigment and rich plexus of blood vessels of the retina glowing like fire in the light of the jack.

We had hardly begun to move silently downstream when things occurred and were seen, quite otherwise than we diurnals know them. The change was casually initiated by the moan of an owl, the last sound before the young "croc" skittered me into the pothole. Beyond the upper part of a tinkling riffle a single glowing eye shone out, which never increased to two. After much peering, being rather nervous from my upset, my companion fired along his light shaft and bagged an unfortunate marine toad, which with a companion had come out upon the moss to seek for stray bits of food. His chagrin was great, for he had hunted with jack light over this region many times.

Ten steps more and I aimed at two points of light moving steadily over the rocks. Rushing over to the spot, and miraculously escaping a whole nest of potholes, I found that I had sunk still lower in the jack hunters' scale, and had dismembered a huge crayfish, which had been sculling along on the surface of a shallow pool.

We squatted silently and waited. The pool held four mancholas, elongated, mottled fish, sleeping quite soundly with wide open eyes, their bodies resting on narrow ledges, often with head and tail unsupported. Creeping over the bottom was a host of small shrimps, each with a glowing pair of stalked eyes, the pool appearing full of restless rubies. Close behind me, a pothole re-echoed with a cheerful *whoooeep! whoooeep!*

and the jack turned gradually upon them caused no disturbance—it was to them merely the moon risen before her time.
Flattened upon the surface were two male whooping frogs
Eupemphix pustulosus, pale brown, and each, physically and
mentally, nothing but a voice. Their bodies were distended
until they looked like translucent, rounded jelly-fish. At
two-second intervals, an enormous vocal sac bladdered out
from the throat, almost equaling the diameter of the entire
distended body, producing in the instant of its expansion
whoooeep! and with its deflation, two or three curious, short,
metallic, creaky sounds as if some part of the mechanism
needed a drop of oil. The *whoooeep!* was the thing, the other
merely an incidental noise. Sitting quietly upon a leaf was
the object of their efforts,—a small female, listening, stolidly,
uninterestedly. In the same hole were many full grown tads
and a single, long-tailed polyfrog. One great tadpole blundered up to the surface for air, upset the female, who leaped
against the pothole wall, the males collapsed and dived and
the batrachian tableau was over.

In other potholes were small piles of froth and gilled tadpoles, but the majority in both rivers were full-grown tadpoles,
ready at the first downpour of the rainy season to take care of
themselves. The frog chorus kept up steadily until the rain
began to fall, when it gradually died down. After a few
minutes in one of my vials the male frogs turned olive green
and showed a sloth-like dorsal mark, a black circle with a
heart of greyish white.

Before we had moved from the first spot we had the thrill
of the trip. What had been in early evening a clear view
upstream, was now only a black well, except when our slender
shafts searched out rocks and riffles. With no warning of wind
or distant thunder, there came five prolonged flashes of lightning. As if we had been expecting it, our eyes were focussed,

our direction was perfect, and there, part way down the riffle were three opossums. One was a smaller edition of our Virginia species, and the other two were what we had come to see—yapocks. The smaller had his back turned toward us and every flash of lightning showed the unmistakable pattern. The larger was in profile, sitting up on her hind legs, eating something, probably a shrimp which she held in her 'possum hands. About the third flash we rose and began creeping upstream, and by the fifth the animals had seen us and started for the bank, swimming a shallow pool beyond the riffle. When we came within shooting distance, nothing was visible except the miserable white opossum, snarling as it backed slowly away over the rocks, avoiding the water, and showing not a fraction of the timidity of those we desired so much. In addition to our disappointment, at the same moment we both fell into separate potholes, and then I climbed out and plumped into another. So we sat down to drip and drain and to post-mortem what we could have done if we had brought a heavy gun.

At the next flash of lightning, I went on upriver alone and found the pools so deep and continuous that I had to make my way along the bank under the overhanging foliage. Reaching out to help myself by swinging on a branch, I was startled to have part of the dry foliage tear out of my grasp and turn into a large lizard which scurried faster up the bark than my flash could follow. Examining the rest of the plant showed five more basilisk lizards, out of reach, all asleep, or just awake and watching me, stretched out along the drooping twigs. Like *polychrus* in Guiana these, and others which I saw during the night, were all lying head up, the long tails indistinguishable from the hanging ends of the twigs, the great expanses of casque, and frill and fins, all merging with the many dried and crinkled brown leaves still lining the branches.

I think the flying dragons of Borneo are the only lizards I have ever seen more spectacular than these basilisks. At night, at least, they were of a rich, warm, leaf brown. From the back of the head rose abruptly a dark, backward curved casque. The body supported a long, fin-like crest, almost as high as the body was deep, and there was a third great dorsal fin on the proximal half of the tail.

The smaller bipedal lizards were wholly different in pattern and colouring, dark grey with yellow and white markings, yet they had the beginnings of the casque and the high fins. I saw none on the lower trees and bushes, except that in the fly-catcher's nest, but they were quite common in daylight, feeding at the edges of the shallows, running, or swimming across the pools.

A bright pair of eyes in a place where none of the rest of the creature was visible was too tempting to resist. I was so close that I fired well to one side and ran up just in time to catch a white-eyed opossum by the tail as she was ambling off into the brush. One shot pellet had slightly lamed her, and made it possible for me to reach her. I bagged her, but not before she had well bitten me, climbing unexpectedly up her tail to reach my finger. The next day she had completely recovered, fed heartily and snarlingly allowed me to examine the two large infants which remained in her pouch. Another had dropped out when I captured her and she refused to reinstate it.

As I squatted by the pothole I was aware of a heavy scent as of jasmine, which spread without any wind, and passed, with the atmosphere still breathless. Next I could distinguish 'possum scent and again the sweet perfume. Twice before during the night I had detected sudden, violent, very individual odors, as from invisible flowers which had just opened aloft.

Back I went to our first stand and then by agreement I started across for the second river. A last glance around showed a pair of shining eyes close to the riffle. No manipulation of my flash revealed any other hint, but the pale pink colour at last made me decide to investigate rather than shoot. For my pride's sake I was glad, for the eyes did not move as I approached and I finally pushed the flash within a few inches of a great grey ctenid spider, flat and with widespread, spiny legs, resting on the moist rocks for heaven knows what provender. Except for the slight difference in color, at a distance of fifty feet, the jack light showed up two of his eight eyes, quite indistinguishable from those of an oppossum at twice the distance.

My companion returned. He had seen another yapock and had overshot it, but had secured, to his disgust, an evil-smelling white opossum, which I kept as consolation. A hundred yards of crawling by intermittent flashes brought us to the second stream. We peered out from a screen of leaves, which in daylight were doubtless green, and close in front another white opossum quartered the rock like a huge albino rat, snuffling here and there for something edible. Beyond were two spiny rats, travelling slowly side by side, their eyes small and less distinct, their dark bodies curiously visible in our low-swung light.

The life of the yapock, *Chironectes panamensis*, is as great a mystery as its haunts are difficult of access. I found three separate entrances to burrows, easily identified by the numerous webbed and fingered tracks. The holes were about six inches in diameter and all well up under overhanging roots and banks, out of reach of all but the very highest floods. They led either straight in or slightly upward, but we could not get to the ends on account of the interfering roots and stones. Between holes, and for some distance up and down stream, the animals had well-trodden runways, which occasionally turned sharply upward and led for a short distance through the undergrowth.

On one side of each entrance was a low mound, pounded hard by the pressure of the animal's feet and body, and scattered about, sticking in the soft mud, were veritable, miniature kitchen middens,—remains of giant crawfish, *Macrobrachium jamaicense*, ancient heads of fish with occasional bits of vertebræ and fins. The flowing stream in front (and the holes invariably faced pools of considerable depth) provided an abundance of food, much of which was apparently brought up and devoured near the burrows. These opossums have as many as nine young in a litter, and even well-grown ones have been seen in close association with one of the parents. It is a mystery how the young can stand immersion, when the mother goes on her fishing excursions, unless during the helplessness of the early period of their existence she confines her hunting to shrimps in the shallows. Few naturalists have ever seen these creatures alive and my glimpse, meagre though it was, made me count my nocturnal adventure a success.

Finally came the wind and in another five minutes the rain. For a time it fell heavily and we saw an end to jacking, but after some glorious lightning and thunder, and another wild burst of wind, came steady, quiet, misty rain which interfered in no way with us. As we left the shelter of the bank, a great tree fell on the other river, and we wondered whether it was the dead giant which had overhung our dining rocks, and upon which we had commented, wondering how many centuries it had stood there, perhaps since Morgan passed, and how many minutes or months or decades it would still resist gravitation.

The shallows of the second river alternated patches of soft green grass with smooth rocks, and seemed singularly free from potholes until my guide got careless and went up to his waist in

a narrow one. We had sat motionless on a soggy bank, flashing occasionally, when we both thought simultaneously of the
two yapocks on the first river, and rose to make our way to
shore. We were in an inch of water on a wide expanse of rock
as flat as a table, when my foot struck against a bit of stick.
Flashing downward perfunctorily, I saw a heavy-built, five-
foot snake in the act of changing its course. My slight blow
had turned it, and it was partly between my feet, winding
slowly around toward its tail. The first glance showed me the
swollen jowls, the x-shaped marks, and I thought fast. My
first impulse was to press my flash down upon its neck, but the
electric had been flickering during the last five minutes and if
the water quenched it I would have to be very certain of
my neck hold. My bags and haversack were far off on the
other river and I remembered the thorns and slippery banks
between; besides we were out for yapocks. So I decided against
toting a living deadly bushmaster snake back, and called
out to my companion who had the gun. He turned ready for a
yapock, and was astonished to see this great yellow and black
form undulating toward him. Potholes did not concern him
as he turned its flank and put two pellets through its neck.
This rendered the snake helpless, but its fangs were a trifle
over an inch in length, and I fastened the head very tightly in
my handkerchief before I wound up and tied its sixty inches
around my hunting belt. And on my way back, whenever
its coils were unusually lively, and I simultaneously ran against
a thorn, the combination tended unduly to excite the imagination, and I was not sorry when I could cache the big master
of the bush farther from my skin. These venomous snakes
are apparently rare in this region, and this was the first which
my companion had seen on the Isthmus in eight years of
hunting.

When we returned we could see nothing but the more

common opossums and rats, although a paca once ran across the shallows. It was half-past two and although the rain-mist still came down, yet the diffused light of the hidden moon showed our figures to one another, which meant that the wild creatures could see us distinctly.

Disappointedly we packed up and squeezing all the water possible out of our clothes, we began the slow, steady, crawl and creep and scramble which is the only gait to be sustained through the long miles of dense thorn jungle. We saw not one mosquito the entire night, but from six to eight o'clock, and for a half hour later in the night, our jacks made a veritable mecca for all the sand flies in the world. During that time we breathed and swallowed and felt nothing but the millions of pests,—eyes, ears, and nostrils were filled with them, and it was difficult to keep constantly from coughing and sneezing and so frightening the animals. When they did go, it was in the fraction of a second; one moment there were untold myriads, the next not one was anywhere near us, nor in the most distant part of our searchlight's path.

Near the end of our journey, after passing the sentry, we awakened our motor boatman, and pushed off. Five minutes later I caught a white-eyed opossum with three young which had also gone to sleep in the boat and was doing a marathon around the gunwale of her suddenly made island. An hour later the yacht loomed up in the mist and as I went down the companionway to my cabin, the first streaks of the new tropical day were spreading over the great canal.

The trip through the Canal means a thousand things to as many minds. To me this fourth crossing brought surprise at the multitude of slender smoke spirals, indicating tiny negro holdings, where the jungle was being cut and burned for cultivation, on islands and on distant mainland slopes. The foreground everywhere was a forest of drowned trees, reaching up whitened, hopeless branches from the areas flooded years ago—havens for woodpeckers, termites and epiphytes. Large turtles basked on sloping tacubas, egrets trod the white water lilies and jacanas cackled amid great hyacinths. I wondered at the total lack of green scum, even in sheltered backwaters. We passed close to Barro Colorado, the four mile island in Gatun Lake, which has been set aside for a biological station— an isolated bit of primitive jungle, said to abound with deer, tapir, ocelots, serpents and monkeys.

At Panama I made the acquaintance of Sindbad, and as I found him fallen among low company I rescued him and added him to our personnel. Sindbad was only an alias—his real name was *Ateles ater*, and less formal acquaintances spoke of him as a black spider monkey. I found him tethered in a narrow street, with a line of Panamanians ten feet distant absorbed in the jolly game of seeing who could spit upon him from that distance. There were two small palms in sight and Sindbad's whole attention was fixed upon these; his dignity, his abstraction from his immediate environment was complete. The native competitors must have had much practice—they were very skilful, and while I am not usually a kill-joy, yet I stopped this out-door sport by purchasing Sindbad then and there. As one long arm went around my neck and he looked inquiringly into my eyes, I almost agreed with Bryan—in fact I am quite sure many human beings have not the slightest kinship with such a banderlog gentleman as Sindbad! He proved to be gentle, very self-contained and with a delightful sense of humour and love of play. Only when suddenly alarmed did he ever threaten to bite, and after punishment never harboured a grudge. He walked upright constantly, balancing with both arms raised high like a gibbon. When walking on all fours the fingers were always doubled under, down to the knuckles. In fact they were primarily hooks for catching hold of branches, even the thumbs having been completely lost in the course of adaptation for a hanging and swinging organ.

Sindbad was the only monkey I have ever known who laughed. Many monkeys grin, like dogs at play, but even when at the climax of a game, they seem never to make a sound. But Sindbad really chuckled when tickled, or when he was wrought up with the excitement of the game he invented, of leaping down upon one from a height, and then trying to escape before being caught and rolled over. The chuckle was a series of jolly sounds which expressed audible mirth in a most astonishingly human way.

When curious about anything which he did not understand, Sindbad uttered a single, high, clear note, and when he saw food approaching, he gave a series of shrill chuckles, quite unlike his hysterical giggle when playing, but obviously related in timbre and expression of satisfaction. He enjoyed teasing the puppy Bonzo, tweaking his ear, or tugging at leg or tail, and then allowing him to get almost out of reach before hauling him back, usually by the tail. When the pup ceased to consider it play, and growled, the monkey at once respected the change of mood and stopped his teasing. This spider monkey showed far more intelligence in regard to his cord than the cebus and other monkeys we had, usually holding it clear of entangling objects by a twist of his tail. When the leash was entwined around a chair leg or stanchion, he would try several ways of freeing it, and eventually follow it back from his belt along the line of the rope until quite clear.

For a time I fumbled in my mind as to the charm of the isthmus, the canal and the wild islands toward which we were headed. It seemed that the great northern city I had left could offer nothing exactly like it, and with this thought came the truth in one word—contrast. Neither marvellous music, nor superlative painting, nor even a great book, heard, seen or read in the city could quite reproduce the emotions of these places. Two things in New York came to mind as perhaps worthy of comparison; the first sight of the five-foot skull of Baluchitherium being slowly chiselled from its matrix,—its second resurrection in an intellectual evolution, renaissance, with the mind of Henry Fairfield Osborn directing and interpreting, as the high priest of some incomparable religion might bring to pass an astounding miracle; or again on another day, when I left the hurry and rush of avenues and watched, in dimness and in quiet, the reconstruction mosaic of the jewel casket of Princess Sat-Hathor-Yunet which had held those superb tears of gold and carnelian. These two had no direct connection, either mental or chronological, with motors and subways, newspapers or food; they stood apart, isolated and startling, like lightning at night, or this cleft between the continents.

On the isthmus there was nothing Victorian, no linking thing, no gradation: all was of early beginnings, primitive, or else stood for the acme of human accomplishment, and as these passed one after another, there arose a peculiar glamour, a new joy of life which touched unusual depths, plumbed realities, dimmed comparatives. And this was to be the key to the whole trip; not gradual immersion in far distant evolutions, with impersonal interests in primitive nature, but make-and-break flashes of past and ultra present, action and reactions of primitive and sophisticative.

Sharks swam around our keel in Colon waters with the still unossified skeletons of early Devonian, the piles of our wharf were furred with tunicates which almost achieved and then lost forever the vertebrate goal, while our ice machine whirred, planes snarled past, and nothing whatever was wrong with the picture of our perfectly appointed and served meals.

PANAMA AND YAPOCKS

Later we left behind the torture chambers of the Inquisition where, in days of olden Spain, men were brought to religious salvation by dismemberment, and we steamed beyond the spillway where ospreys crucified unsuspecting fish to keep their young from starvation, and turned abruptly to a very usual man who by the quick twist of a tiny handle changed an isthmus into a strait and made islands of two continents. I saw human handiwork outfly the cormorants but falter before vultures; at the farther end of the canal I saw Northern Lights creditably imitated by seven score battleships, while at the heart of all was the most primitive, except two, of all instincts.

These are but shadow phases of the many realities which successively arose, impinged and passed, as we went on our way. And now, still another contrast leaped to mind as I thought back. We had spent five days on the Isthmus getting repaired and iced and fed and coaled and watered for our long stay in desert islands. Thousands of casual memories would rush to consciousness if I permitted, but this phase of the trip was to me not the place of wonderful swimming baths, delectable rum punches, quaint coolie shops and splendid hospitality, although right worthy in these and many other ways. But it was the single place on earth where, in tissue paper, in an old pasteboard box, in an ancient chest, a hunter treasured five and twenty skins of the yapock or water opossum, hoping in time to have enough for a coat for his buxom wife. This sounds incredible to the scientific cognoscenti, but it is true. And indeed why not, for I have heard a song of ancient India jazzed in a cabaret, I have seen a Watt's advertising a soap, I have heard Dunsany praised by one who had no belief in fairies, and I have seen a woman of my own race push past the royal guard and thrust a bundle of tracts into the hands of the Dalai Lama!

GALAPAGOS

But after all, Colon and the Canal and Panama were merely taking-off places, and their interests suffered sorely from our absorbing anticipation of what was ahead. But civilization was jealous of our inattention and would not let us steam off into the unknown without a final effort to arouse our wonder and enchain our memory. As we slipped swiftly between the isles of pearls, searchlights from nearly one hundred and fifty ships of war made magic in the black night. A maze of beams of whitest light swung and wavered, winked and curved over the bay, shattering against clouds or stretching up, up, to what our eyes registered as infinity. And yet my mind unconsciously refused for long to pay homage to man's achievement—the first simile that came to mind was of a galaxy of shooting stars. Then as we left them behind they dwindled to a meadowful of fire-flies, and finally the last one flickered—less than the glow of fox-fire from a rotten leaf. Another moment and man and his mightiest engines were non-existent, and only the black water lapped against my port-hole and the darkness of the first Pacific night closed unbroken around us.

Fig. 6

Yapock or Water Opossum

Chironectes panamensis Goldman

One of the rarest animals in America

FIG. 7

FIRST VIEW OF INDEFATIGABLE

The clouds are gathering for the regular morning's rain in the centre of the island

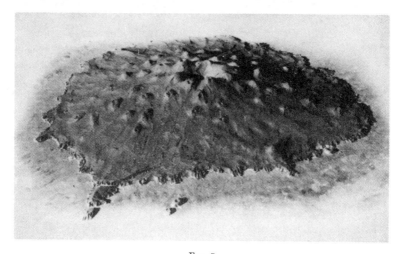

FIG. 8

MODEL OF INDEFATIGABLE

Showing the hypothetical central crater, and Seymour, Daphne and Eden Islets scattered
along the foreshore

CHAPTER III

INDEFATIGABLE—MY FIRST WALK

FAR out in the Pacific, and yet not in the South Seas, is a cluster of cold volcanos which, over three hundred and fifty years ago, was known as the Enchanted Isles. The Seventh Lord Byron has seen them, has stumbled over their rugged lava, and has been astonished at the tameness of the birds; Robinson Crusoe was brought here by his buccaneer rescuers and must have rejoiced that his luck had not cast him upon these inhospitable shores.

As we shall see more in detail later, the historical relation of man with these islands has been through thirst, war, tortoise meat, and mystery. The first hint that we have, coming dimly through the earliest years, is that the Inca Chief Tupac Yupanqui, grandfather of Atahualpa, sailed out into the Pacific long before the first voyage of Columbus, and discovered a mountain of fire which he named Ninachumbi. But if this, as seems probable, was one of the Galápagos, he must have visited other islands as well, for he brought back negroes and a throne of copper. The cold Humboldt current sets strongly northward from Chili, and turning westward at Peru, leaves the vicinity of the mainland and debouches into the Pacific, washing the shores of the Galápagos. Its temperature is from 15° to 20° lower than normal equatorial water, and in times past it undoubtedly bore the penguins and southern seals which are now found on the islands. It was this current, combined with prolonged calms, which also carried the Bishop of Panama against his will to the Archipelago in 1535 and gave us the first definite account of the islands, the tameness of the birds, and tortoises so gigantic that men could ride upon their backs. The Spaniards almost died of thirst but lived by chewing the thick pads of cactus, and finally found a little water cupped in hollows in the rocks.

Then came buccaneers from conquests in Peru, to careen their ships and wait for a favourable opportunity for a fresh exploit. Out of all the generalities comes now and then some definite little incident, as when Diego de Rivadeneira, harassed by thirst and beset with calms, threw a young deck-boy into the sea to seize a great turtle which had been swimming around the ship. At this moment the wind suddenly freshened and the vessel moved off so rapidly that it was impossible to return for the unfortunate Spanish boy, who was last seen clinging to the turtle and waving hopelessly. Woods Rogers buried here an enormous booty which he took from the Viceroy of Peru, and it is believed that the Incas fled hither from the attacks of Pizarro with great quantities of gold and jewels.

With the war of 1812, the Galápagos entered into somewhat more legitimate history and Admiral Porter has left us a long and interesting account both of his adventures and of the natural history of the Archipelago. His interpretation of the name Enchanted Isles was the great difficulty in getting away from them, calms and changeable currents often delaying vessels for weeks within sight of their last anchorage.

The thrill which came to us at dawn on the fourth day out from Panama is indescribable in words, yet lies so lightly and abidingly in memory that it can be recalled by odor, sound or abstract color. From the bridge of the yacht I perceived a low cloud on the southwest horizon undissolving with the other horizontal mists gathered from sea or sky during the night, which at day-break hesitated between allegiance to the sea as rain, or to the sky as clouds.

At this moment of dawn-grey light I forgot the clouds for a mighty school of dolphins, rolling so lazily, breathing so leisurely, that they must all have been sound asleep. Like the shaft of an engine barely turning over the propellers, they came up, curved and dived with a single magnificent motion. All in the same instant they caught the sound of the yacht and if there were two hundred in the school, two hundred leaps came as one, and like the fragments of a bomb they churned off and bounded high in the air, turning at the first touch of water and heading straight for us. They passed just ahead, still leaping skyward, flippers spread, slapping back with cracks like rifle shots.

The upper edge of the strange horizon cloud sloped upward, which is unusual with clouds, and at last the solid mist shredded away from the tenuous peaks, and I saw land— a crater—an island—Galápagos—Indefatigable itself— World's Very End. It was so huge, so unlike any descriptions we had read, that for long we were uncertain of its identity, until we had sighted and checked off the Gordon Rocks, Barrington, North and South Seymour, Daphne, Jervis, James, Duncan, and Albemarle.

On to the northwest of Indefatigable we steamed slowly, wishing for a dozen eyes, so filled was the sea with strange, living things. For three days we had seen hardly a feather, and no aquatic life, and now, from appearances, we might have been near Mount Ararat on the twenty-seventh of a certain second month. Petrels fluttered here and there over the waves like parti-colored butterflies, Shearwaters with steadier wing-beats recalled martins, and two albatrosses floated over the yacht, an unexpected sight as I had not thought to see them at this time of year. Sharks swam alongside, two fur seals turned and looked us over from wireless antennæ to waterline—the only two I saw at the islands. The most spectacular sight was a giant sailfish, at least eight feet in length, which leaped three times in succession, bringing down his mighty blade each time in a different direction. I have since learned that this is a unique record for this part of the world, but there can be no possibility of error. Mr. McKay and I saw this one distinctly; on another day he saw another and I two smaller ones. Pelicans and frigate-birds flew around and around us, and boobies swooped low to quench their curiosity. The rumbling clang of our anchor chain rang out like a desecration, and we came to rest in Conway Bay, the very harbor where the most famous pirates of all time had anchored. I looked around at the island spread out before me, listened in vain for any sound from shore, and let the fact sink deep within me—in my turn I had come to the Galápagos and another dream of my boyhood had become real.

Charles Darwin spent over a month in these islands, and from observations on the varying forms of bird life he derived perhaps the first inspiration for his Origin of Species. From that day to this, the islands had remained almost unchanged. At rare intervals a schooner passed or was wrecked on some outjutting lava reef. But for month after month, and year after year, on most of the islands the reptiles and birds and sea-lions knew only each other's forms and alone watched the sun rise and set. Generations of these creatures came and went without ever seeing a human being. On the twenty-eighth of March of the present year I slipped overbroad from a motor life-boat and waded ashore through crystal clear water. A little duck flew down, paddled and waddled to our very feet, looked up into our faces, and quacked in curiosity and astonishment. I knew it for the fearlessness of the Garden of Eden, the old tales of Cook and Dampier come true again.

And now in a few thousand words I am to try to reproduce some of the beauty and interest of these islands. As well bottle grains of sand and drops of water and show them as a Galápagos shore! A photograph moves one less than a painting, so a word description can hope to succeed only by its very imperfections—with vast gaps and hiatuses to be filled by the imagination of the reader.

As I touched land a curious feeling of hopelessness came over me. When I enter a tropical jungle I feel only elation, and desire to choose some tiny thread of a problem and try to follow it in its devious path through the inexplicably intricate pattern of the whole mass of plant and animal life. There is no chance for a single brain to grasp the inter-relationships of even a square continental yard, and so there is no sense of responsibility, of the possibility of an opportunity slipping past forever. But here there came the feeling of despair, for these islands were isolated, their life was comparatively meagre and, given time, concentration and ability, it was conceivably possible for a human mind to perceive, to unravel and to piece together again: finally perhaps to reweave the simple warp and woof of past years,—the story of the rocks, of the volcanic activity, of the terrestial migrations of a mainland remnant, or of arrivals by currents of wind and water if the Galápagos have risen from the bottom of the sea as single, inconceivably tall, volcanic peaks, eternally isolated, ocean-washed since the beginning.

Scientific expeditions had visited this Archipelago ever since Darwin's trip eighty-eight years ago. One remained six, another seventeen months. Able men had examined, collected and described thousands of specimens, and had arrived at antithetical theories as regards the origin of the islands. I had the advantage of all this accumulated knowledge, although some of it was singularly barren for our present use. A very thorough list of plants has been published, giving exact distribution, elevation, ecological summary and relative abundance, but, except for the technical names, no hint as to the growths themselves. Professor Wheeler and I were very anxious to identify certain plants because of their relationships to animal organisms, but found it impossible from published work on the Galápagos. *Bursera* was one of the most abundant plants, but we correlated it with four entirely unlike growths before we could decide. If explorers would give just a thought to others who might come after, and take the time to write out the simplest kind of a key, it would be a kindness beyond gratitude.

But I knew that my visit must be measured by days, almost by hours. I knew that sleep and meals would claim some of these precious minutes and black despair possessed me. My eye wandered over the vast island in front of me, over the miles of slow-rising volcanic lands, culminating in the unexplored craters in the centre, thousands of feet high, now half hidden by walls of rain, and hiding we know not what wonders. And then I turned and looked back at the sharp ridge of little Eden, at the clustered sculpture of Guy Fawkes Islets, and of the dim, distant little island crater of Daphne Major, and I knew that my only hope of getting anything of clear-cut value would be to concentrate my scanty hours of observation on the lesser islets, almost rocks, which swam in the quiet waters in the lee of great Indefatigable.

A first walk in any new country is one of the things which makes life on this planet worth being grateful for, and in a wonderland such as this it is an event so absorbing, so replete with the impinging of a myriad new impressions that any coherent account of it is almost impossible. That was the case with my first stroll on Indefatigable, but in spite of it a score of things stand out with astonishing clearness. Almost before we reached dry land I believe one of us pointed a camera at the absurdly tame little duck and pressed the button without removing the slide. Then we came down to earth, carried all our dunnage ashore and put up a big canvas fly for general headquarters and shade for the artists.

By sheer good luck we had landed upon a most charming little beach which we named Harrison Cove—a crescent of white sand tipped at either horn with a reef of jet lava boulders, and enclosing a lagoon of clear emerald water. In the sand were rooted two great, dead, bleached, straggly bushes, landmarks for a long distance. Months from now, a taxi-cab driver in New York was to recognize them as marking the place where his pal had died seventeen years before.

The beach was exactly what I always expected a buccaneers' and shipwrecked mariners' beach was like. Henty and Kingston, Stevenson and Conrad had pictured it in words, Howard Pyle had made it live in two dimensions. I appreciated it all the more because of the many times in past travels when I had been disappointed in tropical beaches —finding them as barren of strange shells and as anemic in color as a Coney Island or Margate strand.

Here, generously scattered over and in the soft, fine white sand, were shells which I had not seen since, as a boy, they came to me as treasures in barter for birds' eggs, stamps and cigarette pictures. Here were cones, turrets, conches, glorious murex, chitons and rare colored cowry shells, and I found myself on my knees, picking and choosing, filling pockets and handkerchief and becoming more and more miserly and avaricious by the moment. Then I looked up and there, almost within arm's reach, were four birds, two oyster-catchers and a pair of grey Galápagos lava gulls watching me. I was a sea-lion—that I could almost read in their trust—but I was not an ordinary sea-lion—that was evident in their concentrated interest. The oyster-catchers, although bearing the name *Hematopus galapagensis*, and being residents of the Archipelago, are almost identical with the birds of lower California, and in their black and white plumage and scarlet beaks were colourful inhabitants of our little cove. And the gulls in their beautiful deep greyish brown looked like graceful bits of animate volcanic ash. They were *Larus fuliginosa*, a native, found nowhere else in the world. During all our stay in Conway Bay these birds were our constant companions. Even before we did any seining they trotted up and down the beach after us, watching us with never-flagging interest, as we sketched, pursued butterflies, photographed and botan-ized. Sometimes another pair of gulls would appear, but the first four birds were the real tenants of Harrison Cove. When we began to draw the seine, then their enthusiasm increased to keenest excitement. They fairly hopped up and down as the fish began to leap over the net, and we flipped them small tit-bits which they ate until we began to be alarmed. With the tail of the last one swallowed still protruding from their beaks, they would make futile dabs at another fishlet flopping tantalizingly on the sand at their feet. After a few days if we came ashore with only easels, guns and cameras, they watched silently. But when the long poles of the seine were in the boat, the gulls would recognize them and greet us with rollicking peals of laughter, while the oyster-catchers ran about with half raised wings and cackled with anticipation.

But I left myself kneeling on the sand jealously scanning every inch. An ingredient new to me was innumerable spines of club-spined sea-urchins—millions upon millions of them,

Fig. 9

GALÁPAGOS MOCKINGBIRD

Nesomimus melanotis (Gould)

A young bird in the spotted juvenile plumage

Fig. 10

OYSTER-CATCHERS AND LAVA GULLS

On the beach at Harrison Cove, Indefatigable

Fig. 11

Indefatigable, Showing Distant Mountains and General Aspect of the Coastal Region

Fig. 12

Coastal Vegetation of Indefatigable

Cactus, Bursera and Cordia

purple and yellow, apparently indestructible. Later at Tower Island I found fork-tailed gulls making their nests of several hundred of these short, stick-like structures, and one day on this same beach I came across four of our sailors using them for poker chips, so their utilitarianism is extremely catholic. One of the most beautiful things in death is the giant thorny lobster of the tropics. Each part of the great carapace and the sheaths of the legs, the eye-stalks, the huge thorns and cornices of the head are marvels of delicate carving and colour, and when death comes to this crustacean and the fishes and the scavenger mollusks and worms have made away with all his muscles and flesh, then the empty shell, as wonderful in carving as the Taj Mahal, is washed up and pounded to pieces upon the lava, and all the fragments scattered through the sand—a myriad mosaics of the most exquisite sculpture and with pigments faded into unnamably delicate tones and hues. As I casually unearthed some jewel of a leg-joint, well worthy of a setting in platinum, a slender rod splashed with mauve and crimson, with a galaxy of blue stars wound in a spiral about it, I realized more than ever what a casual thing is man upon the earth. For untold ages since thorny lobsters first crawled about in the waters of the upper chalk, perhaps sixty million years ago, beautiful detritus such as this has littered the tropical sands. Only by the merest accident had I stumbled upon it, and its shape become beautiful in my brain, its pigment colour in my eye. Bryan and his anthropomorphic isolation, the fear and the egotism of his aloofness from our lesser blood brothers, —how silly and childish this all seemed. Once we were taught that the earth was the centre of the universe; then that man was the raison d'être of earthly evolution. Now I was thankful to realize that I was here at all, and that I had the great honour of being one with all about me, and in however small a way to have at least an understanding part. I knew very well that my carnelian and sapphire lobster joint was not made for my special delectation, nor that of Bryan or any man, but that in spite of the millions of years which had passed since the crustacean's ancestors and mine were one, our ways had by accident touched again at last, the tropical crustacean to bring beauty of form and colour, I an humble appreciation.

For a few moments longer I kneeled and thought of the other trinkets hidden beneath me in the depths of the island sands—trinkets of gold and real jewels less beautiful but holding more of intrinsic wealth than my lobster-leg—coins hacked into eighths, with trade and water-worn imprints of kings and queens these many years dead, filigree earrings from Peruvian beauties whose very race has vanished, and the bones of men who died as only Howard Pyle has shown us.

I left the beach and walked up over the sandy ridge, and at once found myself in a wholly new land. Every grain of sand was left behind, and in its place was a charred and burned-out world. The first hour's walk gave me a panoramic view of a type of country and vegetation which, in general, never changed in any island. The exact abundance of plants varied, but there was little variation in the coastal lowlands, the semi-arid zone surrounding the central, high, humid area.

My first half hour's walk showed me why commercial man had left Indefatigable severely alone, why only buccaneers careening their heavy barnacled keels, and shipwrecked taxi drivers and hungry whalers and irresponsible scientists had been the only tenants—and these as transient as possible.

My first quarter hour of walking showed why the interior of Indefatigable was still unexplored. Like most of the other islands, its lower zone was one solid cinder, dotted with scores of dead, cold craters, with sparse vegetation springing from cracks and crevices and existing only by a never-ending fight for water. Every step must be tested, else a four-foot sheet of sliding clinker, clanging like solid metal, would precipitate one into a cactus or other equally thorny plant. A careless scrape of a shoe and the sharp lava edges cut through the leather like razors. Here and there a semblance of meadow offered a temporary haven. The rich, red, pasty earth supported a rather dense growth of coarse grass, *Cenchrus platyacanthus*, but after a dozen steps I shifted back, in preference, to the terrible piles of shifting lava disks, for each clump of pseudo-grass gave off at a touch a host of seeds, barbed and rebarbed, and the effect on clothes and skin was like a hundred fishhooks. When one of these seeds had worked inside the clothing, it meant blood from fingers and body to pull it clear, while one or more spines usually broke off, to work their vengeance at some later time, Never have I known a worse country for forced marches.

Indefatigable is rounded in contour, about twenty-five miles across, and rises very gradually to a height of twenty-three hundred feet or more at the great central crater. Rumour has it that there is a lake of fresh water in this crater, but no explorer has ever mounted above a thousand feet or approached the centre of this island, and lacking time to arrange relays of food and water, five miles was as far as I succeeded in getting. As I walked inland, I was surrounded by piles and hills, slopes and gullies, all fashioned of great sheets and disks of clinker, like thousands of misshapen manholes balanced on edge or thrown together as the last upheaval or earthquake left them. Huge cacti raised their oval pads aloft, angular and posed like Javanese dancers, and lower growths found somehow space for roots in jagged crevices, and nourishment from scant volcanic dust and

ash.   Two qualities stood out in the flora, the predomi-
nance of spiny and thorny plants and of those with thick,
fleshy leaves or stems.   All the coast lands were semi-arid,
rain falling only during two or three months, and then but
sparingly.   So it was easy to account for the plants which
hoarded water within their tissues, not for a rainy, but for a
rainless day, such as the cactus, *Opuntia* and *Cereus*.   But the
thorny plants, although most of them had been here so long
that they were peculiar to the Galápagos, yet had not lost
their guardian coat of spines.

The aridity of the coast lands had been so emphasized
in the various accounts I had read, that I was astonished to
find all this country surprisingly green, and almost every
plant in full flower.   Yet when I came closer and examined
the growths more in detail I saw at once the superficiality of
this verdant mask.   Only the terminal twigs, the end branches,
were in leaf, but as most of the bushes had a rounded, wide-
spread manner of growth the leaves gave an impression of
almost luxuriance.   Parting the green skin, one saw at once
the bare bones underneath—the wooded stems which were
whitened in the months of rainless sunshine, dead to all
seeming, sheltering deep beneath the armour bark the spark
of life, which another rainy season, perhaps seven or eight
months away, would arouse.

The guard against the heat and dryness of all the other
months had to be so complete, so hermetically perfect that it
could not be broken even by the softening rains.   Only the
tips of the branches and twigs were susceptible, and all the
plant's energy shot forth at these points.   So a veritable
Roman candle burst of leaves and bloom appeared at the
terminal joints, leaving all the rest of the plant apparently
as dead as a faggot.   When I walked among taller growths
I seemed to be passing through a maze of whitened dead

branches which bore tufts of parasite-like foliage overhead. The weedy annuals of course were green throughout their length, and these grew with tremendous rapidity. Revisiting Conway Bay eleven days after I first landed, I found that many lesser growths had in the meantime flowered, fruited and wilted.

January to April is the chief period when rain falls and then only irregularly. The season was late this year, and now in early April it seemed at its apex. Yellow dominated in the colour of flowers, the pale, strangely thick blossoms of the cactus and acacias, the delicate, scarlet-hearted flowers of the wild cotton and the *Cordia* whose tree-like bushes were often a mass of bright yellow. Curiously enough these were among the largest flowers on the island, the great majority of the inflorescence being very small and inconspicuous.

Even before I reached the top of the sandy beach I was greeted by the first land birds of Indefatigable—three mockingbirds which ran rapidly down toward me—one singing as his feet flew over the sand. They stopped a yard away and looked me over. Here I was in a strange foreign land, hundreds of miles out in the Pacific Ocean and yet anyone who had ever watched the mockers in a Virginia garden could never have mistaken these even at first sight; the same jaunty pose, tail slightly raised, the identical greyish brown colours, with dull white underparts and dark wings and tail with conspicuous white spots. The white in the wing was lacking and this islander had tried to disguise himself by drawing a black mask over his eyes and face, but even if he had clad himself in blue or gold, the moment he opened his mouth he would have given himself away. He was more original in his song, less of a *mimus* than our bird of the north, perhaps because of the poor quality of other songsters in these islands, but the spirit, the timbre, the relaxed mélange of all the notes in the world—liquid and harsh inextricably mingled—this was sheer mocker medley.

First, as always in days to come, I was surprised and charmed by their tameness. There were two old birds and a youngster in spotted baby plumage and when they stopped he came on and picked off a grain of wet sand from my shoe.

I went on slowly, and the two kept me company, and mounted the first bush to have a close look at my face. It was a bush worthy of these strange islands, a bush of ten thousand green thorns, growing close together in lieu of leaves. A human finger could not touch it anywhere without being pricked, yet the mockingbirds perched and hopped about as if it were velvet leaf.

Nesomimus melanotis is the name of this black-eared mocker and one of the first things I observed about it was the remarkable speed it could attain on the ground. As I look back, I remember these birds far more on the ground than in bushes. They walked and ran, they chased flies, and often leaped over obstacles without opening their wings.

With this in mind I was interested to discover that in relative proportions they differed from our mockers in a most significant way. They were considerably smaller than birds of the United States, with shorter wings and tail, while the bills of the Galápagos birds and their legs and feet were larger. On these islands, without tall trees, and with only a scanty growth of low bushes and shrubs, with most of the insects on or near the ground and berries often developing on the sand or lava itself, with a steady wind blowing and the ocean so near, there was no inducement for extended or frequent flight and all the reason in the world why they should do much and rapid running. The irregularities of the lava combined with the scanty insect fauna would make the securing of food a matter of greater difficulty and create a necessity for putting the bill to all sorts of unusual uses, such as prying into crevices, splitting small twigs, cracking open the shells of young crabs and mollusks and of the eggs of fellow birds, and even pulling the tails off small lizards—all of which I have seen them doing. So without going into the theory of cause and effect, or natural selection, or which comes first —long legs and terrestrial speed, or rapid running and long legs—the fact remains that these birds show a very significant structural change from continental mockers, and are very well adapted to a successful pursuit of food and happiness on these islands. Although from a wholly different angle and to an infinitely less extent, the change is comparable with that which I was to find in the flightless cormorants of Albemarle. To visualize the change in the wings and legs of the mockers we may compare it with human beings who if they should alter to a corresponding extent would lose two inches in the reach of their arms and gain five in leg length.

As an example of what has occurred among some other groups of birds on the Galápagos, I will continue with the mockers. They have been found on every island of any size in the archipelago and have been measured and examined by many ornithologists. The results are as follows:

Rothschild,—5 species divisible into 11 subspecies.

Hopkins–Stanford Expedition,—5 species divisible into 11 subspecies.

Ridgeway,—11 species.

California Academy Expedition,—4 species.

Knowing that no more material was necessary to decide this question (the California Academy Expedition alone took six thousand skins of land birds which have not yet been worked up), I collected only the individuals which were of especial interest to me, thirteen all told, together with nestlings and embryos. These came from Indefatigable, Albemarle,

James and Tower, and showed very plainly the differences upon which three of the forms had been based. The characters in question are the relative length of bill, wings and leg, the lighter or darker colour of the upper parts, the presence or absence of a moustachial streak and spots on the breast.

As regards the exact nomenclature of these forms, I am wholly in accord with Ridgeway who calls them species. Twice I saw mockers at considerable distances from the islands. Once a bird flew aboard when we were five miles off Indefatigable, but I saw none attempting to bridge the twenty to fifty miles which separate most of the group, and I imagine that it is seldom that any cross. This being the case, I see no reason for adopting subspecific names when there can be little or no crossing of the forms: especially as no linear classification, whether bi-, tri-, or quadri-nomial, can reveal the actual interrelations of these birds.

The most interesting aspect of this as with some of the other groups of Galápagos birds is the confusion arising in descriptions of the various insular forms—a confusion which is perfectly natural when we attempt to consider any linear relationship under conditions such as these. Without doubt we have here birds which once were living on a single land mass —connected long ago with the mainland, probably along the Costa Rican-Panamanian latitude, and later insulated on a single large Galápagos island. This in turn has, through subsidence, been reduced to a few volcanic peaks, on all of which *Nesomimus* still survive, and which have begun to differ slightly among themselves. This is due probably to causes far other than resulted in the general, generic wing reduction and leg increase, and causes of relatively less importance, whether initiated as internal variations or in response to external conditions we know not. But however these were stimulated and preserved, we have a splendid example of evolution in the three planes of space. Hence the facts so annoying to the taxonomist of a mocker in Tower Island, far in the northeast of the archipelago, which has a black moustache like the birds in Chatham, seventy miles to the southeast, a black back like those in Bindloe, thirty miles due west, a large beak corresponding to the beaks of the Abingdon birds, forty-five miles northwest, and yet which in all characters is almost identical with the *Nesomimus* of little Culpepper, a speck nearly a hundred and fifty miles to the northwest.

To give just one instance, the islet of Wenman is just southeast of Culpepper, and the mockingbirds which inhabit it are quite indistinguishable from those of the slightly larger island of Barrington which lies on the windward side of Indefatigable, a hundred and seventy-five miles southeast of Wenman, with Abingdon, Bindloe, James and Indefatigable lying between. It is important to know and note all these small differences, but their dominant interest is as examples of generalized variation in all directions, unrestrained by any important controlling factors of humidity or aridity, enemies or sex. If every individual of the none too numerous remaining Galápagos mockingbirds was shot and examined, I doubt whether we should be any nearer knowing whether the Tower bird is *Nesomimus personatus bauri* or *Nesomimus melanotis melanotis*, for it is very likely both. By which I mean that if we conceive that the ancestors of the birds on Wenman, Tower and Indefatigable were identical, and gradual subsidence separated the three, casual variation would go on with a slightly greater emphasis on one character or another, and with no possibility of indication of specifically separate relationship. A glance at the map will show what I mean without the need of additional instances, such as the moustacheless Tower individual which I shot, and the other which I found breeding in juvenile plumage, with more spots on the breast than the most maculated Hood Island bird.

Again I return, after having said my say about *Nesomimus* in general, to my particular first trio. During parts of the next two days I watched these birds and observed them most astonishingly to haunt the outer lava reefs within wave spray, and there to feed on flies and on small crabs and other crustaceans. And then I observed a thing of the greatest interest— the fact that very often the birds thus engaged teetered and tipped. All my life I have wondered why sandpipers went through this motion, every few seconds tipping up the whole body, as though it were hinged at the thighs. The head and neck rise and then the whole body bobs instantly forward, the tail rising in its turn. This habit gains in significance when we see the water-thrushes doing exactly the same thing—birds which are actually wood warblers, taken to a littoral life. The ouzels and smaller herons also bob, but less enthusiastically. Here were mockingbirds with a littoral habit developed probably only a short time ago, who occasionally teetered with their whole being in a fashion identical with that of the sandpipers. Two others of my party observed this very thing in mockers feeding along the edge of a land-locked lagoon. We have not the slightest clue as to the reason for this, it seems only in some strange way connected with a life of walking along shallow waters, whether mountain brook, millpond or the shore of the open sea.

I threw my mockers a few crumbs and they flew down at once to my very feet and ate them as though every buccaneer and visitor of past centuries had done likewise, and inherited memory was a real and operative thing.

Another family joined me as I walked slowly inland, keeping closely parallel in branches or on the ground, and sending to each other, now and then, a loud, high, shrill *peeent!*

the call or flock note I took it to be. Each individual came close at first and his bright, intelligent eye studied me—soon dropping behind or close beside. Then came the first note of fear or suspicion I had heard, and at the same moment something struck my helmet a slight blow, and clung. I put up my hand and touched a feathered leg, and then whatever it was flew and I saw that a short-eared owl had attempted to alight on my hat. The mockers had retired to a short distance, and the old birds watched the newcomer intently, and with wavering crest feathers uttered a short, harsh, scolding note. I never heard this again, for I never again saw the island mockers suspicious or afraid, and even now they did not act as though in great dread of this bird of prey. At another time I saw a small ground finch strike its head against a branch, in such haste was it to leave the vicinity of an owl. Those I secured were feeding on grasshoppers and crabs, but in dried pellets which I picked up under roosting places, I found the skulls of mice, small ground finches, two ground lizards, and the carapaces of giant centipedes. This owl is a native, *Asio galapagensis*, differing but slightly from the typical species which is found in almost all parts of the world. It varied almost exactly in the same way that the mockers did, having a larger bill and much larger feet, but shorter wings.

My owl flew a few feet away, perched on a slab of lava and looked with great, round, yellow eyes, unblinking even in the full light of day. I walked up within four feet, when it flew, softly and gently as a bird of cotton. Again it perched, and a big grasshopper immediately blundering against its breast, it reached down, caught the unfortunate insect, transferred it to its feet and ate it inconcernedly. When the bird flew again, a mocker at once scaled down and began eating a wing of the insect which the owl had dropped. These sound like trivial happenings, but the effect was tremendous, they appeared all so in keeping with the spirit of the islands,— a spirit of unexpectedness, of bizarre phenomena on every hand. I felt that if a diminutive mud-pie of a volcano should erupt at my feet, or if I should find a web-fingered pterodactyl or a real pirate's cache, I would not be more astonished than to see owls that alighted on hats, grasshoppers committing suicide, mockers playing sandpipers.

No sooner had I passed beyond the open sandy beach than little lizards began to dart along my path. As the mockers looked a trifle askance at the owl, so the small *Tropidurus* lizards watched the mockers warily. They were more afraid of these small birds than of me, and when I went on hands and knees I could almost get within touch.

These lizards, from four to eight inches in length, were marked with colours which I was beginning to expect in these islands, grey and black and scarlet-ash, and lava and flame— appropriate for a land where every hill was a volcano, every path a flow of lava, and with the plants growing from tufa beds and ash heaps. The male lizards were grey and brown above, mottled and banded with black, with the throat and underparts a mixture of pink, red and contrasting black. The females were usually more of a monochrome brown, with a brilliant slash of fiery scarlet over face, shoulders and sides.

They ran and frolicked about, running in and out of the lava crevices, with always a lookout for marauding birds. Of course they were absurdly tame and investigated all our luggage. When pursued, they would impudently pause until almost within reach, at the last moment going through a great show of intimidation, nodding the whole head and body violently up and down, and expanding the scarlet and black throat pouch to its fullest. When caught, they accepted confinement philosophically, and spent their time catching flies or taking

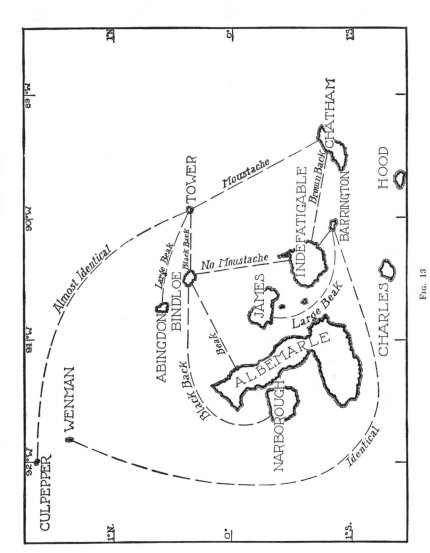

Fig. 13

Chart Showing Confused Distribution of Certain Characters of Galápagos Mockingbirds

FIG. 14

MALE TROPIDURUS LIZARD

On the sand, showing xerophytic vegetation

FIG. 15

FEMALE TROPIDURUS LIZARD

Wholly unlike the male in colouring
She is standing high, about to execute her bowing challenge

them from our fingers. At night, even in preference to taking shelter beneath an overarching stone, they burrowed as deeply as possible beneath the sand. In a small cage a half dozen would thus completely efface themselves, leaving no hint of disturbed surface. When dug up, they were always very sleepy, and it was some time before their eyes would open wide, and with a wild struggle and rush they would realize that this was not a dream.

The species found on Indefatigable is known as *Tropidurus albemarlensis,* and seven other forms have been described from other islands, the distribution closely paralleling that of the mockingbirds. In spite of my increasing belief in the original connection of the Galápagos with the mainland I was interested to observe several instances of possible redistribution of these lizards. Twice I found *Tropidurus* which had crept into our row-boats on the beach, one of which escaped as we pushed off, the second was on board when we reached the yacht. Again, three of our captive lizards escaped from their cage on board, and two did not appear until twelve days later when we were well out in the Atlantic Ocean. This would make their occasional transportation from island to island on floating materials a not impossible thing. As long as the lizards are fed a few flies each day they do well in confinement.

I crept up to the first one I saw, anxious to get a photograph, and, while looking into my Graflex, almost trod upon it, so tame was it. While waiting for it to turn sideways, a big male crawled between my feet and nodded frantically to a scarlet-throated female sunning herself on a bit of lava. He crept a little nearer, nodded again, whereupon the lady lizard rose as high as possible upon all four legs, making them look like straight little sticks, arched her body, blew herself up with air until she lost all semblance to a lizard, and turning

77

her head slowly, spat upon her admirer. He turned, non-chalantly caught a fly, and sadly made his way elsewhere. Never have I seen such a sudden transformation or a more unmistakable indication of disposition.

Again there came vividly to mind the delight of work on volcanic islands, the relief from the complexity of scores of nearly related species and larger groups. Here within a few yards, I had seen a mockingbird and a lizard, yonder were several small black ground finches, and now a hawk soared overhead and a dove whistled past. These five organisms were the only ones of their several families or even larger assemblies to be found anywhere on the Galápagos. The niche which the mockingbird occupied took the place of all the brown thrashers, catbirds, wrens, creepers, bluebirds and thrushes of our northern fauna; the finches had no near relatives, no grosbeaks, tanagers or buntings, they had to compete with no blackbirds, larks or crows. The only other lizards on the islands were tiny geckos, and iguanas, which I was soon to encounter. Strangely enough the great iguanas were both to prove vegetarians, while these nervous, quick-moving little *Tropidurus* chaps fed principally on animal food. In their little world, beetles and spiders seemed the chief items of diet, along with small grasshoppers, land shells, seeds and moths. Twenty-five per cent. of those I examined had included the last in their list of edibles.

I circled through the half-dead vegetation, making my way up steep lava ridges, slipping into crevasses, leaping wide cracks, never, except when standing still, able to pay attention to anything but sliding clinkers and terrible cactus spines.

At one spot I saw, at a distance of thirty feet, a tall, peculiarly bent, Bursera bush, and made my way toward it, counting my steps and roughly gauging the distance as I proceeded. Had the land and vegetation been as land and vegetation is wont to be, I could have strode to my bush in ten steps and half as many seconds. As it was I found an impassable chevaux-de-frise of cactus, two volcanic crevasses, and one elbow ridge, which forced me half a turn to the south before I struggled up the last slope, sending an avalanche of tufa dust and lava to the bottom. I had accomplished my journey in eighty-three steps, including leaps and backward slips, I had covered about one hundred and fifty feet, and was not a second less than two minutes. This particular part of the island was, I must admit, worse than the average, but I realized vividly why much of the interior is still untrodden.

Continuing to the west I reached the shore and came out on a lava reef, where an old black marine lizard was sunning himself, and within a yard of him was a great blue heron in full breeding plumage, to outward appearances exactly like our northern bird, the "blue crane" of rural nomenclature, but he had shed all his wildness, and took to wing only when I approached within ten feet. This particular bird had a limited range of less than a quarter of a mile up and down the shore and in the lagoon. Day after day we saw him there. Before I gave orders to the sailors not to kill or injure any island life, a boatload went ashore just east of Harrison Cove and one of them walked up to this same heron, knocked him over with a stick and brought him on board ship. One wing seemed slightly strained, but the bird took up his position on the stern awning, coming to the midships rail for water and fish. After two days it got so tame that it pursued our pet monkey everywhere, and got under our feet so that we had to build a barrier. It maintained its position when we steamed away in search of water, and with us visited James, Albemarle, and Chatham. When we returned to Conway Bay, Vincent as we called him, suddenly spread his wings and slowly and gracefully flapped his way over the water back to his old accustomed haunts.

The soul-satisfying inclusiveness of desert-island natural history was evident just after my first great blue heron flew. He alighted, gently as a shadow, twenty yards away on the sand of a lava-strewn beach, and there, scattered along the sand or perched on a low mangrove clump, were all the species of herons which are ever to be seen in this archipelago—a yellow-crowned night heron, *Nyctinassa*, in young nondescript plumage, a Galápagos green heron, *Butorides*, only it was blue, and a pair of American egrets, *Herodias egretta*. These last flew from the mangroves, and I watched them as they slowly vibrated across the water, becoming mere specks of white before they vanished along the coast of Eden. I was interested in seeing them, but did not realize that this would be the last time, nor that they had not been recorded from Indefatigable before. Their wariness was in marked contrast to the tameness of the other three species. The small blue herons were constantly flying aboard the yacht, and twice we almost caught them by hand.

Striking inland again beyond the barrier of mangroves I followed a big, meandering pond or lagoon, green with bottom scum. Ducks were seen here on another day, but now the waters were barren of life. Suddenly, like a burst of vocal rockets, four Hudsonian curlew and a pair of black-necked stilts shot up and with loud outcries circled around several times. The curlew with their sickle bills and unforgettable, plaintive cries, would soon be off on their brave journey to their summer home in the heart of Alaska, six thousand long miles to the north. The stilts were unchanged from their fellows of the continents, but were residents, and the female of this pair, pretending a broken wing, very likely had a quartet of sand-colored youngsters hidden somewhere near.

But the striking thing—the character which instantly set both of these species apart from the more essentially Galápagos birds was their fear, their timidity. Unworried by their frantic rush of wings and outcries, all the native birds in sight, instead of being warned and alarmed at our presence, came straight to the nearest bushes or actually hovered about our heads in mid-air, striving to make us out. Throughout the whole of my visit, there was continually impressed upon me the value of psychological as well as physical characters. I believe we could plot the birds of the Galápagos in two groups; those which fled at sight, or allowed no near approach, and those which approached of their own accord and called us friends. We would then have a fairly accurate delimitation of the seasonal migrants, plus comparatively recent arrivals which had not appreciatively altered in form or mentality, and on the other hand, the birds which probably harked back, not only to the unification of the Galápagos, but to its junction with Central America.

A tiny meadow of red hematite was sandwiched between the lagoon and the lava reef, and in the centre of this was an oval wall of basalt stones, two tiers high, marking some grave of long ago, a ship-wrecked sailor, or some New Bedford whaler who died far from home, or perhaps a swarthy buccaneer thus escaping the hangman's noose.

Excited by the curlews, a Galápagos flycatcher, *Myiarchus magnirostris*, came fluttering from a distant clump of wild cotton. He belonged to the same genus as our own great crested flycatcher, and although only two-thirds as large, still preserved all the familiar characters; above he was plain brown, the throat and breast ash grey, and the rest of the under parts primrose yellow, while he still clung to a bit of rich rufus on the inner margins of his tail-feathers. As he came flying he also gave the whüt-tweé—call of the clan of *Myiarchus*.

GALAPAGOS

If the tameness of the Galápagos birds in general was astonishing, the fearlessness of this little flycatcher was unbelievable. He alighted three feet from me, and when I drew back a little to focus my Graflex camera, the mirror was suddenly obscured, and looking up I saw the bird clinging to the lens, pecking at the brass fittings. Wherever I went, on every island, although they were not as abundant as the mockers, *Myiarchus* came straight and close, perched within arm's reach and with crest feathers rising and falling gave me stare for stare with those bright, intelligent eyes of his. For half an hour this first individual neglected his fly-catching to watch every movement, twice interrupting his concentration to do me a good turn by seizing and devouring two little green horseflies which were bothering me. These beasts, *Tabanus vittiger*, were small of size and on this equineless island ill-named, but at times on the beaches were very troublesome.

I squeaked like a young bird, and two mockers arrived at once, flying close together and perching on a low stone long enough for a photograph. A small pile of grass straw lay at the foot of a cactus, and when the birds saw this they lost interest in me, and gave another exhibition of their really remarkable cursorial habits. They had begun a nest in a rugged old Bursera and to this they now made trip after trip, first seizing a bit of grass, then racing with closed wings along the sand and line of red earth, finally slanting upward in a half leap, half flip of the wings to the crotch ten feet above the ground. The return was made in similar fashion. Once a bird made a side excursion, and for no reason that I could see, used its wings, flying easily low along the ground. But on all the trips back and forth, ten or more in number, to their lumber pile, they might rather have been road-runners than mockers. Whenever they stopped for breath they wasted it in little phrases of joy, the commonest being *cher-o-hee! cher-o-hee!* sometimes followed by liquid single notes, with now and then the inevitable harsh chord of their kind.

It seemed treachery to take advantage of any of these confiding birds, but we hardened our hearts and the three which we caught without any trouble, accepted their roomy cages and excellent provender with no interim of moping, and today, after many months in the Zoological Park, they are as tame and in as perfect health as they could be on their native heath, or rather thorns.

I walked back along the shore to Harrison Cove where the tide was coming gently in, and with patient fingers, obliterating the scratchy furrows of my unlovely grubbings after shells. A pelican soared over my head, almost brushing my cap and fanning me with a swirl of air from his great wings. He alighted a short distance away and burying his beak in his neck or vice versa, he watched what was perhaps the first human being of his young life. A giant grasshopper flopped against the tent-fly which we had erected, and I caught it. It was very large, its name was *Schistocerca melanocera* and it was splashed with red and yellow and blue pigments, a much more brilliant insect than any of the butterflies which I saw on the islands. Climbing over a narrow reef of lava I surprised a big red and black centipede just disappearing into an inaccessible crevice. Wherever I went on the sands, the oyster-catchers and gulls trotted after me, just as though man were not the inventor of guns and sport for sport's sake.

In the lee of the reef I saw a large dark patch on the water. I thought at first it was bottom reflection, and then cloud shadow, but when I waded out a few feet, the mass became troubled and moved slowly away. I discovered that this surface film was alive, a solid sheet of thousands of blue and black water-striders. This was of the greatest interest because it was salt water, and marine insects are rare beyond estimate.

With a swoop of my net I captured a whole fleet and found, to my delight, that they were *Halobates*, almost the only group of strictly marine insects. They were variegated above with pale blue and black, and beneath were tinged with terra-cotta, which would become deeper as they grew older. Would that every scientific name was as apt as *Halobates*—Treaders of the Sea! (Plate IV.)

Pioneers are always worth cultivating, and from the first time I ever heard about them, these insects have always excited my admiration. They alone have left their half million fellow species, who find life agreeable on mountains and plains, in air or underground, in fresh waters or as para-sites, and have pushed out into the open oceans all over the globe, where they must pass their lives with no contact except the surface of the water and whatever floating sea-weed or other jetsam may drift across their path.

In every ocean in the world, far out at sea, I have seen these little creatures skating ahead of the bow of vessels, leap-ing frantically at the last moment to escape the wash of the ship. Twice only have I ever succeeded in capturing them, by a lucky scoop of the net as I hung from the anchor chains. In a basin of water I have watched them struggle for a few minutes, then sink to the bottom and die. This was as as-tonishing as the invariable tendency of albatrosses to become sea-sick on the deck of a ship, or the old joke about the death of a pet trout by drowning. The second time I caught a *Halobates* was in the Indian Ocean on a P. & O. steamer, and I remembered the former tragedy and attempted to avoid it. I placed the insect on a rough towel where it could not move about. When it was thoroughly dry I let it slip on to a piece of glass, whence I gently lowered it into water. It floated off buoyantly and skated about for two days. The insects cannot endure getting their backs wet, and apparently sink and die as a result. This handicap appears to an inquiring scientist about as reasonable as an Arctic creature susceptible to cold, or a mountain sheep with a weak heart. It is inconceivable how in a storm or heavy rain these insects can keep dry.

So it was wholly unexpected to find them in such numbers and near land. It is evident that they do not count oceanic islands as part of the forbidden mainland territory. I repeated the water experiment with these young Galápagos *Halobates* and, when thoroughly soaked, three out of five failed to reright themselves and again float.

Little is known about the habits of these insects. A floating feather has been found far out at sea covered with their eggs, and each female is said to deposit about twenty-five. Here in this cove, close under the lee of a tumbled reef of lava were, I estimated, about four thousand immature sea-striders. When undisturbed they were motionless and so close together as to appear a solid mass. But when my approach and my net stirred them up, each began to move in its own small orbit, and the whole turned into a boiling, blurry maze.

In a basin I saw that the insects were supported only by the surface tension of the water, each foot resting in a very slight depression or liquid dimple. Imagine spending your entire life skating over a thin, yielding skim of ice, the breaking of which would mean certain death. When the basin was placed in sunlight, the extent of the dimple was evident in the six large circular shadows cast on the bottom.

Halobates has six oars, but he is actually a two-oared shell, as the bow oars are chiefly passive supports and the stern pair do most of the steering. The legs amidships have a free sweep from level with the head clear back to the abdominal thwarts. They are long, slender and the blades are composed of a line of very fine but stiff hairs, a feathering which any oarsman would envy. The speed and control of direction which can be attained is amazing.

As soon as I realized what I had found, I searched carefully for any younger individuals or trace of adults or eggs, but found none. Whether the ever-flowing currents had herded the eggs from mid-ocean into this little sheltered cove, or whether many females had united in depositing their ova close to the shore I could not learn. The fact remained that here were several thousand immature *Halobates*, gathered in a dense swarm within a few inches of dry land. When I scooped a net full out on the wet sand, they skated about on the film of water, or where there was not sufficient liquid to row upon, became crippled and helpless at once; they had nothing of the versatility of Captain Shard of the bad ship *Desperate Lark*.

I saw the insects on only one other day, when the sun was very warm, and they were correspondingly more active. The movement on the whole was much less zigzagging than the erratic darts of our mill-pond water-striders. Sometimes fifty or more would shoot steadily along in one direction for several yards, as if they were rowing trial heats in anticipation of the skill needed in the terrible storms which they must learn to ride out.

And so ends the tale of small happenings of my first walk in the Galápagos—all unimportant in the telling, but of vital interest to me. There were no large, dangerous animals, or poisonous snakes, no thrilling adventures, no incident of great import, and yet if I have failed to give the strange, mysterious atmosphere of the place, the sensing of unreasonable happenings, of unearthliness, then there is nothing in my walk but my own memories.

We gathered at last, all telling at once of the new and

unexpected things seen and collected, all sensible of the spirit of weird isolation of this land. At dusk we climbed the companion-way to our home of orderliness and electricity, a land of linen and cut glass, of delicious food and drink, easy chairs, music and absorbing books, and the contrast with my day came almost like a blow. Each made the other so much more worth while.

I went forward and watched the swift tropic night close down over the distant beach and bushes. I now knew something about that beach and the life on lava and leaves, I had seen and heard and smelled it—and I begrudged my night of sleep.

CHAPTER IV

EDEN AND GUY FAWKES

WE found our anchorage in Conway Bay, off the northwest coast of Indefatigable, to be ideal in every respect. To the north rose the jagged outline of James, from this distance appearing even more barren and rugged than the other islands. It was on James that the last active volcano had been reported and we watched it carefully from day to day for any signs of smoke or fire. Dimly to the west Albemarle was seen, but our immediate surroundings interested us more at present; several miles to the northeast the three Guy Fawkes Islets, rising almost sheer, and due west, only half a mile away, the little island of Eden. This last was well named and many of our earlier vivid impressions were formed on its shores and slopes. It was a mere speck on the map, a miniature islet half a mile across, but true to its archipelago in possessing several perfectly good craters. In fact the whole islet was a third of a volcano's rim, with lesser craterlets scattered along its shoulders and the shore.

Three of us started out in the little dinghy puff-boat early in the morning of our second day and landed on the shingle of Eden. So protected was all this inner part of Conway Bay that only a slight ripple ran up and down the bed of rounded lava pebbles. Straight up from the shore high cliffs rose,

basaltic columns and angles in strong bas-relief, splitting off
at a touch into mighty ax-heads with sharpest of edges ready
made for any passing cave-man. Just above high-tide mark
the pebbles gave way to huge black lava boulders, some
smooth, others fretted with a surface which tore skin or rubber
soles at a touch. I kneeled down to watch the progress of a
great cone-shaped mollusk. The snail flowed slowly along
the rock, trailing its twisted shell of pale yellow and purple,
now bathed by the clear water, now left dripping in the air.

Suddenly I heard a snort, and a raucous roar sounded in
my very ears. I started up and there within six feet was my
first southern sea-lion. He was unique in character and
behaved as none of his kind ever did again. He presumably
thought I was another sea-lion, but crippled and deformed,
and he launched attack after attack upon me. The first
one or two were so startling that I was really alarmed, and
made certain of a line of retreat. Then, like "Gen'al O'lando
Jackson" I remembered that bravery was only a conviction
of safety, and I became brave. The lion of the sea would
weave back and forth a few yards off shore, splashing and
grumbling, arousing himself to the point of attack—then,
with a magnificent round-backed dive, he dipped down and
toward me, and shot up into the air, eight, five or four feet
away, measured by his exact conviction of safety. I rose from
my pinniped posture, stood up on my hind legs and watched
the courage of the great beast ooze away, and his terrific roar
die out in echoes against the cliffs, as he sank back into his
element. Finally, by subtle retreats I led him to think that I
was on the point of flight, and he made a dive which stranded
him on the pebbles. The moment he appeared, I charged
in my turn and before he could go into reverse I had his
hind flippers in my hand. Pebbles flew in all directions, water
foamed in masses as he flippered and caterpillared on the loose

shingle, in abject terror. If he again appeared above the surface it was very far from the beach of Eden. I saw and played with dozens of his fellows and never again was charged or threatened.

This adventure with my premier Galápagos sea-lion showed how erroneous first impressions or experiences often are. I remember that when a boy the first time I ever took a croquet mallet in my hand, I went clear from stake to stake without stopping; the first time I ever shot over dogs, I dropped two partridges with the left and one with the right barrel. These were first experiences; what happened the second time is best left unrecorded. And now, if we steamed away from these islands today, I should carry the impression that albatrosses and fur seals were as common as boobies and sea-lions, and that the latter were ferocious and attacked without provocation; the truth being that the individuals of the two first mentioned were the only ones seen throughout the trip and the sea-lion must have been an outcast rogue judging from the good humour of all its fellows. In fact just beyond the next headland, we surprised a sleepy pup on a half-submerged rock, and he came at once to inspect us, and followed us along shore, playing and frolicking and striving to learn what manner of creatures we were.

As we walked along, hosts of scarlet crabs scuttled away from our path—crabs which we were to know as the most conspicuous and ever-present feature of Galápagos shores. They were old friends, and I had seen them years ago in Cuba and Mexico, and scampering over the rocks of St. Thomas and Jamaica. In the latter island they call them Sally Lightfoot, a name much more apt than that by which the carcinologist knows them—*Grapsus grapsus grapsus*,— the old Greek word for crab thrice repeated.

Catesby, writing of the natural history of Florida and the

EDEN FROM HARRISON COVE ON INDEFATIGABLE

As the ship-wrecked taxi-driver said, "Nights was the worst. We'd lie there and think about things, and wonder how much longer we could drink blood instead of water; and then we'd get up and look at the sea, and think where we was, and suppose a ship didn't ever come. Nights was the time we felt it." (Page 302.)

Fig. 17

A Big Bull Sea-Lion Rounded Up by the Scientific Artist

Tower Island

Bahamas one hundred and eighty years ago, gives a vivid paragraph of them in smooth and sonorous English. "These crabs inhabit the rocks overhanging the sea; they are the nimblest of all other crabs; they run with surprising agility along the upright side of a rock and even under rocks that hang horizontally over the sea; this they are often necessitated to do for escaping the assaults of rapacious birds which pursue them. These crabs, so far as I could observe, never go to land, but frequent mostly those parts of the promontories and islands of rocks in and near the sea, where by the continual and violent agitation of the waves against the rocks they are always wet, continually receiving the spray of the sea, which often washes them into it, but they instantly return to the rock again, not being able to live under water and yet requiring more of that element than any of the crustaceous kinds that are not fish."

Nowhere had I ever seen them of so vivid a scarlet as against these sombre lava rocks, and nowhere as abundant. Lying flat on the gentle slope of a huge cube, twenty feet each way, I watched the waving tentacles of anemones far below me. Suddenly a scarlet curtain swept across the whole face of the rock, as an army of crabs skittered into view. Their armour clattered as they ran against or over one another, in their fright they blew strings of little bubbles out of their comic faces and in their haste a number scurried over me. Then came the cause of their fright,—a little bob-tailed, slaty-blue heron, a copy of our own northern green heron, except for the dull, lava-like hues and stouter build, *Butorides sundevalli*. Whether the easy life led by these small herons on the islands has induced corpulency I do not know, but it is a fact that they weigh more than two-thirds as much again as their continental relatives.

My cancrivorous heron watched me suspiciously for a

moment, flicking his tail up and down, then dashed at the crabs and seized a leg which its owner promptly discarded. Flying to the top of a neighbouring rock, the heron hammered off the terminal joints, crushed the largest and swallowed them. After a short rest the bird made another foray, again secured a stray leg and feasted upon part of it. During coming days I observed this performance on the part of at least four individuals on two islands,—a source of food as certain as it was remarkable, for the discarded limbs are almost always renewed, the crabs being little the worse for the loss of one of their eight feet.

As Catesby truly says, these crabs never go inland, but are always found within reach of the water, and yet they are indifferent swimmers. This was the first of the many confusing facts with which the Galápagos teemed,—that this species could occur on the Pacific coast with fifty good miles of dry land, or equally inimical fresh water between them and their Atlantic fellows, and here, seven hundred miles or more away, were thousands forever hurrying about the cooled lava. No change had taken place in their anatomy, although thirty-five millions of years may have elapsed since the Oligocene period when the waters of the Atlantic and the Pacific were mingled.

These hosts of Sally Lightfoots were the most brilliant spots of colour above water in these islands, putting to shame the dull drab hues of the terrestrial organisms and hinting of the glories of colourful animal life beneath the surface of the sea. When such an outburst of crabs occurred as I have described, darting out of all possible and impossible cracks and crevices of the lava, they appeared to the imagination as organic reminders of the sparks and flames which once reddened these great beaches and these plains and mountains of lava.

Flying about, close to the precipitous cliffs, just this side of a cave which marked the end of the beach, was a flock of Galápagos purple martins, *Progne modesta*. They are smaller than the martins of the mainland and have less white upon their plumage, and have established themselves in small colonies on many of the islands. Darwin observed them on James Island, and long before his visit, old buccaneers wrote in their journals of the "swallows" of these islands, birds which took their minds back for a moment to English downs or the eaves of France.

There were seven birds in sight, one of which was immature, all swooping about and twittering in the usual martin way, and by their fluttering visits to the side of the cliff I located six nests. These were in deep sheltered ledges or in angular tunnels made by the splitting off of the rocks along the weak lines of crystalline deposit. One nest was not more than twenty feet up and by the aid of acrobatic efforts on a long improvised ladder I managed to reach the old bird on her nest, and saw a full-fledged youngster scuttle up a narrow chute out of sight. On my next ascent I was able to dislodge a mass of the cliff which almost buried me as it fell. I just managed to leap out of the way of the hurtling rocks and looking up saw the young fledgling flutter out. It dropped nearly into my hand, recovered itself and flew with remarkably steady flight out over the water and around the adjoining promontory. Climbing over the jagged summit I was in time to see the little bird drop exhausted into the water, and flap itself ashore, only to be promptly seized by a huge scarlet crab. Holding its prey aloft in the great pincers, the crab climbed the lava slope with others of its bright-hued envious fellows following at a respectful distance. I thought the little swallow had had about enough for one day, and I relentlessly pursued the crab and rescued the bird quite unhurt.

Returning to the martins' beach I watched the birds and

within five minutes saw a most interesting thing. A yellow butterfly fluttered slowly down over the cliff toward us, and at once a martin set off in pursuit. It was a long zigzag chase with the "sulphur" trying to dodge, now down to the water, back to shore, and around in spirals—a veritable whirling bit of yellow tissue. At last an unlucky turn fairly shot the insect into the mouth of the martin and the bird flew about for a full minute before the wings disappeared, either dropped to the ground or swallowed.

Urged by Professor Poulton I have for many years kept on the watch for instances of birds attacking butterflies, as considerable weight of certain mimicry and colour theories depends upon butterflies having aerial enemies. That lizards often devour these insects is well known, but a bird as assailant is a rarer event. In Ceylon and in Burma, in the high Himalayas, and in central China I have seen such pursuits, but they were few and far between and seldom successful, often appearing to be mere half-hearted, sporting activities, a pitting of wing-power against a worthy opponent, as birds will pursue each other in mid-air. I have seen many thousands of opportunities neglected, where migrating butterflies were passing, scores to the second in sight, and flycatchers and swallows hawking about, wholly indifferent to this abundant but fuzzy source of food.

Seventeen years ago E. W. Gifford made four notes on this subject, writing of the martins of Tagus Cove, Albermarle. He says:

"I saw one with a butterfly in its mouth being pursued by two others."

"I saw one enter its nest with a medium-sized yellow butterfly in its mouth."

"I saw one make a dozen or so unsuccessful attempts to catch a yellow butterfly which was crossing the cove."

"On April ninth I noted one chasing a sphinx moth over Tagus Cove; the moth finally dropped into the water and the bird left it."

Stimulated by the observation which I had made so early in my visit to the islands, I kept on the watch, and for the first time in my life I found aerial birds which fed largely on butterflies and moths. Within five minutes after my first butterfly-martin incident, I saw others chasing a red butterfly which they failed to capture. The first butterfly and at least two of those mentioned by Gifford were the cloudless sulphur, *Callidryas eubele*, almost identical with our northern form, and the reddish one was the fritillery, *Agraulis vanillæ*, which I have already mentioned. During the ensuing twenty days which I spent on the islands I made notes of thirteen additional instances of the same character, twelve of the victims of which were sulphurs, and the other a fritillery.

Not only this, but when I returned to the *Noma* from the first trip to Eden and examined the food of the martins I had taken, I found that both the young fledgling and its male parent had been feeding almost entirely upon small moths. Two wings were still recognizable as a new species, *Melipotis harrisoni* Schaus. At another island, as I shall describe in greater detail, I saw the same species of bird pursuing and feeding on a small diurnal sphinx moth.

It is a usual thing for cuckoos of various species to feed upon hairy caterpillars and other unpleasant appearing provender, but it is not common for diurnal birds to be willing to devour such fuzzy creatures as are these millers. I remember in Garhwal, high up in the Himalayas, half round the world, I once shot white-crested kaleege pheasants with their crops stuffed with two or three dozen small moths, all swallowed whole and quite identifiable. As I shall mention again, both the mockingbirds and flycatchers of the Galápagos were expert and willing butterfly catchers. All this is in very decided contrast to what obtains elsewhere, for in my experience, the relation between birds and butterflies is quite a negligible factor in any theory of lepidopterous evolution of pattern, colour, form or activity. With winged grasshoppers of all sizes so abundant everywhere in these islands, the diet of butterflies became all the more inexplicable.

I was not through with the martins, however, and indeed in their fascinating ways and habits they were typical of all this insular life. On every hand new problems offered themselves, in strange forms and under unexpected conditions, appallingly hopeless of solution to a naturalist who must eat and sleep and do so many other things which distract concentrated attention.

When I secured the martin sitting on the nest which contained the fledgling, I supposed it was the female, although the other parent appeared exceedingly dull as it fluttered close about me. Dissection showed the bird to be a male in full breeding condition, and yet in not even second year plumage. Beneath it was quite sooty grey with only a hint of metallic on the upper plumage, and much of this, curiously enough, on the exposed portions of the flight feathers. The under tail coverts were pale edged, and in all the general proportions, except in size, length of tail and in the newly sprouting wing feathers and soft gape there was little to choose between it and its own offspring.

Two weeks later on Daphne Major I shot another breeding male in juvenile dress. This was the first hint of unusual conditions of life on the Galápagos, but I soon came to consider the absence of such abnormal characteristics as rare and improbable. Correlated, perhaps, with this breeding in immature dress was the fact that the nesting season was spread over a longer period than that of their northern relatives, and too they had no annual migration, being found in the same place every month in the year. The number of eggs laid was considerably smaller than with our northern purple martin. I have data of my own and of other observers on nine nests; two, two, two, three, one, one, three, one, two,—an average of two and a half eggs to a nest.

Here we have a most interesting series of data concerning a bird which at first glance seems in no respect different from those of our northern native woods and fields. And although the birds have been on the islands long enough to change specifically in size and colour, yet their newly acquired insular diet has as yet had little or no appreciable effect on either mode of flight, colour or size of the yellow butterflies.

The most significant point is the possibility of a more or less sudden change in the birds' diet, so radical that if it should spread to the mainland it must be a potent factor in the development and control of new adaptations on the part of both eater and eatee. An analogous case is the inexplicably widespread fish-catching habit of the kiskadee tyrant flycatcher. [1]

We went on around the eastern promontory into a wholly different land. Here were deep caves and fissures, and, a few yards off shore, circular water-worn cinders, ten feet in diameter on which were perched smug pelicans and boobies with feet so blue that they reflected an azure sheen over all the immaculate breast feathers. Here and there, sprawled amid the worn black pinnacles and galleries, were great marine iguanas. The rollers came in majestically and, breaking against the iron battlements, were dashed high into spray. This foamed and raced through the crevices, spurting out of little lava faucets, and dripped from the beaks of the birds and the dangling tails of the lizards. There was no chance of landing here, but a hundred yards farther on we came to a

[1] "Mangrove Mysteries," *Harper's Magazine,* October, 1922.

delightful little semi-circular bay, guarded on each side by rugged lava dragons who received the full shock of the curving breakers, and protected the white sand from all disturbance. Even before we landed we named this Amblyrhynchus Cove, for on the sand, on little trails in the bank behind, on black rounded boulders, looking out from paths worn in the rank grass were dozens and dozens of these great lizards. Our boat grounded gently and we walked up through the colony, the iguanas hardly moving out of our way. For a long time we watched these strange beings, we saw them challenge one another, and pay court to indescribably ugly black lady lizards. And then I saw something which I shall probably mention in another chapter, but it will bear repetition. An iguana about three feet long was resting partly on the sand, and partly on a water-smoothed stone, when a large scarlet crab approached slowly, crawled over the whole length of the reptile, and picked off two ticks as it went. If my companions had not been watching at the same time I should hesitate to record such a remarkable occurrence.

Minute by minute we were ever more impressed by the tameness of the animal life of the Galápagos, but the crabs were an exception. I could creep close to them, but their eyes would watch me closely and they would always sidle out of reach, slowly if I approached gently, or like a scarlet flash if I grabbed quickly. Here however in this magic cove of Eden there were three crabs which in point of fearlessness might have been the Three Musketeers. When I walked up from the boat one came slowly to meet me, one of the biggest, his carapace fairly aflame in the sunlight. On and on he came, his eyes twiddling comically on their stalks, his thick, spoon-shaped claws making queer cabalistic signs, intelligible only to the race of Cancer. After waving them about regardlessly for a time, he would go through a series of about a dozen motions, absurdly like an American private saluting with both hands at once, the claws being raised to the stalked eyes and brought down simultaneously with a quick jerk. Then followed more irregular signs, and later a quivering as of a drummer's hands beating a tattoo.

I stood quietly until he came within a foot, then leaned down, slowly reached out and rubbed his shell. He sank down upon the sand, lowered his eyes into their sockets, and wiggled his maxillipeds ecstatically. I took all manner of liberties, lifting one leg after another, raising him from the ground and replacing him, standing him upon his head, and tapping gently upon his hard back. Surely this must be a very ill crab, or an idiot crustacean, or somehow abnormal. I walked slowly away, and to my amazement he turned and followed me. Another crab joined the first and I felt like a walrus or a carpenter. I took a few steps inland, rounded up a half dozen ordinarily wary crabs and rushed them in the direction of the tame pair. The stampede was infectious and as I splashed along the water's edge, all fled at full speed. My tame crabs were inextricably mingled with the crowd, and although I tried to stalk them again, they had lost their fearlessness and psychically as well as physically had merged into the mean of the race.

Just back of the beach rose a terrace of reddish sandy soil filled with the burrows of the iguanas and net-worked with their trails and steep paths. Beyond this the vegetation began, typical of much of the islands. Scattered plants of cactus reared high their oval pads, and low, meadow-like growths alternated with thickets of spicy *Bursera*, and the *Maytena* shrubs, with their dense foliage of small rounded leaves. As on the main island, to step from the sharp, fretted lava on to the soft-appearing grassy slopes was from frying pan to fire. The soft leaves concealed long panicles of seeds, ready at a touch to cling to one's person or clothing, and then with every movement to "walk" where they wished, and they usually crept beneath shoes and clothing, where at the least touch they drew blood. Without high leather puttees, any continuous activity was impossible. Small white convolvulus blossoms were abundant, and the flowers and pods of a small legume, about which were flying very beautiful little pink, black and white moths, *Utetheisa ornatrix*. As I have observed in Guiana with the identical species, the young caterpillars had a very ingenious hiding place. After they had eaten their fill all night and the dangers of the daylight approached, they retreated each to a special pod, into whose hollow interior thay had previously chewed their way. Even when their nightly foraging took them down to the ground and up another plant, I have traced three marked individuals back to their particular pod sanctuary for several days in succession. But like the hermit crabs the caterpillars soon outgrow their shelter, and either shift to a larger pod, or cling to the under side of leaves.

Just above the cliff dwellings of the iguanas, the vegetation was very dense and here I found many interesting things. One of the first was a harmless serpent, small and at first glance much like our garter snake. It was only two feet in length and of a general yellowish brown with longitudinal stripes. Like all of its less lowly neighbours it was quite tame, and made no objection to my picking it up, curling contentedly at the bottom of a snake bag. I found later that it had been feeding on beetles, two grasshoppers and a small moth. Of this single genus *Dromicus*, seven forms, rather than species, of snakes have been distinguished, and as with the mockingbirds and Tropidurus lizards the differentiating interest lies chiefly in the almost inextricably mingled characters. Many features which on the mainland serve to distinguish well-defined species, here are found in snakes from the same island, and occasionally combined in individuals themselves. Such are the number of rows of scales, the pattern, whether striped or spotted, temporals and scale pits. The latter character for example, may be absent, or with one or two pits present, or a few scales on a particular individual may show pits, while all the rest are free from them. So again we have irregular, uncontrolled variation, of the most significant kind—a striving toward species as it were, without definite enough environmental stress to produce very definite distinctions. One specimen which I captured later on Indefatigable was striped, but in alcohol, there appeared very definite, pale, interspaces, which were quite absent in the living reptile,—an ontogenetic tendency toward the breaking up of the stripes into spots. One snake contained two oblong eggs which were almost ready for the deposition of the lime shells. If this, as seemed probable, was the full complement, then the snakes of the Galápagos produce a much smaller number of eggs than their nearest relatives in Central or South America, where eight is more nearly the usual laying.

There were many loose stones lying about among the grass and weeds and under these I found a scanty but very interesting fauna, which was equally characteristic of all the other islands which we visited. Spiders, earth-coloured, were here, a few scorpions of medium size and a small, pale-coloured centipede, and under one small bit of lava there was a very limited colony of termites or white ants. They had driven their tunnels several inches into the soil, but search as I would I could find no queen. The soldiers were very much like the workers, unusually elongated, and moving very slowly, with almost no attempt to avoid the light. Beetles were rare, small, brown Pantomorus weevils crouched in the angles of half-curled leaves, but all others, including big, green, carnivorous Calosomas, hidden under stones.

Yellow flowers, Cordia, cotton blossoms, and Cassia, were abundant, together with a large, white morning glory. Small black skipper butterflies were fluttering about and many small "blues," while grasshoppers and day-flying moths were everywhere. I caught several small, green, long-horned katydids, while pillbugs rolled up at a touch, just as they do under more familiar stones.

A strange, new aspect came to consciousness; the presence of so many flowers, such abundant vegetation, a hot sun, and yet no continuous undertone of insect voices. I once caught a low, distant fiddling as if the first cricket in the world was trying out his first song. Even when an occasional large, black bee appeared, it flew and drew nectar from the flowers silently, with muffled wings, and a bird's chirp caught the ear at once from its complete isolation. The trade winds made no sound among the thin foliage, as if they had blown so long and so regularly that no sough was left in this part of earth.

A pair of mockingbirds haunted the beach and ran down almost into the water in pursuit of some small food. The song of another came faintly from far up the slope, and nearer at hand a new note was heard. I climbed fifty feet and saw the new bird. In this country of lack-fear I needed no careful approach. I pushed through the branches and clattered through dead leaves until within six feet of the strange singer, then watched it as it fed a full-sized fledgling, singing a simple little ditty between mouthfuls of spider gathered for the young bird. The youngster was fluttering after its parent, and constantly uttered a sharp cheep. It was Darwin's Certhidea, which is its generic name, and for the present I use this advisably as its common name, for it has been successively

Fig. 18

GUY FAWKES ISLANDS

Cliffs of stratified layers of Volcanic Tuff
The home of many families of Sea-lions

Fig. 19

Galápagos Sea-lion and her Pup

So tame that one can approach and pick up the young animal at will

classed as a sugar-bird, finch, and finally is perched, but none
too securely, among the wood warblers. I shall have more
to say of this bird later—a cheerful, little person with the
general actions of a sugar-bird, but, at least in worn breeding
plumage, clad in a most sad, simple garb, dull olive brown above
and greyish white below.

Eden was rich (as Galápagos riches are counted) not
only in land creatures, but on the south shore with strange
lives of the sea filling beautiful great tide-pools, and here, as
I shall narrate, we spent a full day of our precious three weeks.
And finally when the last chance came to gather living sea
lizards, it was to Amblyrhynchus Cove on this same island
that we returned. When we wanted the best moving picture
panorama possible of Indefatigable, it was to the top of Eden's
ridge that we toiled, struggling over the lose shale with camera
and tripod, and the view that it gave us will make the scene
live forever in our memories.

GUY FAWKES ISLETS

Not very far from Eden, on the very largest scale maps,
are two or three tiny specks—pin-points of ink known as Guy
Fawkes Islets. From the deck of the *Noma* at anchor in
Conway Bay these showed through powerful glasses as sheer,
majestic cliffs with numerous black creatures moving slowly
about the slopes. Why are they called Guy Fawkes, or, for
that matter, why not? The same question might be asked of
America or Shakespeare or any of us. I felt certain that I
was anchored near Indefatigable; an Ecuadorian would correct me and say it was Santa Cruz, the sailors of the doughty
Admiral cruising here one hundred and ten years ago, would
dispute us both and call it Porter's Island. Guy Fawkes was
as good a name as any for me; my chief concern was to get

a boat ready as early as possible on the third day. I went due northeast and stopped at a small, low island well this side of the main group. In a quiet cove we stepped out and splashed ashore, and around a rock-sheltered pathway I came upon a hollow boulder where five doves were drinking,— sleek little birds in pale, wine-coloured plumage, with black and white wings, face of cerulean blue and scarlet feet. They continued to drink until I came close, then rose, clapped their wings and flew swiftly away.

I went on and found myself among tide-pools near the surf. A very large and brilliant chiton shell caught my eye and I stepped down into three feet of water and began to pry the great mollusk loose. I could not see what I was doing while my hands were busy, and it was only by turning my head sideways that I could keep my mouth above the surface. Suddenly I got a tremendous shock as I felt a soft, warm, rubber-like substance press against my hands. I leaped back and at that instant a baby seal rose directly in front of me, treading water with his hind flippers, while his front mittened fingers were folded funnily across his breast. He looked at me with all his soul, and forthwith broke into a loud, raucous wail. A deep roar sounded from the other side of a barrier of huge boulders, and instantly there appeared, swiftly swimming and banking sharply on the turn, a mother sea-lion and two more infants. She saw me at once and her fear died so instantly that it was not wholly complimentary. She might have explained it, "That thing, whatever it is, is not a shark, so it's all right!" She barked a something out to the youngster and swam back and forth watching me both above and below water. I went on with my chiton prying, greatly to the edification of the four young seals, who, gathered in a circle not more than six feet away, never missed a movement of mine. Again and again one would swim forward under water and nuzzle my fingers to find out what I was trying to do.

At last, having overcome the powerful suction of the shell, I walked along the series of pools with the young seals keeping alongside, eager to see what would happen next. On an out-jutting flat rock which seemed the day nursery hereabouts, were several mother seals suckling their young.

A huge male milled about in the surf outside and expressed his disapproval of me in mighty roars, but the dozing mothers vouchsafed me but a single glance and went promptly to sleep again, with an occasional protest to their disturbing spouse. Another female lay half in and half out of the water in a shallow, tidal lagoon. Every incoming wave would gently lift her and her baby, hold them suspended for a moment, and then quite as softly replace them on the sea-weed-padded rock. When this baby spied me his curiosity overcame his hunger and he dived straight for me. Instantly two of his companions raced for the vacant cradle or lounge or whatever one may call it, but their thieving anxiety defeated their purpose,—they bumped into the mother, thoroughly aroused her, and with a roar of anger at these changelings she chased them away. The rightful pup had by this time satisfied his curiosity, and with a rush returned for dessert.

I killed a red crab and threw it to my particular quartet. They dived after it and nosed it, but apparently were not yet weaned to such food. In fact I believe that crabs seldom if ever enter into their diet, for these crustaceans show absolutely no fear of the sea-lions, sometimes actually scampering over their sleeping bodies. In the stomachs of those we secured I found only remains of fish. The sleek brown animals and the black rocks, the scarlet crabs, the pure emerald of the shallows and the dashing white foam and spray taxed the extreme gamut of colours of our marine artist.

GALAPAGOS

From here we went on to the group of three **Guy Fawkes** Islets, tall, impressive cliffs of finely stratified rock, deeply etched away below by centuries of storms, and overhung with glorious masses of basalt, the whole grey, and olive and pink, with traces of pale yellow which made us all exclaim at the resemblance to a distant view of the Grand Canyon. On the lee side there was no surf, only a great swell which rose and fell as if the Pacific were breathing quietly and regularly. A broad base left by the waves sloped down to the water's edge, and below the tide mark, great tufts of olive sea-weed formed natural buffers for our landing dory. We backed in slowly, watching our chance, and on the beginning of an upward heave, pushed close. It was exhilarating, exciting, at the very summit of the lift, at the moment when every particle of movement had ceased, to leap lightly out, push the boat clear, cling with fingers and toes to the crevices and sea-weed and watch the boat sink down, down, leaving a swirling fringe of white foam along the whole front.

Digging in firmly, with a good foot grip so that the hands were free, I would wait for the return. The extreme of rise and fall was not more than six or eight feet, but from my position it looked thrice that distance. As the boat rose again, a school of brilliantly coloured angel fish would rise with it, half sheltering under its keel, or a grey, lithe shark would weave slowly upward, a perfect symbol of undulating grace, balance and power. At the same favorable moment, with the boat lifted to my very hands and the clear water swirling around my knees, I seized camera or bag or gun, or one of my companions and held quiet until the great ocean had again exhaled, when I clambered up to where even the mighty Pacific has never reached.

Again there was realized the joy of the small, the simple, the unexpected. When, following one's finger on the map, one lands on the shore of a new continent, capacity for surprise is forewarned and limited; of course there must be untold beauties and mysteries in such an enormous portion of the earth's surface, so many in fact that two eyes and a single mind can grasp only the merest fraction of them. But to resolve a speck of ink on the map into islets such as these, is to combine all the joys of discoverer, explorer, and owner in one.

After a short reconnaissance I seated myself on a marvellous throne—a mighty segment of cliff with easy arms and a back fitted for a giant. I was directly beneath a great breach in the high wall, and the end of the ridge towered overhead like the prow of some majestic vessel. A breeze seeped down through the break behind me and brought the penetrating, unmistakable musty odor of an age-old sea-lion rookery. Between the boulders wound strange paths, trodden hard and worn smooth by centuries of pounding flippers and the dragging of sleek, rounded bodies.

In front was a gently sloping monolith which stretched downward thirty feet and flared outward until it was the largest desk in the world. The cliff above me reached up two hundred feet in great billowy rolls and strata, recording every separate outburst and overflow of countless years past, thousands before pirates and yachts and scientists ever were. At one side, on a third great slab, almost within arm's reach, were four pup seals, watching me with their great, expressive eyes, with a straining concentration which hinted at adaptation rather to watery depths than to clear air and sunlight. They were silent, and now and then lost interest in me to the extent of nodding sleepily or of scratching their half-dry, rich brown fur. A newcomer began frolicking with one of the four and they raced all round in their hunched, caterpillar method, sending down a shower of fine sand which fell on my writing and dried it, as used the sand boxes of my grandfather.

Everywhere were the crabs—the same great scarlet fellows, scurrying out of the way of the playful pups, then following them up without fear. One climbed over the rim of my desk, as the one on Eden had done, and twiddled his stalked eyes at me in a most disconcerting way. These eyes were a rich lavender and his shoulders a deep violet blue, while here and there the scarlet of his back reappeared in small monograms and hieroglyphics—a palimpsest telling more of the past history of Grapsus than I shall ever decipher. There must have been five hundred of these crabs in sight, spotting the black lava and pink strata with violent splashes of colour.

A few yards away, the parents of the baby seals were stretched out in sound slumber. I could go up and push and slap them without awakening them. A grandfather of an Amblyrhynchus crawled out behind my seat, and I leaned back with my fountain pen and scrawled A. C., his own initials, on his greyish scales, before he shifted beyond reach. The pups watched him go past and then one picked up a gull's feather and rolling over on his back gave himself up to the joy of chewing it. Thus passed ten minutes of sea-lions' life until they spied Ruth Rose climbing along the water's edge and all hunched excitedly down to see what this new sealy person might be. They all played together in an algæ-padded pool, while fortunately the moving picture camera was unlimbered and recorded it all.

Out beyond the froth of rock-broken water I saw grey shapes eternally slipping past, and knew why the pup seals when alone always kept close to the rocks. This is their only danger but a very terrible one,—the sleepless patrol of the sharks. On shore nothing but man would ever disturb them. I watched a great yellow sea-turtle come up and dive, and as I stood up, a blue-footed booby alighted nearby, and a pair of fork-tailed gulls flew past, calling.

Two oyster-catchers whirled by and alighted on the ledge near the water, then a gaily-coloured turnstone, an old friend of our northern shores. The first birds trotted up to our camera and ran between the legs of the tripod, but the turnstone, after a suspicious glance, rose at my first movement and fled headlong around the end of the island; no native this, but a migrant from the land of men and fear.

The sun was swinging low and I had already leaped into the boat to return to the launch when a loud, sweet, long-drawn-out cry arose, a call wholly new to me, and I quickly returned. I saw the bird, a grey, long-legged sandpiper, which I could not identify. At long range I fired but the bird did not fly away, and then followed a terrific chase. I threw down my gun, and with Bob McKay in the boat and I on the shore we pursued the bird over and around promontories, in and out of the water, within an inch of our lives. It was a better swimmer and almost as good a runner as I, and everyone was near the point of exhaustion when I overhauled it far out of my depth, and we were both dragged into the boat, just in time to miss impalement on a cluster of steel-hard lava spikes. It was a worthy acquisition, with the rather nice name of wandering tatler, a stranger to us on the Atlantic coast, a hardy breeder in the far north, migrating from Alaska to these distant equatorial islands.

As I looked back from the launch, I saw the last rays of the sun falling obliquely on the great barren slope of the second island, a slope dotted almost to the summit with groups of sea-lions. McKay had been fishing while we were ashore, and small sharks had taken most of his tackle, but two gorgeous fish which he had caught were brilliant gold with five, broad, longitudinal lines of blue, each bordered with black. Why it should be burdened with the specific part of its name, *Evoplites viridis*, is a secret which has died with its christener,

M. Valenciennes, who took a specimen in these very Galápagos on the voyage of the *Venus*, seventy-eight years ago.

Again I looked at Guy Fawkes from the deck of the *Noma*, this time with a certain feeling of possession, and hereafter when I see a tiny speck on a map, with or without a name, I shall become discontented, and remain so until I have explored it.

THREE SEA-LION PUPS AND THE EXPEDITION'S HISTORIAN ON GUY FAWKES ROCKS

The Sea-lions were wholly without fear of human beings

Fig. 21

Indefatigable from Amblyrhynchus Cove, Eden

Giant Marine Iguanas are in the foreground, waiting for the tide to uncover their sea-weed food

CHAPTER V

BLACK LIZARDS OF THE SURF

A LTHOUGH the Galápagos were first discovered by Europeans in 1535, it was exactly three hundred years later when Darwin visited them, found the great sea iguanas, and gave us the only good account of their habits ever written. Ten years before, in 1825, Bell had described them under the appropriate name of *Amblyrhynchus cristatus*, and sea or marine iguana is as good a name as could be desired.

We had just put up a tent on Harrison Cove, Indefatigable, when the Artist suddenly called our attention to a small seal or snake swimming across the quiet waters. We all ran down to intercept it and there clambered out our first black sea iguana. I once saw some crocodiles in brackish water, but for a lizard to climb out of the sea was as surprising as dolphins in fresh water or song-birds in a brook, both of which I have seen. The big reptile slipped down a deep crevice of the rocks, and we had given it up and turned campward when another rushed out from under foot and crept beneath a flat lava rock. My fingers just reached its tail, and for five minutes all my strength availed nothing against the twenty claws of the lizard. Little by little he gave way, but when I had acquired about fifteen inches of tail, I had to yield my place. We won in the end, but our first lesson was a thorough one in the tremendous grip of these talons.

111

When an Amblyrhynchus once entrenched himself in an ill-fitting crevice, he blew up his body with air, thus pressing all the myriad scales against the rough lava, and then with the grip of his score of long, curved claws, offered a resistance that had probably never been overcome except by the occasional muscles of pirates and scientists, all one to Ambly in such a crisis.

We soon found that it was quite unnecessary to go to such trouble, however, for when I saw the Amblyrhynchus first, and was willing to creep on all fours on the rock, in the eyes of the lizard I became a harmless sea-lion, and could approach closely and with care even stroke the flabby shagreen skin without causing fear. If any field character was needed to identify these lizards it was that of the tail which I had first in my hand, long, crested and flattened, decidedly a swimming tail, for this is the only marine lizard in the world.

Within two days we realized that these islands were still in the age of reptiles, or rather of reptiles and birds; amphibians and indigenous mammals being wholly absent, and fishes above the water negligible,—although sail-fish and mullets leaped high and blennies climbed out and flicked here and there upon tide-soaked rocks.

Giant tortoises and land iguanas dominated the upper parts of the islands while the jolly little Tropidurus lizards ran everywhere under foot. But the shores were held by the big, black iguanas, who, more than any creature I have ever seen, except the hoatzin, brought the far distant past vividly into the present.

Iguanas have been recorded as reaching fifty-three inches in length, and a weight of twenty pounds. I saw several which I am sure were four feet long, but the two largest captured were thirty-five and forty-one inches respectively, the latter weighing thirteen pounds. Young ones a foot in length weigh only a quarter of a pound.

BLACK LIZARDS OF THE SURF

Superficial observation would class the lizards as black when they are wet, and dull greyish, lighter on the head, when dry. This holds good on close inspection except for the dirty white or pinkish underparts, and for a multitude of dull fawn or reddish brown spots and blotches which cover the back and sides in infinite variety. In three-fourths of the individuals of all ages, these take the form of more or less regular cross bands, extending up to and including the tall spines of the dorsal crest, and occasionally reaching down quite to the belly. The scales above the eyes and around the mouth are often strongly tinged with green, but these lizards never show the brilliancy of colour seen in their land relations.

We found sea lizards in abundance at widely separated localities, as at Tagus Cove, Albemarle, and all around Indefatigable. Collectors have recorded them from twenty-one islands, and I can add to the list Eden and Guy Fawkes, northwest of Indefatigable, the fact important only as being the locality of many of my observations.

Although several forms have received specific names, yet recent study of nearly two hundred individuals[1] from various islands indicates that only one species is valid. The comparative ease with which they can pass from island to island under stress of circumstance would account for this archipelagic identity as compared with the multi-specific non-natatory Tropidurus, Conolophus and Testudo.

I gave as much time as possible to these remarkable iguanas and found them superlative in the two dominant qualities of the Galápagan fauna,—strangeness and tameness. A bird's-eye diagram of their haunts would be a narrow outline of the islands, the slenderest of hair lines along the coast, for high tide marks the equator of their few yards of terrestrial and aquatic wanderings. In this narrow zone they spend their entire lives,

[1] Proceedings of the California Academy of Sciences, (4) 11, 1913, pp. 192–194.

finding food, safety and mates within its confines. Neither drought nor seasons nor food-supply require any migration,—a burrow, a flat rock, a tuft of seaweed, and Amblyrhynchus is content.

Darwin quotes Captain Colnett as saying that "They go to sea in herds a-fishing," and aside from the obviously erroneous reference to their occupation I am inclined to doubt the first part of the statement, for the sociability of these reptiles does not extend to their feeding, which is purely an individual affair. Darwin himself negatives his "occasionally seen some hundreds yards from the shore," with his widely quoted experience of failing to drive the lizards into the water.

The character of the shore on which they live is chiefly jagged dykes and out-jutting masses of the roughest lava, much of it as sharp and evenly pitted as if centuries of breakers had not been forever pounding upon it. Here and there, between the lava, were small sandy beaches, and on the beaches, still more rarely, appeared isolated patches of mangroves. The trade-winds blew continuously from the southeast, and only in sheltered coves and behind lava breakwaters were the waves gentle and landing easy. Like Darwin, I tried in vain to chase Amblyrhynchus into the breakers, but like the great crabs which live with them, they refused because all their swimming ability would avail nothing against the smashing power of these great waves. The sharks which were thought to be the explanation of this reluctance are not to be found at the edge of such breakers, but when they do come close to shore it is in quiet waters, off deep sandy beaches, or just beneath the ledge of low cliffs beloved of sea-lions. And in such places it is far from easy to drive iguanas into the water.

As with the first one which we saw, I frequently saw them swimming across the calm waters of the wide but shallow coves, from one promontory of lava to another, where, a short time

FIG. 22

MARINE IGUANA IN THE SURF, FEEDING ON A SPECIES OF SARGASSUM ALGA

Fig. 23

Giant Marine Iguanas

Amblyrhynchus cristatus (Bell)

They have strong claws but are good tempered and never attempt to bite

before, my water-glass had showed occasional sharks, but of so small a size that an Amblyrhynchus would have nothing to fear from them. When I found that they were abundant and tame all around our camp, I took pity on the big fellow who had given me such a tussle to get out from his rocky den,—I untied him high on the beach and faced him inland. He turned in his tracks and waddled dignifiedly down the sand and into the water. Here he floated for a minute and then dived three times and swam along near the bottom, gaining twenty or thirty feet each time. He then struck out, or rather undulated, for the feet are never used in swimming, aiming for the farthest point of lava one hundred and fifty yards away, and swam steadily, with head, and much of the back halfway to the thighs, above water. In this unhurried swimming the sideways motion was hardly noticeable, the lateral undulations of the tail being absorbed or counteracted somewhere aft of the mid-body. The legs were allowed to hang back, dangling helplessly alongside the body, except at times when swimming close to the bottom when a push with a claw helped now and then.

Their abundance on almost every type of shore and their absurd lack of fear indicated a life unusually free from danger. They were decidedly gregarious and definitely marked colonies were distinguishable. Yet a walk of two or three miles along shore always revealed a solitary lizard now and then, sunning himself or feeding in complete isolation. I only once detected a family association such as one finds among crocodiles, whereas a well-placed boulder would often be covered by a solid mat of iguanas, usually all of the same size and age, either very large or all youngsters. In one place I noticed for several days in succession a pair of unusually large lizards sheltering under a protecting ledge of volcanic clinkers, while out in the sunshine a foot away four young ones were piled three deep. I watched these and saw the spirit of play show itself. Two seized one another and rolled over and over, pretending to bite, like puppies pouncing upon each other. One of them reared high upon his hind legs and tail, with forearms bent and claws outspread, and balanced for a moment before dropping upon his playmate. I was crouching almost flat so that the players were silhouetted against the sky, and they might have been any size in the world. The pose I have described was exactly that of the attacking Tyrannosaurus of Knight's splendid painting, clothing the great skeleton in the American Museum of Natural History, and again, as so often in these islands, I had the conviction of ancient days come once more to earth. Whatever the type of shore they choose, the one necessity was seaweed, and wherever I found one, or a whole colony, low water was certain to reveal tufts of the olive-green Sargassum appearing out of the water.

The daily round of life of the sea iguanas was very simple. They spent the night in their burrows in the earth, or deep down in lava crevices. About eight or nine o'clock in the morning, if the sun was shining, they came out and waited for low tide, then, making their way slowly to the edge of the surf, they fed on the short, glutinous algæ. Afterward they sometimes basked all day in the sun on some favourite rock, out of reach of the water, individuals going back day after day to the same spot. I never saw a single animal out at night, not even in full moonlight at low tide. In spite of the opportunity of the six o'clock tides, they feed but once a day.

Young animals were more eager than older ones and usually arrived on the feeding grounds first. They walked slowly down and began to feed immediately, biting and biting before the algæ gave way, and then swallowing it at once. They bite sometimes with the front, again with the side of the mouth, and often beat up considerable froth between the lips. Only the tips were eaten and even in the midst of a large colony there seemed always to be an abundance of new sprouts. Digestion is very slow and the residue of the food may not pass through the body for four or five days.

I never saw the iguanas dive for food, and indeed there would seem to be no need for it, for at ordinary times an abundance of the weed was always exposed. As this growth thrived only where there was active surf, so the feeding reptiles were often completely covered, three or four feet deep, by an incoming wave. At such times they and the big scarlet crabs all about gripped tight with their claws and, limpet-like, were immovable. Never did I see one dislodged, and from my experience of trying to drag them out of crevices, I count any such danger as negligible. Twice I tried Darwin's experiment of throwing lizards out into heavy surf. They dived at once and walked back toward the shore on the bottom, clinging fast at each backwash and working their way into deep, underwater crevices which led up into half-tide caverns, draped with algæ and tenanted by club-spined sea-urchins, sponges and anemones. There was no haste to return and as I have said, at such a place the danger from sharks was slight.

During all the period of my observation I detected no enemy of Amblyrhynchus, and abstract consideration of all the members of the Galápagos fauna revealed none. The indigenous hawk, which is a Buteo about the size of our Swainson's, was abundant on Eden, and twice I have seen these birds swoop at a Tropidurus within six feet of several iguanas, causing no emotion except a passing interest; indeed a hawk of any size would suffer severely from the long, curved, powerful talons of the reptiles. Martins, mockingbirds, flycatchers, and Tropidurus all hold the hawk in fear and showed it plainly, while none but very young Amblyrhynchus ever took to cover when one flew over. I have already spoken of sharks, and I have no doubt that both sharks and any barracuta-like fish occasionally take toll of swimming lizards, but if the actions of the iguanas were any indication, this danger is a very casual one. I never saw an imperfect tail among iguanas, and in fact they can be dragged by main force out of their hiding-places by a grip on even the tail-tip without dismemberment, in very marked contrast to Tropidurus and the continental arboreal iguanas.

They associated with sea-lions fearlessly, sometimes crawling over the seal's body and showing no disposition to move when these pinnipeds caterpillared their way close to where the reptiles were sprawled in the sun. When a young seal leaped out of the water to get a better look at me, an iguana at my elbow dived into a crevice, but this flight was only momentary and only to save its body from being actually crushed. Sea-lions were certainly not a source of danger.

Wherever I found sea lizards, the effect of my appearance was the same. They turned their heads in my direction and watched me with interest, and as I approached, slowly climbed out of my path. Many times I almost stepped on them as they moved. By sitting down and slowly reaching forward, it was almost always possible to touch and stroke them. We caught lizard after lizard this way and let them go again. One of the first we saw was on a little promontory of rock and we kept after it, taking moving pictures until finally it took refuge in a deep crevice. We then saw a still larger one which had been following us about, two or three feet behind, and which finally went to sleep between the legs of the tripod. As we collected marine specimens in tide-pools, numbers would creep out and watch us. When I wanted thirty or forty within a few hours I used a tarpon rod with a flowing loop at the end and with this I never missed one. Where three were sunning themselves side by side I noosed one after the other, the only emotion aroused being an added curiosity as they saw their comrades swinging off through the air. The larger ones would fix their claws wherever they might be and resisted all efforts at being pulled free.

To test the acquisition of fear, I caught an iguana of medium size, jerked him into the air, played with him for a few minutes and then loosened the noose and set him free. He ran off a few feet, turned and looked at me and offered no resistance to being again caught and swung through space. Six times I repeated this, and if anything he was tamer after the rough treatment than before, in the face of a series of experiences which would have driven any ordinary wild creature insane with fright.

One of their most curious habits was revealed on a late afternoon when I lay flat on the sand watching the ageless surf pounding on the lava boulders. Over the jagged, tortured summits there climbed the largest iguana I saw on the islands. It was a full four feet in length—appearing forty to my lowly viewpoint. His head was clad in rugged scales, black and charred, looking like the clinker piles of the island; along his back extended a line of long spines, as if to skin of lava he had added a semblance of cactus. He saw me and stopped, looking long and earnestly with curiosity, not fear; then with his smug lizard smile unchanging, he dismissed me with an emotional feat as strange as his appearance; he twice solemnly nodded his whole massive head, he sniffed and sent a thin shower of water vapour into the air through his nostrils and clambered past me on down toward the water. If only a spurt of flame had followed the smoky puff of vapour, we should have had a real old-fashioned dragon! He had come from whatever an Ambly finds to do inland at midday, and was headed seaward at high tide on whatever errand calls such a being into activity at such a time. His nth great-grandfather had perhaps looked at the first Inca or Spaniard with the same dignity and nonchalance. As far as appearances went he himself might have been as old as the lava.

Again and again I saw these lizards sniff like this when they were feeding or when they challenged each other, or even when simply watching me. The two streams of nostril vapour were very distinct and visible momentarily for a considerable distance. During this or afterwards a mass of bubbly froth would often become visible along the lips.

The only other attempt at defence, if it could be dignified by such a term, was the inflation of the body with air, the belly swelling until the skin was tense, and held so for a considerable time. Two or three times I have seen them bite one another in apparent irritation, without causing damage, but in spite of the roughest handling I have never induced them to bite me. I have received very severe and painful scratches from their talon-like claws but only through involuntary actions, when they were struggling to escape. Here are great lizards with mail of scales, which become like solid masonry around the head, with a formidable saw-like ridge of horny teeth down the back and tail, with many small but efficient teeth and powerful jaws, with twenty long curved talons, backed by incredibly strong muscles and yet with no desire or power of defence that I could discover, no biting, or clawing or lashing with the tail, —much less dangerous to pick up than the big scarlet crabs which lived with them.

Like most lizards, these iguanas were very sound sleepers and would seldom awaken at the sound of my footsteps when sprawled in the sun. Indeed I have known them to remain undisturbed at the sound of a gun not far away, and I believe the sense of hearing is extremely undeveloped. The pineal eye was well marked by a specialized scale, and there was occasionally an instantaneous reaction in sleep to my hand passing back and forth between their heads and the sun. When walking about they never used the tongue as an external tactile organ. Sight was their most acute sense and they detected me yards away, even though they were to show no fear of me at arm's length.

Challenge and courtship were indistinguishable in external manifestation and of a majestic simplicity. The lizard reared high up on his front legs and nodded his head vigorously up and down a few times—that was all. When two large males passed close to one another, they stopped, went through this intimidation formula, waited with the statuesque patience which only a lizard can achieve, then, honour satisfied, both passed on. One could imagine a mental battle in which there was no vulgar external advertisement of the victor. A male would approach a female with amorous intent, stopping every few steps to send forth his little steam exhaust, and solemnly to nod. This was all he could do to express the passion within his scaly cuirass, but judging by the size of the colonies and the abundance of the young, it was satisfactory.

At Amblyrhynchus Cove on Eden Island I walked to within four feet of a lizard with my graflex camera. It was half on sand and half on lava, and I wished a photograph in just that position as showing the two general types of habitat. As I took the first picture I saw a large red crab approaching on the lava. When I had changed plates, the crab had reached the head of the iguana, and instead of turning aside, crawled straight ahead, the lizard closing its eyes to avoid the sharp legs of the crustacean. On and on the crab went, slowly descending the whole length of the lizard. Three times it stopped and picked a tick from the skin beneath it, the black tissue being pulled high as the crab tugged away. I took another picture as the crab reached the fore-legs. I could not see whether the crab even attempted to eat the ticks, but when it had gone on its way over the sand, I put down my camera, crawled forward, caught the lizard, and with my lens found two places where ticks had been. There was no sign of a third having been pulled off, but there were sixteen ticks still remaining on the skin. The whole thing had come as an absolute surprise, for while I had several times seen crabs walking on the iguanas, I had not noticed any deliberate attempt at tick-catching. These ticks are *Amblyomma darwini* and are closely related to a new species which I discovered on the land lizards.

Of nests, eggs, or recently hatched young I observed almost nothing. One of our party found an old eggshell of a lizard of this species, as there were no land iguanas within fifteen miles, near the shore of Eden Island, about twenty-five feet above the water. Heller, writing twenty years ago, gives the following facts in regard to the breeding habits, "the eggs are deposited at the end of the rainy season in the sand near the beach, usually in boulder-strewn places. A nest found at Iguana Cove, Albemarle, and from which the female was driven, was situated in the sand which partially filled a fissure in the lava rocks bordering the beach. The eggs in this nest were six in number, soft-shelled, elliptical and measured approximately three inches in length, by one and a half in diameter."

The most unexpected deviation from this nesting locality was on Eden where Ruth Rose found burrows and from which she brought shrivelled eggshells, up to an elevation of three hundred feet on the slope of the ridge. Other nests have been described as shallow holes in the shore, which on Eden and elsewhere describes the shelters in which the twelve-inch young ones spent the nights and the overcast days. Each iguana had its individual crevice or burrow, although several often shared burrows with communal openings. The upper stretches of the lava afforded deep, narrow shelters. I do not remember seeing any lizards living in sand. In Amblyrhynchus Cove was a large colony of cliff-dwellers, which had excavated deep burrows on ledges along a high bank of reddish soil and rock. On each side, well-worn paths led upward twenty to forty feet on the hillside to burrows beneath isolated lava boulders or driven directly into the bare earth sheltered only by an overhanging growth of grass and weeds. In front was a sandy beach, so that to find food the iguanas had to clamber down from their burrows and over the rocks on each side of the cove before they could reach the algæ, twenty or thirty yards away.

Ten- or twelve-inch iguanas were the youngest which I saw and their habits appeared to be identical with those of the adults. But in all my travels on these islands I never saw small ones on the main shores, only on the outlying islets. On Eden for example, there were several times as many young as old iguanas and on the southeast side, which we called Martin Cliffs, there appeared to be a pure culture of the twelve-inch individuals, nurseries of six to eight occurring here and there on the great out-jutting boulders.

Although there would seem to be so little reason for it, these reptiles were protectively coloured to the highest degree, and only with the greatest difficulty could one distinguish more than a fraction of those within sight. I often began creeping toward an individual which I wished to obtain, and two or three equally large ones would crawl aside from my path, individuals which my eyes had passed over unseeingly.

On the last day when we prepared for a big haul of Amblyrhynchus, I had a boat cage made—a sloping wire screen on a small dinghy, which we towed to the beach where they abounded, Amblyrhynchus Cove on Eden Island. Into this we placed them as fast as captured, and not only brought them to New York in it, but even carried it to the lawn party at the

Zoological Park on the day after arrival. From Indefatigable Island, five hundred miles out in the Pacific, to the Bronx, and for two months later, these lizards lived and apparently thrived on salt water and air. No variety of seaweed or terrestrial vegetable tempted them to break their fast. Individuals were killed from time to time as material for the big group to be built in the American Museum, but after one hundred days of complete abstinence from food, the remainder appeared as active and as strong as when first taken from among their native lava.

In Chapter XVI many references will be found to these lizards written in past years by various buccaneers and whalers. One hundred and ten years ago, Captain Porter wrote the following about Albemarle:

"I took my boat and proceeded for the aforesaid point, where I arrived in about two hours after leaving the ship, and found in a small bay, behind some rocks which terminate the point, a very good landing, where we went ashore, and to our great surprise and no little alarm, on entering the bushes, found myriads of guanas, of an enormous size and the most hideous appearance imaginable. The rocks forming the cove were also covered with them, and, from their taking to the water very readily, we were induced to believe them a distinct species from those found among the keys of the West Indies. In some spots a half acre of ground would be so completely covered with them as to appear as though it was impossible for another to get in the space; they would all keep their eyes fixed constantly on us, and we at first supposed them prepared to attack us. We soon, however, discovered them to be the most timid of animals, and in a few moments knocked down hundreds of them with our clubs, some of which we brought on board and found to be excellent eating, and many preferred them greatly to the turtle."

Long after this Mr. Beck found very large assemblages on Narborough, a photograph of which I have reproduced, but

FIG. 24

HEAD OF A GIANT MARINE IGUANA

Slightly enlarged

From a painting by Isabel Cooper

[Note: This figure has been reproduced in the Dover edition at 71 percent of its size in the original edition. The caption has been reprinted unchanged.]

Fig. 25

Marine Iguana

This is the only lizard in the world which is marine.   It reaches a length of four feet and never
leaves the shore to go inland.   It refuses all food in captivity

while we found hundreds of Amblyrhynchus on various islands, we saw no such hordes as Beck describes. They have no enemies of which we know, but they are slowly but certainly decreasing. No other living inhabitant of these islands seemed so thoroughly a part of its environment as Amblyrhynchus. In colour, in rough contour, in the scales of its head standing up like volcanic cones, in its intimacy with lava and surf, it seemed an organic embodiment of the shores of these desert islands. Its swimming ability has either enabled it from time to time, even in spite of sharks, to pass from island to island to such an extent that there are no well-marked separate insular forms, or perhaps its limited environment has made for absence of variation. It has very remarkable powers of orientation as I proved on throwing an individual overboard when we were anchored two miles from land. It splashed into the water on the seaward side of the yacht, swam around the stern and started straight for the nearest land off the port bow, although the lava shore could hardly have been visible from the lizard's viewpoint, especially in the dim light of late afternoon. It made five long, deep dives before I could recapture it.

CHAPTER VI

THE WATER LIFE OF INDEFATIGABLE

THE life beneath the surface of the water about the Galápagos presents as great a contrast to the creatures of the land, as the fauna of a tropical jungle differs from that of the Arctic regions. The terrestrial organisms are relatively few in number and of sombre shades; the submarine creatures form an unnumbered host, many equalling the gayest butterflies and birds.

Before we reached the Galápagos almost all our thought was given to the creatures we should find on the islands themselves. It was not the ocean around them, it was the specks of dry land, and their inhabitants, and the problem of how the insects and reptiles and birds had reached these, that occupied our reading and conversation. But once in the archipelago the sea and its tenants forced themselves upon our attention and interest. It was startling to walk past the forecastle and see a grimy stoker, with wide grin on his black face, unhooking a living thing of gold, silver and turquoise. By day strange fish and sharks gathered about the yacht on the lookout for scraps, jelly-fish drifted past, and crabs strove for foothold along the water-line. At night the water life was even more distinct, glowing in phosphorescent silhouette.

It was only by utilizing every possible moment of our precious three weeks that we were able to accomplish what we did, and we soon discovered that we need not sit impotently idle while going from the yacht to the shore. Robert McKay made this discovery and thereafter was the master rod-and-reelman of the expedition. As he tells so well in his chapter, trolling usually resulted in great excitement and valuable specimens. My own reactions always went through a very definite sequence. When I had trolled for ten minutes with no result I became impatient and often yielded my place to see if I could not scoop up some interesting organisms with a net as we went along. But when that high tension electric shock came along the line, no fishermen who ever lived enjoyed the fight more than I. As the fish came closer, and his battling weakened, my sportsman's thrill gave way to the scientific, and the size and weight and fighting ability became of less importance than the species. Although the fishermen groaned at the sight, I welcomed a shark as much as any game fish, if it yielded interesting parasites.

Not until I began to study the birds of Indefatigable did I realize the importance of the sea in the archipelago. Exactly three-fifths of the birds of this island were wholly dependent on water life for food, while a number of others included fish and crustaceans as a part of their diet.

We found that our work with marine organisms resolved itself into several phases of activity; hand-line fishing from the yacht and trolling from moving launches; sitting on the lower platform of the companionway and scooping up whatever drifted past with the tide; dredging on the bottom and towing for plankton along the surface and at intermediate depths; seining along the sandy beaches and capture by search under the lava boulders and in the tide pools.

I have found it impossible in the short space of time since my return to have many of the groups of marine invertebrates worked up, but thanks to Dr. Nichols I have more definite information in regard to the fish. The approximate number of shore fishes recorded from the Galápagos is one hundred and fifty-six. In our limited time we were able to collect or observe sixty species, roughly forty per cent of the whole number. Among these were two new species, *Eupomacentrus beebei* and *Runula albolinea*, and twenty fish new to the Galápagos Archipelago.

Tide-pools among these islands are not very numerous. Where the shores are low, they are formed of lava blocks which allow the water to run out between them with the tide, leaving only the bare, rough surface. Or when old craters occur at sea level along the shore, sandy beaches are formed. Occasionally, as on the south shore of Eden, conditions are ideal, unexcelled by any pools of Fundy or colourful caves of Antigua. Here, among the Octopus Pools, as we called them, we spent parts of two days.

The rugged cliffs came down several hundred feet almost sheer, and near the base were etched deeply by centuries of pounding breakers, curving inward into shallow caves, with boobies and pelicans perched in niches high overhead. At low tide a great table of algæ-padded rock was exposed, extending twenty or thirty feet out to where it dropped abruptly to blue invisible depths, beloved of sharks and angel fish. Some of the pools were shallow—two or three inches deep and six feet square. Here we could make a clean sweep, but where great boulders were clustered together it was a different matter. Creeping slowly up I would see a fairyland of life,—fish so brilliant that they seemed to furnish their own illumination from within, crabs of impossible shape and sculpture, shrimps so transparent that only their faint shadow on the bottom betrayed them. Then, at the first incautious movement, all would slowly fade into safety among the shadows under the boulders, some of which were as large as a bungalow. Years of intensive study could be spent on these ever-renewed treasure pools; I can hope to give but one or two shadowy vignettes.

The first fishes one notices along any tropical shore are the gobies and blennies, principally because they refuse to stay in the water, but go skipping from pool to pool. They are small, rather elongated fish with elaborate fins and a great variety of markings. Some have the pelvic fins united in the middle line to form a leg-like support or sucker. They are capable of most unfish-like activities. I still retain a vivid memory of the mud-skippers of Java, flipping about, lying on their side or back in the sun, and waving their fins to keep away troublesome flies. Here were several species, all draped in pool colours and patterns, bits of green and brown algæ on their sides mingled with light sun-glints and dark lava points. They could change their colour at will and no two were quite alike. At rest they chose to perch in tiny niches on the sides of the pools, exactly as the pelicans and boobies were doing in the caves high overhead. One big brown-banded blenny had marvellous eyes—two orange-framed crescents of emerald on each side of the pupil, set in dark, mottled sepia. It was necessary to put them in covered jars, for in a pail they walked up the sides by successive leaps and flipped over the rim. Two closely related species had diametrically opposite escape reactions; one was found on wet rocks or seaweed out of the water, and when alarmed dived to the pool bottoms, the other perched on stones or fronds beneath the surface and instantly leaped clear of the water on being disturbed.

The seaweeds hung curtain-like before scores of small pools, and when these olive portières were lifted, there were exposed the wonderful colours and the indescribable forms of sea things to which no words can do justice. Hundreds of club-spined sea-urchins were here, each of which rested in a more or less complete chamber of rock. Some lay in a shallow depression, others were all but hidden in lava cells out of which it would be impossible to extract them. We are not certain of the method of excavation, whether by direct action of some dissolving chemical which the animal secretes, or by mechanical attrition. A sea-urchin usually turns away from bright light, so by continual revolution from the exposed light side, it might be that it would sink deeper and deeper. But this is hard to believe in the case of this basalt which resists so stoutly even the eternal pounding of waves. Other urchins with very long, needle-like, purple spines crept slowly along, carried by their ambulatory spines and tube feet.

Crabs, hydroids, sea-fans, sponges, serpent stars and weird worms filled every grotto, while naked mollusks moved about or clung motionless, mimicking everything else in the pool, both animal and vegetable. In one small cup were three close together, all less than an inch in length, all as unlike each other as the elements. One was oval, brilliant orange, and only when I saw it flow gently along a frond did I know it was not a bit of sponge. Six irregular pale brown spots were in two rows on the back, the front of the mantle was turned up in folds, exposing the rounded head, while a tuft of gill foliage sprang from the posterior part of the back. A second was slender and elongate, pale translucent grey like some of the tunicates, with two hydroid-like, frosted tentacles, while four more sprang from the head, long, thin and in constant motion like the common round worms. The third was an anemone to all appearances, the base rich lemon colour paling to grey behind, the neck tentacles deeply ringed, those on the back in four, close, clustered patches, grey with blood red bases.

When a stone was lifted, there would often be seen a whole nest of eels or sea-worms, writhing and twisting down out of sight, but when at last I was able to balk this retreat by putting my hand under another stone, I discovered that all were one organism, the medusa-like arms of a serpent starfish, ringed conspicuously with successive annuli of black, green and white.

The crabs should have a whole chapter to themselves, the ever more wonderful scarlet chaps who came down from the upper caves to see what we were doing, appearing and disappearing like flickering flames among the smoky seaweed. Bristly crabs inhabited small individual lava cells, which were so like the hollows worn by sea-urchins that it was certain that they were worn only by the drawing and quartering of the rightful architects. Yet they seemed to fit strangely well into their niches, and could not be pried out. They were ever ready to pinch with their claws, or to draw blood by shutting down on a finger with any of their walking legs.

My favourite crabs were rich crimson little fellows, the largest not of a penny's diameter, who had no height, only length and breadth. Everything about them was broad and flat, and thin as tissue, and it needed all one's attention to catch one, for they could slip between closed fingers and into the merest crack of a crevice like magic. When cornered, they gave up all attempt at escape and waved their claws valiantly and even rushed forward to attack a finger.

The jewels of the pools were little fish, less than an inch over all. Each pool had one or more, so gorgeous in colour that no artist's pigment can do them justice. They were fish—but they were colour before they were fish. Looking at them they hardly seemed real, and one thought of colour long before any suggestion of fins or scales came to mind. I have seen enthusiasm personified, and cruelty too, just as here were tints and shades harmonized into organic motion.

As they slipped in and out of the sunlight they were like rainbow fragments, electric blue, blazing red, glowing yellow, cool violet. After revelling in the violence of their hues, the eyes sought the restfulness of the cool green algal background. Three of us working together felt well repaid if we caught one after a half-hour of constant effort, for we had systematically to stop up all cracks with seaweed caulking, sink our hand-nets low down and then move slowly, a half inch at a time, until we had the tiny creature cornered. One which we called the azure fishlet was azurite blue in general, with shadings of dark violet; another was violet blue on the upper half of the body, cadmium orange on the lower half and tail, ventral and anal fins flame scarlet, and the cheeks grey blue; still another was dead black with electric flashes of blue in certain lights, while the posterior fins were apricot yellow. The red-backed azure fishlet was new to science, *Eupomacentrus beebei*, and had the upper third of the body scarlet, and the remainder violet blue, shading into soft grey below. A chubby black and yellow banded fish no larger in size rejoiced in the name of *Abudefduf saxatilis*. It has persisted unchanged in the Pacific for some thirty-five million years, from before the geological closing of the isthmus, for identical individuals are found in the Atlantic as far north as Florida. On our second and last visit to the pools we captured one of each of these little fish and brought them alive to the yacht, where Isabel Cooper painted them before death erased their brilliant colours.

This final visit was memorable for reasons besides the specimens we secured. We reached the pools at what we thought was dead low tide and made the most of every moment. We had been working about an hour, when I straightened up to ease an aching back. Almost at my side I saw what will be ever to me the most remarkable sight in the animal world. Frightened by our long continued splashing and tramping, a big octopus had crept quietly out of a crevice just behind me and was making his way as rapidly as possible over the seaweed shelf down to deep water. Nothing animate is comparable to this sight. The bulging mass of the head or body or both, the round staring eyes, as perfect and expressive as those of a mammal, and the horrible absence of all other bodily parts which such an eyed creature should have,— nothing more but eight horrid, cup-covered, snaky tentacles, reaching out in front, splaying sideways, and pushing behind, while one or more always waved in the air in the direction of suspected danger, as if in some sort of infernal adieu. I have seen them before, but I have always a struggle before I can make my hands do their duty and seize a tentacle, at the same time protecting them from being drawn into the parrot-beaked mouth. This octopus was over two feet across, jet black when I first saw him, but turning to a mottled grey when we engaged in our struggle. When I headed him off, he stood on defence and did not retreat. After much feinting and slipping and unpleasant pulling away from the myriad suckers, I got the beast into a snake-bag and tied it firmly.

I looked about and saw two of the party standing almost on their heads in frantic attempts to round up three new little fishes. Just beyond, the marine artist was absorbed in trying to fix on his canvas the spectrum of colours in the water and weed and caves in front of him. Around me were cameras, jars, pails and a large white basin with our best treasures. Down on my knees I dropped for a renewed attempt at a fish whose eyes were duplicated by a pair of pigment ocelli near his tail fin. Suddenly a warning shout came from an upper shoulder of the island where our Snake Hunter was looking down on us. A glance showed the whole Pacific Ocean rolling in—or so it appeared to my startled eyes, a seething, boiling wave coming over the ledge rim. I let out my wildest yell, grabbed at the big basin, held it over my head, and was swept against the side of a giant boulder, losing half the water and specimens. Harry Hoffman the artist had acted instantly and instinctively, and stood up, seizing paint box and canvas and holding them aloft, while easel, seat, palette, coat and hat washed merrily in and against the cliff wall. One by one, dripping and astonished faces rose above the foaming flood, and hands grasping something more or less of value. At once the wave retreated, all but washing us out over the ledge into sharkland. We looked about, saw kodaks and other things at the bottom of the deepest pool, while Hoffman made frantic attempts to kneel or step on his seaward-headed easel. He missed it, but he soon had another chance, for a second wave came in waist high, and after that a third. This was shoulder high, but by now, like Noah's distant friends, and the third-plus individual animals of an earlier day, we had climbed up, partly out of the reach of the water. Diving and pulling ourselves down by seaweed strands, we salvaged the limp cameras and other jetsam, and now observed that the water did not recede. The tide had begun to come in with remarkable rapidity, in trios of waves, and if we remained we would soon be covered, with no chance of further retreat. No boat was in sight, and in fact it was not due for another quarter of an hour, so we began to creep and crawl along shore, inch-worming over sheer, smooth slopes, I always hugging my residue of fish in my basin. Never have I known before the most awkward thing in the world to carry, but now I am certain,—an oval basin, half full of water, with a small handle at each end. The lava cut our feet and knees, the incoming waves played pitch and toss with us, but when we reached the last possible promontory, we heard the welcome put-put of the dinghy's engine, and our adventure was over. Looking back we saw the waves rolling over the highest boulder of the Octopus Pools, and we wondered what the octopus in the snake-bag was doing.

Our preoccupation with the creatures of the land gave us but scant time for trawling, but every scrape of the bottom, even in ten fathoms, yielded a rich harvest. There was little use dredging over lava, for the sharp points either tore the net to pieces or anchored us firmly. Our method was to chug slowly along until our water glass showed a sandy bottom.

On my first return from the Octopus Pools of Eden the sun was bright and just at the right altitude for correct refraction on the bottom. I could see every detail; a long, deep furrow, at the end of which was a great mollusk, plowing sturdily along like a submarine tractor. And as the gulls of our western states follow the plow for worms, so here was a cloud of tiny fish, swimming just behind, now and then diving down to the sand as some small edible creature was exposed. Giant starfish were scattered about, palms down and fingers wide-stretched on the white sand. In one spot I saw the Big Dipper done perfectly in living starfish, as if the fairy tale of our youth had come true, and here in this equatorial latitude the nocturnal reflection of the low-swung constellation of the Dipper had tipped, filled with water, and sank to the bottom. When we brought up these starfish, they proved to be brilliant scarlet above and blood red beneath. Fawn-coloured sand dollars also came up with the net, small stones covered with bright sponges, interesting shells with still more interesting and less known tenants, shrimps and small fish, the latter mostly young and not to be identified, so slight is our knowledge of the fish life of the equator.

The same sunshine that lighted the bottom thirty feet down so clearly, showed nothing floating in the water between, so that it was with scant hope of success that I paid out the rope of the surface net. For thirty seconds I drew it slowly through the water, just beneath the top in the full sun of noonday. When I lifted it into the boat I saw on the bottom mesh a handful of brown ooze. Gently washed into a pail, this resolved into a tremendously rich haul, and in a narrow, high aquarium in the yacht's laboratory it was indescribably beautiful. There were hosts of shrimps and larval crabs, zoëas, bearing no resemblance to their ambulatory parents, but consisting chiefly of great fore and aft spines, with huge eyes and a very diminutive body, well protected against the maw of any tiny enemies of their small cosmos. Copepods—minute crustaceans with very long antennæ which they use as oars—formed the majority of the mass of plankton. They are useful scavengers of the sea, and in turn compose the principle food of many fish. They are delicate little beings, mostly about one-twentieth of an inch in length, of various colours, orange, white, pink and blue. The pink ones so dominated in numbers that they gave a tinge to the entire mass. Some appeared brownish in the net, but with the light behind them, they showed as brilliant ultramarine. Crystalline, arrow or torpedo-shaped beings, a half inch in length, shot with incredible rapidity in all directions through the mass of life. They were transparent worms of sorts called *Sagitta*, and well merited their name, for they had a many-barbed, or rather jawed head, and even guiding feathers of fins. They are very voracious and feed on copepods, diatoms, small shrimps and even larval fish. When they shoot to their mark, the teeth and bristles never fail to transfix their small prey. While they are thus engaged, like the rest of us, with the search for food and the avoidance of enemies, courtship and marriage form no concern of theirs, for they are hermaphrodite, and are sufficient unto himself or herself, or rather themself, producing and fertilizing their own eggs and depositing them freely in the sea-water. My haul brought hundreds of fish-eggs, so transparent that every detail of the tiny curled-up embryos could be made out, and there were besides, eggs of many unknown organisms, green, brown, blue and yellow. Shrimps of all shapes and sizes swam about, some on their backs, either with minute bodies

Fig. 26

Tide-pools of Eden

The home of octopus and wonderful fishes

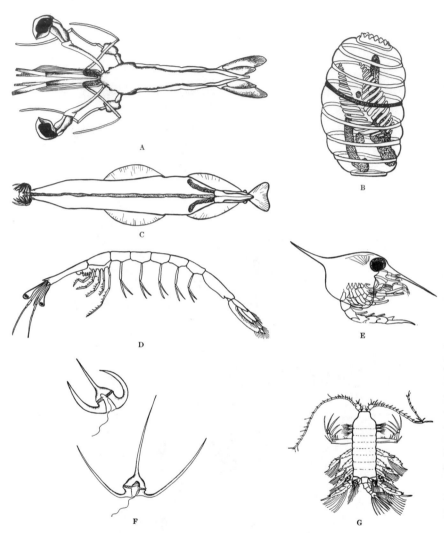

FIG. 27

COMMON ANIMALS IN SURFACE HAULS

A—Stalk-eyed Shrimp   D—Leucifer, Long-necked Shrimp
B—Doliolum,—A Pelagic Ascidian   E—Zoëa stage of a Crab
C—Sagitta, a Sea Worm   F—Cerithium—a Single-celled Animal
              G—Shrimp with Asymmetrical Antennæ

and enormous legs, or with eyes elevated on slender stalks, half as long as the entire animal.

The average human being is seldom profoundly affected or impressed with animal life, either common or bizarre, because he does not actually perceive it. Our subconscious recording of every creature, through eye, ear, or touch goes steadily on, filling that hollow thing at the end of our necks with untold details and facts and marvels, which to the very death of the untrained layman are to lie dormant, unremembered and unused. His consciousness sees little, hears little of all this,—his human field of interest registers the monstrous advertisements in the scenery, of cocoa, or watch, or turkish bath, and misses the birds, the startled fox, the colour of new-sprouted willows. He looks over the steamer's rail and records actually nothing, allowing the passing water to glaze his eyes with a hypnotic film, or he peers ahead and thinks of the next day ashore. On the other hand we scientists tend to become blinded by myopic consciousness of detailed facts, like Darwin, who, with his marvellous introspection, sadly recorded his declining appreciation of music, of painting, of sculpture as his days were filled more and more with the concentrated recording of facts. Yet now and then the most technical scientist is startled from his factual coma, to sheer fervour of unreasonable, illogical amazement at the habitations in which Nature can evolve and sustain life. I doubt if the most abstract Ichthyologist breathing, if suddenly, and for the first time, confronted by a young moonfish, would calmly pronounce it, in quiet voice and with steady pulse, merely an immature Carangid, in particular *Selene vomer*. At one of the hauls of the seine near the shore of Harrison Cove, six of these little beings were entangled. Seeing them all lying on their sides, apparently dead from the shock, the Artist piled them one on top of the other like the half dozen silver dollars which they resembled, and put them in the fish can of unusual specimens. They promptly came to life and showed remarkable viability, long outliving all the other twelve or fifteen species caught the same day. Moonfish is their pleasantest name, whether we say it in English or use their generic *Selene*. Fishermen with impressionism rather than poetry call them horseheads or lookdowns, while Linnæus' mind, one hundred and sixty-five years ago, worked along very mythological lines, and he dubbed them *Zeus*. This however fell in some taxonomic passage-at-arms and Cuvier's more appropriate name has persisted. Somehow or someway, long before the Panama Canal was planned or executed, this little fish had established itself in both the Atlantic and Pacific, and the identical moonfish which we seined here off Indefatigable Island is occasionally caught by Cape Cod fishermen, three thousand miles to the northeast, as the swallow flies.

Look up its picture and forgive my failure at detailed description. It was like nothing else in the world, and yet its profile reminded me with terrible persistency of several of my good friends. To those of us whose physical spareness is our most emphasized gift in life, the sight of a young moonfish makes us feel positively obese. In depth my fish was exactly two inches, while its absurdly long-drawn-out, high foreheaded face lacked only a quarter of an inch of this. From the side it glowed like a freshly opened shell of mother-of-pearl, or unoxidized silver-foil—or as only a moonfish can. Its wide, surprised eye rolled inquiringly and its downward-turned, pessimistic mouth forever opened and shut, engulfing the current which meant life. As it slowly rotated and appeared head on, the moonfish went into eclipse and showed nothing but a narrow sliver of silver, with tiny mouth and two slight black dots marking the eyes. At its thickest it was only a quarter of an inch, and yet in this vertical line of a fish were compressed all the bones, organs, emotions and life processes which characterize the plumpest trout or grouper. It happened that in the same haul we secured a half dozen small flounders and when these were all put in the same aquarium, more diverse types of piscine life could not be imagined. Both seemed to be living in a land of two dimensions, the moonfish with no breadth, and the flounders without height! And the colours were equally different, those of the flounder absolute representations of the sand and shells on the floor of the cove, the moonfish painted with aquamarine, with silvery spray, with the sheen of foam and sky reflections.

The silvery sides of the young moonfish were oxidized with a few spots of delicate bronze, and back from the eye streamed a stain of the same colour, like the tail of a comet. Aside from this there were only three areas of dark colour, first the round pupil, which was not a colour but a hole, second, the two long streamers from the dorsal fin, extending back quite twice the length of the whole fish, and finally the very long slender ventral fins which were darkly pigmented. In the course of time a thing will occur as strange as the form of the moonfish. As it grows older and larger, the long filaments and the fan-like fins grow shorter, are either worn away or, like a tadpole's tail, actually absorbed. Why this is we do not know, only it is interesting that the larval characteristics which will be lost are already outlined in black, marked for amputation. With all their intense verticality, their short fins, their downward looks, they had themselves under most excellent control, and to catch one in a net in a small aquarium required continued swift, well-judged efforts. The moonfish and flounders lived their lives at right angles to one another, yet seemed to get out of life food, safety, and a mate, and no fish could ask for more.

Harrison Cove was one of our favourite seining places and James Curtis who took charge of this phase of work, secured some very interesting specimens. Mullets were the most abundant species, and sometimes a school of two hundred would be caught. They flopped about the sand and shallows until incoming waves salvaged them, and they swam swiftly away. A great ray swept back and forth on the bottom at high tide, but was too canny to be caught. We saw no flounders over four inches in length, but these were of two species, one round, and one more oval in form. The small larval mullets known as *Querimana*, were ordinary in appearance, but had many interesting habits in aquariums. I shall detail these elsewhere, and here only mention two from the first haul we made. Eight of these little fish were put in an aquarium together, two of which instantly detached themselves, and swam about close together, alongside and one slightly in advance of the other. No matter how abrupt the turn or unexpected the movement even in intensest fear, they maintained the same relative positions,—both moving as one. The next morning all were dead, scattered about on the bottom of the aquarium, while the two inseparables were lying on their sides, quite stiff and dead, but in the same position in which they swam about when alive. Although we missed the ray we secured a two-foot baby hammerhead shark, and innumerable puffers which squeaked and blew themselves up with air, until the gulls could do nothing but poke at them as they bobbed on the surface of the water. Then, without warning, with a final gurgle, they would collapse, sink and swim away. There were trumpet-fish with mouths drawn into unbelievably long tubes, and half-beaks with sword-like lower jaw, and no upper lips at all.

The current and tide swept past the *Noma* as she lay at anchor, and time after time some strange creature would be seen to drift past, escaping our frantic efforts at capture from the companionway. One evening the phosphorescence was unusually beautiful, the sea a mass of lambent fire and sparks. The next morning while looking over the rail I was astonished to see a number of small, rounded specks, of the most delicate, turquoise blue, phosphorescent light in the water, drifting slowly past. I got out in a boat and after several attempts I caught one. The moment it was disturbed it sank and the light vanished. I saw nothing in my net, but rinsed it carefully into a small aquarium. After all had quieted down, the pale blue glow again came and in bright sunlight glowed with all the energy of phosphorescence. It was the copepod *Sapphirina*, and its light, in spite of appearances, is due only to interference colours such as we see in soap bubbles or oil spreading upon water, and not at all to phosphorescence. Later we saw many more,—beautiful drifting globes of most marvellous shades of blue. In this group of copepods the eyes of the female are unusually well developed, and correlated with this, the males show all sorts of strange decorations,—brilliant colours, and elaborate plumes radiating from the various appendages. It gave an added wonder to the drifting specks of turquoise to realize that even in their small life there might be competition, rivalry, courtship and choice.

The commonest species of *Sapphirina* hereabouts was over five millimetres in length, an enormous size for a copepod, as a man thirty feet tall would compare with one of usual height. The body was broad; it had short antennæ, and two rod-like appendages. It could move very rapidly by rowing with its feet, or glide gently along, apparently by means of invisible cilia.

At high tide, if an abundance of sea life was desired, we would go to one of the narrow-mouthed lagoons, some of which extended a mile or more inland from the north shore of Indefatigable. Paddling slowly, or drifting quietly along on the tide, the water seemed often aboil with living organisms,—

great yellow sea turtles barging along, watchful and rather wary, although what they had to fear from enemies hereabouts was difficult to imagine. Sea-lions popped up, looked, barked and sank; sharks sometimes rubbed their backs against the boat, mostly under seven feet, and thousands of fish of various kinds swam about, usually in great schools, and of carefully graduated sizes, all of a size in a school by themselves, perhaps a mute reflection of a tendency to cannibalism. A mist of silvery fry would drift past, suddenly to vanish at some unseen threat, and their places would be instantly filled with half a thousand angel fish, all a foot long, black, with bright yellow tails and an astonishingly conspicuous white slash from the shoulder down each side of the body. Mockingbirds and little black finches alighted on the nearest bushes or the gunwale of the boat itself, and sang; crabs wandered along the bank, herons came close and squawked, hawks dived down, an osprey picked and chose among the piscine multitudes, and for an hour the lagoon completely refuted the paucity of Galápagan life.

CHAPTER VII

RAINBOW CHASING

By William Beebe and Ruth Rose

THE preference for person, animal or object possessed of a past, a pedigree or a history, is a characteristic common to most human beings. The man of adventures, known or suspected, is sure of an audience, even though it is an enviously sceptical one; the animal, bipedal or otherwise, whose ancestry is one of sounding titles commands a respect that is independent of his individual worth; and who would not exchange a houseful of furniture for, let us say, the battered stool from which the Maid of Orleans faced her judges?

So who would not prefer to put to sea in a yacht whose forward stack bore two service stripes, won in the Bay of Biscay? Ordinarily it is only after trial by voyage that one endows a ship with a personality, but when you see on her after stack two stars, symbols of submarines met and overcome, your feeling for her is no longer akin to that absence of interest with which you regard a Pullman, but the emotion of one invited to share perils with a tried warrior of the deep.

If I could imagine myself in a position of having to choose between several yachts, the choice would have surely fallen on the *Noma*, of staunch and doughty record. She is symbolical of the change that the war brought into many lives, a well-groomed, luxurious lady, born to a life of ease, suddenly precip-

143

itated into hardship and danger and toil, but bearing herself as gallantly as ever she did in former days of elegance.

Vincent Astor came recently to my studio and brought the diary of a young seaman who was aboard the *Noma* when she was a patrol in stormy sub-infested waters, and it has been interesting to recognize familiar spots in the narrative. That sheltered nook on the boat deck where our pet monkeys used to discuss the voyage, was the only place where the war-time crew could cling during a night of black and raging storm. We boasted that we rolled 34° crossing the Caribbean, but before the 1917 record of 47° in the Bay of Biscay, we were abashed. It was to my cabin that the dying sailor was carried after his rescue from a shelled ship. And there is something more than faintly reminiscent in the seaman's description of the leaks over his bunk that let in generous quantities of ocean.

One sad thing about a hero is the difficulty of being continuously heroic. People expect it of him and are bitterly disillusioned when they discover that he has perhaps a sulky temper, or an ineradicable tendency to eat with his knife. So we had focussed our admiring attention on the *Noma's* past as a war-vessel, and we were shocked to find that she was not also an inexhaustible reservoir of fresh water and a floating coal-mine.

In our innocence, we depended on charts and the pilot-book for information about obtaining water on the Galápagos. We had no other reliable source of knowledge. In fact, we did not even have that one, had we only known it. Here and there on charts of the archipelago were cheery words,—"fresh water," "fresh water," sprinkled about in a liberal way, and always on or very near the coast. With such optimistic tidings, anyone who could have cherished forebodings that all was not well would have been a super-Cassandra.

RAINBOW CHASING

Since rainbow chasing is traditionally a futile and a profitless pursuit, it is the title of this chapter, but not with entire justice. Had fresh water been the sole object of our explorations, we would have been indeed empty handed, but every time we sallied forth, the resulting treasure of strange beauty and interest almost made us forget that we were thirsty. Spend three or four hours under an equatorial sun, scrambling and sliding over never-ending heaps of corrugated broken rock, which support a liberal crop of spines and thorns; postpone again and again tasting the meagre supply of fresh water that gurgles so enticingly in the canteen; and then at last take one swallow of the lukewarm tinny liquid contained in that canteen, and you will know what nectar and ambrosia were really like.

When we left Panama we realized that the *Noma* could not carry a sufficient quantity of water for the whole cruise. For her boilers alone forty tons were required, and we agreed to be very saving in our use of the liquid. Of course we were not worried: our implicit faith in the charts prevented that. At our first anchorage in Conway Bay, off Indefatigable, we had no hope of finding water, and that was not yet the absorbing topic it later on became.

The thrill of one's first desert island is a quite indescribable thing and the fascination of these barren Galápagos is inexplicable. A land pitted by countless craters and heaped into a myriad mounds of fragmentary rock; gnarled trees bleached by salt spray, with twisted, stunted branches; grotesque cactus stiffly outlined against black lava boulders,—this picture, truthful in its bare statement of facts, cannot convey the feeling of mystery that led us on and on, wondering always what lay just beyond the next barrier of sombre rock.

The impression of strangeness, of difference, was augmented by the silence. Now and then the brief sweet song of a mocking bird or finch sounded from a thorny tangle; the scream of a gull served merely to italicize the stillness. At night even these sounds were missing. There was no hum of insects, no chorus of frogs, and though the fresh wind blew continuously, it made no sough or rustle in the sparse, brittle foliage. The swish of gentle ripples on the stony shores soon became only a part of silence.

If we, living in comfort, felt all this, what must those castaways have felt, who from time to time have found themselves on one of these lonely islands? The Ecuadorian convict, who, for extra punishment, was marooned on this very island of Indefatigable, lived for three years in utter solitude, encompassed in a stillness like a vacuum.

After a week at Conway Bay, during which we visited numerous islets in the vicinity, the water question became one requiring an answer. On the opposite side of Indefatigable was Academy Bay, and here, the chart told us, was fresh water. A detachment went round in the large motorboat to investigate this source and found it wholly mythical. The bay was large, with high cliffs all around, and at its head was a sandy beach. Here was brackish water, but none fit or sufficient for boiler use. There were many yellow sea-turtles to be seen and at the entrance to the bay was a little island covered with sea-lions, old and young, where pelicans and sea-birds were nesting.

This party brought back a story which, in our then comparatively well-watered state, was far more exciting than reports of rivers and cascades. Before they reached Academy Bay, they passed a wreck, which looked like a recent one. It was a schooner stranded high on a reef made unapproachable by heavy surf; one mast was broken short and the hull was being rapidly battered to pieces. No sign of life was to be seen.

RAINBOW CHASING

In Academy Bay they encountered the first and only ship we were to see in the Galápagos,—a small steamer sent out by the Ecuadorian government to search for survivors from the wreck. From this rescuing party it was learned that the schooner had struck the reef during the preceding week, and that twenty-two men, women and children were missing. Two survivors in a small boat had been picked up some distance from the islands by a passing ship, and thus the loss reported at Guayaquil. The search party had visited Academy Bay in the hope of finding the castaways there, since there is the only water, scanty and brackish but drinkable, on the island. They had found no traces of human beings and were giving up the search. For many days thereafter we were haunted by visions of suffering perhaps only a few miles away, and more than once a fancied resemblance in tree-stump or boulder to a human figure made us hurry to reach a possible exhausted survivor. Later on we found a small boat washed up high and dry, and a heavy cushioned seat floated in to the beach, but so far as our knowledge goes the fate of those twenty-two is an unfinished story.

One day we were so fortunate as to be rained on. No one accustomed to regard rain merely as a deterrent to pleasure trips or a spoiler of best hats, can know what a luxury it is to stand in a drenching shower and let the blessed drops trickle down one's back and off the end of one's nose. Even though we lived in bathing suits and were constantly in and out of the clear water of the bay, we now spread our hands to the rain and realized the feeling of parched plants after a long drought. Members of the crew ran about the deck, holding buckets solicitously under every tiny stream that ran from stanchion and davit, and almost a ton of water was caught in scuppers and spread awnings. That was the only Galápagos rainstorm that the *Noma* encountered.

GALAPAGOS

From the time when we first approached Indefatigable and saw dimly far to starboard the rugged outlines of James Island, we looked at it with longing, for one of the two active volcanoes was said to be there, and our imagination easily transformed the drifting clouds into steam, and the glints of sunset into glowing lava. When the question of fresh water began to be serious, our Captain discovered a chart on which James Bay was marked "fresh water." Four of our party took a motorboat and went on a scouting excursion in search of the alleged H_2O. They brought back glowing accounts of fresh water, but what they had found came from heaven and not the lava and brimstone where the chart located it. It had rained hard for four hours, sufficient for the *Noma's* wants for a week, if we had only been there to catch it on the decks and awnings. The topsy-turviness of these islands affects more than the habits of their animal denizens. A group returning from a day's outing does not usually beam with excitement in describing how it rained and rained and they were all thoroughly soaked for hours, nor is it customary for their hearers to exclaim enviously, "How wonderful!"

On the chance of encountering a similar storm, we left Conway Bay on April 4th, and steamed the fifteen miles that separated Indefatigable and James. As we rounded the western side of the latter island, we fancied that likely-looking clouds were gathering near the coast, so the course was changed to bring us nearer shore in case the longed-for rain should fall. The spectacle of a large steam yacht hurrying to and fro in the Pacific, hoping to overtake some rain, has its unusual and comic aspects. Luckily for us, it never assumed a tragic one. The clouds, however, provided us with nothing more substantial than a thrill of hope and some æsthetic satisfaction in their beauty as they piled and dissolved in lovely kaleidoscopic changes.

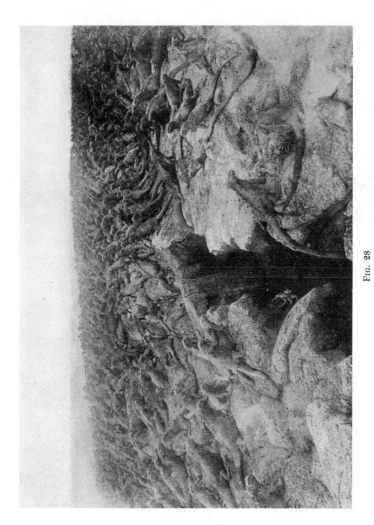

FIG. 28

AN ENORMOUS HERD OF AMBLYRHYNCHUS IGUANAS

Photographed many years ago by R. H. Beck

FIG. 29

THE "NOMA" AT SUNSET IN JAMES BAY

Showing the surf which upset our boat and made landing very dangerous

FIG. 30

THE BRACKISH POOLS OF JAMES

Near here lived wild Donkeys and Pigs, Flamingos and Vermilion Flycatchers

RAINBOW CHASING

We passed swiftly between Jervis and James, the former island looming high and broad-shouldered, with lofty cliffs and a low shrubby skin of vegetation. James was sinister with suggestion of recent volcanic outbreaks. The lofty central craters had evidently long been dead, but the lesser ones along the outer shoulders had spilled over recently, and crucified the island with layer after layer of lava, dead and black at present, rugged and jagged, with crevices and crevasses in every direction. Close to the shore were two sugar loaves with perfect, circular crater rim, and steep slopes which challenged climbing. Inside there might be hot earth stuff, or a fresh water lake, or more probably merely thirsty lava which sucked up every drop of rain.

The last lava flow, probably that of 1907, formed a belt about one hundred yards in width, for a long distance along the coast, with a straight line of vegetation behind. On higher slopes the new and the old lava flows often cut each other into rectangular shapes so that parts of the island had a superficial resemblance to the sugar-cane planted slopes of St. Kitts.

In less than three hours we covered the twenty-eight miles to James Bay, on the northwest side of the island, and anchored two miles offshore. Albany Island, a bit of land perhaps half a mile long, sheltered us on the east. No sea-birds were in sight, nor were there any small islets or reefs where they might breed or even perch.

Looking ashore we saw that the black lava came down to the shore on the right, then a tortured, half-crushed hill of reddish slag, and then vegetation. In front was a beach of dark brown volcanic sand, an astonishing contrast to the crescents of dazzling whiteness that scalloped the edge of Indefatigable. The shore led at once into dense growth, and this into relatively high trees and almost jungle on the slopes of the mountain. The proximity of the mountains to the water brought the rain nearer, hence transition flora and humid vegetation within a mile. No cactus was visible on all the mountain side, and only a solitary plant near the beach on a peculiar castle of lava.

This aspect of the island was encouraging and we hurried ashore in the lifeboat and a dinghy with an outboard motor. The waves seemed only moderately strong though there was considerable surf. We found, however, that the great smooth rollers over which the boats slid so easily, smashed with terrific force on the sand. The brown beach dropped off steeply at the water's very edge, and a heavy undertow and a strong cross-current made landing an unexpectedly difficult and athletic affair. The receding waves washed back almost at right angles to their onrush, and the boats swung sideways, the next breaker crashing with a Wagnerian suddenness that threatened to destroy both boats. By leaping out and rushing them as high as possible, we landed without mishap. But looking back and seeing how the high-arching, emerald masses completely eclipsed the *Noma* in the distance, we wondered how the return would work out.

When I had regained my breath and looked around, I realized that I was on classic ground, for Charles Darwin had spent a whole week on shore near this very spot, and had made some of his most interesting notes here.

James was a favourite resort of the buccaneers in the seventeenth century and they named it in honour of James I of England. Both Captain Colnett and Captain Porter found here such souvenirs of their visits as broken wine-jars, old daggers and benches that they had fashioned of earth and stone.

The onrushing tide swiftly erased all evidence of our struggle to land, but from high water up there was an almost unbroken series of strange, stolid tracks,—deep scallops cut one after the other into the dark sand. And at the crest of the beach, often under the outlying bushes, were many 'deep craters,—each the rifled nest of some sea turtle, filled with the withered bits of shells. Scattered along the shore, standing out against the dark sand like golf balls on a green, and of exactly the same shape, size and colour, were a few turtle eggs, all with a dent or two in the side—all dead and very ancient. Something had dug up all the others and devoured them. What that something might be, we perfectly surmised when we saw on the inshore mud the cloven footprints of wild pigs— hundreds of them.

I started to enter the barrier of green a few yards to the north and there I found a turtle tragedy. Long ago, an aged Chelonian of the sea had flippered her way laboriously up the sands, and intent on finding a safe place for her nest, had lowered her wrinkled old head and pushed under the spiny branches. Under the first and over the second she went, and then a jagged end snapped around her shell, a crotch received her neck and for effectiveness she might as well have stepped into a steel trap. These turtles live very slowly, so this one must have died slowly, and unless all our knowledge is at fault, probably suffered little in her lingering death, as we understand suffering. Then the elements began their work,—sun and rain, heat and moisture—and the katabolic bacteria, who know no distance or isolation, to whom a far distant island is as much home as the heart of a city. All, gradually working together, loosed and scattered the atoms of the old turtle's body,—into the sand, to the far winds and the swirling currents, and today a mosaic of whitened bones, the skull still in the crotch, the bent branch yet around the big shell, was all that was left.

Bagpipes in a mosque would be no more unexpected and comically out of place than the first sound that greeted us here.

Far up in the hills, echoing above the pound of the surf, we heard a long-drawn-out "Hee-haw! Hee-haw!" as a donkey, long since reverted to the wild, saw, for probably the first time in his life, that species of animal that had once upon a time captured and enslaved his ancestors.

The presence on the Galápagos of the descendants of once domestic animals is susceptible of many explanations. Buccaneers deliberately stocked some of the islands with goats and pigs as a provision for future visits. Attempts at colonization account for the numbers of wild cattle that roam on Albermarle and Charles, while such creatures as wild dogs and cats, said to be present in large numbers, are probably the offspring of deserters from visiting ships, castaways from wrecks and strays from abandoned colonies. On James alone there are said to be upward of twenty thousand wild donkeys, though it is hard to fancy how the imaginative person who made this estimate set about taking his census.

Just inside the heavy screen of bushes the donkey trails began. Everywhere that we went on James we found a network of these trails, crossing and recrossing each other in a complete and complicated traffic system. Underfoot they were clear and well-trodden and in the days to come of scrambling over other islands, we often remembered with regret those paths so well made by a myriad little hooves. Our only faultfinding was with the donkeys' height. If they had only by taking a little thought, added a cubit to their stature, we need not have had to bend double in order to follow in their footsteps.

Crouching along in this fashion, I came almost at once upon two shallow salt ponds, not more than a hundred feet from the beach, their muddy shores etched with the trails of various creatures. The heart-shaped spoor of pig and the neat round impression of donkeys' hooves were large and clear, while around and over them was a maze of fine scratchings and grooves where fiddler crabs of divers sizes had sidled through the mud, and everywhere were the branching prints of birds' claws.

I walked slowly into the shallow, more than tepid, water, and at once there came a slither and spatter, and a pair of Galápagos pintail ducks alighted within eight feet. Up they went again, circled, and fairly spattered me with water as they again alighted. They whirled around, watching me first with one, then with the other eye, the head feathers rose with excitement, but no trace of fear. They were trim little things in drab brown, much mottled with darker, a bright green mirror in the wings, white throat and cheeks and a carnelian patch at the base of the beak, this coloured area small and dull in the female. The female of this pair always took the lead, and was even more fearless than her mate. They swam off after a few minutes' scrutiny of me.

Then a low, musical whistling came to my ears, and close behind me was a second female with six downy ducklings in the rear, and a seventh far behind, paddling like mad to catch up and see what he was missing. The mother brought them close, and then individually they came nearer still. As I walked slowly along through the shallows, she kept pace with me, three or four yards farther out, talking in low tones all the time.

On a little muddy spit at the farther end of the first pool was a greater yellow-legs, two black-necked stilts, and five turnstone, the last gay in black, reddish and white. They did not fly until I was very close to them and then not until I had shot a ground finch.

Between the ponds was a narrow strip of land where the ivy-clad ruin of a castle tower stood out above the trees. When we reached it, it was a strangely-shaped rock of lava, possibly forty feet high, with tiny plants growing in every

153

crevice, giving a picturesque imitation of the wreck of a
mediæval fortress.

There were woods here, real trees of moderate height, and
we revelled in the luxury of shade and of walking almost free
from broken lava. Mockingbirds, flycatchers and finches
were numerous and even tamer than those we encountered
elsewhere. They were curious too and followed us at arm's
length, cocking their heads and inspecting us carefully. Now
and then a scuttling sound as of some large animal in the
carpet of dead leaves drew attention to a giant hermit crab
hurrying away through the forest in the house that was his only
by right of seizure. Fiddler crabs skittered about more
quietly. I followed a donkey's track which had been made
only a few hours before, and heard two of my party in an
excited discussion. They had found a rather scattered
skeleton of a donkey, and close by had picked up a big jaw
bone with large tusks. The argument was whether donkeys,
when they became feral, ever developed carnivorous tastes
and evolved tusks! Evidently a big tusker's skull had become
mixed with a donkey's bones.

Turning to one side after a huge-billed black finch, I saw
something white glistening through the thick Bursera bushes.
In an oblong glade on the far side of the second pool was a
solid mass of white donkey bones, some very recent, others
with weedy stems growing through them. I counted sixteen
lower jaw bones, and there were probably others which I
missed. During my succeeding walk of a mile inland I found
three big skulls widely scattered, but the donkeys seem to have
made a cemetery of this glade, and, as elephants and guanacos
are reported to do, to have come here from all over the island
when they felt the approach of death.

I went on and on along the narrow trails, bending almost
double to avoid the low swung branches which were polished

from having scraped the backs of innumerable donkeys, or else the backs of some donkeys innumerable times. Another hundred and fifty feet brought me out into open country with tall weeds and scattered trees of good height. Here butterflies and day-flying moths were abundant. Land birds were more common than I have seen them anywhere. Yellow warblers and several vermilion flycatchers made bright spots of colour, a novelty to our eyes that had grown accustomed to the dull plumage of the other native birds. Until I saw these, crabs, grasshoppers and land iguanas had been the only really colourful organisms.

Mockingbirds were far in the lead in numbers, with black finches next. Nothing could be less striking than these latter birds—the black, full-plumaged males and the black-headed or mottled drab females. The sooty ground finch, *Geospiza fuliginosa*, was the most abundant. One black male sat near his nest and darted fiercely at every passing bird. He was the species with a moderate sized bill, and he sat on the top of a moderate tree, and sang continuously a very moderate song, consisting of the syllables *cheèsee-cheèsee! cheèsee-cheèsee!* with apparently no intention of ever ceasing.

One hundred feet away, another black finch with bill of monstrous size, perched in the top of a larger tree, sang a clear variation, *witchee-witchee! witchee-witchee!* This was *Geospiza magnirostris*, the great-billed ground finch. He did not sing continuously, but courteously stopped now and then to give opportunity to a parrot-billed tree finch, *Camarhynchus psittaculus*, lower down in the same tree. This bird with its curious, short, rounded beak, was a male with the black plumage only on head and shoulders and a light breast. As soon as his great-billed cousin stopped, he came in with *chee-chee-chee-chee-chee-chee!* uttered much more rapidly than the human imitation can give it. The strange thing was that in the course of my walk I saw and heard three different pairs of birds of these two species doing the identical thing, keeping up regularly alternating duets, first the great-billed and then the parrot-billed.

In the whole afternoon I heard complete mockingbirds' songs only twice, while every thicket rang with the modest, but indefatigable attempts of the finches. Hardly recorded by the ear, so familiar a sound of all my life was it, the yellow warbler sang out sweetly now and then.

At last I reached the end of the short level coastal belt and began to climb. The going was easy, for the donkeys' trails in the open were as comfortable for me as for their makers. I stopped suddenly when I heard a sound which stirred the very depths of memory. When I returned to New York I discovered a strange coincidence. In my journal, written in the wilds of western Mexico twenty years ago to the very day, I found a description of the *coloradita* or vermilion flycatcher, "a male in full plumage seeming to outburn every other object in Nature, with the back, wings and tail dark grey, as if the scarlet had burnt itself out in these portions."

In the air overhead I saw a fluttering speck of red, a Galápagos vermilion flycatcher in the midst of his flight song—the identical notes which two decades ago had come to me from the arid deserts outside of Guadalajara, and now over the sparse vegetation of these lava slopes of James. My account of the Mexican bird holds good for that which I saw and heard today[1]:

"Half hidden on a spray of thorn-bush perched a shy *coloradita*, her dull striped breast and darker back merging with the grey environment. Ardent admirers had found her out and one after the other had tried their utmost to out-do each other in the little performance which nature had taught

[1] "Two Bird-lovers in Mexico," p. 92.

them. All thoughts of pursuing the gnats and grey-winged flies which swarmed about the cactus blossoms was gone, and quivering with eagerness, the brilliant little fellows put their whole heart into their aërial dance.

"Up shoots one from a mesquite tree, with full rounded crest, and breast puffed out until it seems a floating ball of vermilion—buoyed up on vibrating wings. Slowly, by successive upward throbs, the bird ascends, at each point of vibrating rest uttering his little love-song, — a cheerful *ching-tink-a-le-tink! ching-tink-a-le-tink!* which is the utmost he can do. When at the very limit of his flight, fifty or seventy-five feet above our heads, he redoubles his efforts and the *chings* and the *tinks* rapidly succeed each other. Suddenly, his strength exhausted, the suitor drops to earth almost vertically in a series of downward swoops, and alights near the wee grey form for which at present he exists."

Although a different species, the Galápagos bird was superficially indistinguishable, the differences being the smaller size and, like so many of the island birds, larger feet. The song was the same, and oceanic life had diminished the enthusiasm no whit.

The finches were either beginning to build or else sitting on eggs, and their round balls of nests were visible in all directions —covered, and with the openings directed toward the northwest—away from the winds and rains. I found a good many nests with four eggs, so as far as number goes island life has made no difference in the ontogeny of these birds. But a very suggestive thing was apparent when I came to blow the eggs. In the sets where embryos were formed I noticed that two of the four eggs were clear, and this struck me as so interesting that when I came to eggs which were almost fresh, I blew the eggs separately and made certain with a hand lens that the same thing was true in these cases; that fifty per cent of the eggs were quite infertile. In two sets of three eggs, one and two respectively would never have hatched. This observation extended to four species, *Geospiza fuliginosa, fortis, magnirostris* and *Camarhynchus psittaculus.*

This is important as revealing a condition where the adaptive diminution of numbers in reproduction is inaugurated as an internal, obscure phenomenon, not certainly referable to either one sex or the other. The full complement of eggs is still produced, but the failure of fertilization may be due to a defect in either sex. The absence of enemies, or the effect of some other environmental insular relaxation has apparently called forth this subtle but quick response,—a conserved correlation of offspring average.

In connection with this it is interesting to record the number of young of three mice from Indefatigable, *Nesoryzomys indefessus.* I collected one female with four young about a week old, and trapped two others each of which contained three embryos. This average of three and a half young is far below the number in litters of closely related mainland species, which range from four to eight, with an average of about six.

The mockingbirds showed a much greater temporal breeding latitude than the finches, and both fresh eggs and full-grown young were found. The average of infertile eggs was only about twenty-five per cent, the elimination being brought about in some other way, for it was rarely that more than two young birds were seen following their parents, and a single one was a very usual sight.

A typical nest was found to the east of the first pool by Dr. Mitchell, about nine feet up a Bursera tree, and was exceptional in containing four naked young about three days old. They showed, even at this early age, distinct evidence of the pattern of the adult bird. At my approach, these raised their heads and opened their beaks wide, but uttered no sound whatever. The nest was roughly globular, with the entrance one-third down from the arched top, and the young exactly in the centre of the mass. It was made of twigs, and lined with grass and donkey hair.

The young birds do not seem to have been described. They were flesh colour, with the closed eyes and the sprouting flights leaden blue; the beak was horn colour, the lower mandible lighter, and the gape bright yellow, orange within the mouth. The down was very sparse, four filaments over each eye, six in a transverse nuchal band, widely divided in the centre, two on the scapulars, and three large, brown downs on each femoral tract. In the smallest, the flight feathers were barely visible above the surface of the skin, in the largest they projected two millimetres. There were nine primaries and nine secondaries followed by two very small tertiaries, the greater and median secondary coverts pale and very well developed, an incipient down on each. Greater primary coverts dark, rather less developed than the secondary coverts, and bearing no down. An extra outer greater secondary covert lacked any attendant primary feather. The four nestlings were exactly graduated in size, the extremes weighing 4.5 and 6.8 grams.

Toward sunset, after I had signalled a return to the boats, I went back for a last visit to the pools. From the mountain high above me came again that sound which is one of the strangest in the world—weird and uncanny when heard from a barnyard, but from the jungle heights of an uninhabited island in the midst of the Pacific, it might have emanated from the throat of any fabulous beast of myth; a wild donkey was gasping and shrieking in preparation for that indescribable hee-haw, that trumpeting, rasping series of double outbursts which is known around the world. While the emotion of this was upon me, a great black animal walked out from the trees on the opposite side of the mirror pool, just this side of the strange lava castle. It stood still, cocked its ears and looked at me. Later a milk white donkey followed—both giants and with coats sleek as if just groomed. The sun had disappeared behind the mighty, northernmost volcano of Albemarle, and the pool and the wild donkeys were softened and indistinct in the early dusk. As I looked up four flamingos flew over, blazing in the last of the upper sunlight, four visions of scarlet; each, a pair of black-tipped, blood-red wings beating on a slender axis of neck in front and legs behind.

If there were ever watering places on this coast they had dried up when we were there, although it was at the height of the rainy season. Not so much as a drop of dew rewarded our anxious search, and toward sunset we gathered on the beach to compare our specimens, and to exchange gloomy views on the water situation. Two miles offshore the *Noma* was silhouetted against a golden sky, and a dazzling path marked our way straight to her side. High overhead the four flamingos again appeared, flapping slowly homeward, glowing in the slanting sunrays like pink sunset clouds. The surf was heavier than when we arrived, so we stowed guns, cameras and bags out of the way of a probable wetting, and prepared for a difficult launching. McKay had trekked far up the mountain in search of feral game, and as he had not yet appeared, two volunteered to wait for him with the motor dinghy. Wading out waist-deep into the incoming tide, we waited for the right wave. Eleven rollers piled and broke, each lifting us off our feet before the moment came. I saw a broad area of momentarily smooth water, and seven pairs of arms shoved mightily, and to an accompaniment of meaningless shouts, everyone flung himself madly over the gunwale and struggled to get to the oars before another wave should dash us back again. Jim Curtis did not fling quite fast enough, and suddenly found himself in deep water, holding to the boat by his chin, with nothing visible but his pith helmet and no longer able to spring aboard. Clutched by some portion of his attire by the Chief Hunter, he presently came in head first with a loud splash, we rowed like mad, and the succeeding wave, which was a mighty one, rolled harmlessly beneath our keel and broke with a roar just beyond.

It transpired later that McKay had stalked and shot a large sow and was struggling down the mountain with it, before darkness should overtake him. Galápagos mosquitoes work on a strict schedule, and precisely at sunset they appear in innumerable swarms, taking over the night shift from the biting flies that are very industrious during the daylight hours. We had discovered this arrangement very soon after our arrival in the archipelago and always hurried off to the *Noma* before twilight. Tee-Van and Broking, waiting on the beach, put in an active hour or so in fighting off the pests before the missing hunter appeared, staggering under his load of fresh pork.

We on the ship, busy with the arrangement of our specimens, gave no thought to the shore party until it occurred to someone that it was pitch dark and that three men were missing. When the big searchlight was turned on the distant shore, three tiny figures could be made out through the binoculars. They ran up and down the beach and waved their arms frantically, but as we could also see the dinghy, apparently unharmed, beside them, we attributed their wild gestures to the usual pitched battle with ravenous mosquitoes. A more careful scrutiny revealed the fact that two of them were wholly without clothes, and this aroused our suspicions that some accident had befallen them. The lifeboat was despatched and presently returned with the survivors, exhibiting among them a bruised head, a cut leg and various minor injuries. They had come to grief in the surf, for just as they succeeded in launching the dinghy, the outboard motor stalled after the manner of its kind, and a huge roller turned the boat neatly upside down. The dinghy was a complete wreck, and as their clothing had been cached under the thwarts to keep it dry, it was all lost in the catastrophe. They came aboard in costumes reminiscent of Private Mulvaney's at the Taking of Lung-tung-pen. Next morning a rifle was miraculously cast up by the sea, but the pig, first shot, then drowned, was lost forever.

Fresh water was now the chief topic of conversation. The boiler supply was getting low, and the health of the boilers had become to us more precious than the health of our nearest and dearest. We had long ago given up fresh water for any external use, and had found that toothpaste in combination with salt water gives the flavour that lasts. At table a meagre portion of fizzy bottled water was doled out three times a day; on other occasions apples and oranges satisfied our thirst.

On shore the red and black fritillery butterflies, *Agraulis*, were about as abundant as the black skippers, and in the same high, open, weedy places. When I was on shore I saw no concerted movement, the *Agraulis* flying slowly from flower to flower, or resting on leaves waving their wings in the sun. About seven o'clock the following morning, April 5th, an interesting migration of these butterflies took place, visible from the deck of the *Noma* two miles offshore. As far as my eyes or my glasses could reach in all directions, *Agraulis* appeared flying steadily and swiftly from southeast to northwest. Although they flew well apart, in a short time I counted three hundred and sixteen and caught six on the deck. When we left James at nine o'clock the migration was going on as strongly as ever. If maintained, this line of flight would take the butterflies parallel to Albemarle, out into the Pacific, with only the specks of Culpepper and Wenman far to the northwest, islets from which these butterflies have never been recorded.

Interesting as James Island was in its difference from Indefatigable, we dared not linger there, lest, with water exhausted, we should be forced to linger forever. Great things were promised by the chart on Albemarle, largest island of the group, so next morning we got under way for Tagus Cove. This was formerly a meeting place for whaling ships, and was called by them Rendezvous Bay. It is on the western side of Albemarle and to reach it we crossed the equator twice that day.

As we passed along the north and northwest coast of Albemarle, the views were magnificent. A geologist would have waxed most enthusiastic. Great mountainous crater slopes showed black streaks of lava alternating with almost equally black cloud shadows. Jet black cliffs at the water's edge were forever whitened with boiling breakers and faint cloud-like spray. Once a huge throne appeared at the edge of the cliff, as if waiting for some stray god; vast fissures developed and passed, dykes winding over the horizon like the great wall of China, and fumaroles everywhere: I counted twenty-seven on one small lava slope. Narborough came in sight to the south, a long, steeply sloped island, and as we turned the corner, the cold Antarctic current and air met us full force. The water dropped from 81° to 73°, and the air from 86° to 76°. Porpoises in large schools and shearwaters in greater numbers appeared, boobies single—white or grey—all these kept us company, while to the west we looked out toward the horizon leading unbroken to the farthest islands of the South Seas. When the lofty crater of Narborough was opposite and the wind was shut off from the narrow strait, the temperature again rose abruptly.

Albemarle is as rugged and barren as any of the Galápagos;

everywhere over the second great volcano the sparsely green slopes were streaked with patches and furrows of lava, spreading out here to fill a hollow and narrowing to a line there as the molten stuff found a channel in which to flow. Wonderful cloud effects formed round the crater tops, the highest of which reaches a five thousand foot elevation, and knowing that there are here vents from which steam and gas are constantly issuing, we strained our eyes in the attempt to make out smoke clouds over the misty mountains. Innumerable small craters pitted the flanks of the large ones, and even close to the water's edge there were little truncated cones that once upon a time had heaved themselves up and spit fire in imitation of their betters.

Tagus Cove looked so little like its portrait on the chart that we refused to recognize it, overshot it several miles, steaming further south in an endeavour to find something we liked better. We saw nothing but surf-drenched lava shores with no sign of shelter, so somewhat shamefacedly we returned. As we steamed slowly inshore, Tagus opened out before us into a perfect anchorage,—a long, narrow cove, with sheer black walls coming almost straight down to the sea all around, and with deep water to the very foot of the cliffs. Opposite the entrance rose the ravaged sides of Narborough Island; dark and grim at the head of the cove was a ravine which cleft the surrounding heights and offered the only possible landing place, though even here the surf broke heavily on an abrupt face of rock. A few shearwaters and one pelican were the only signs of life except the alien *Noma* creeping cautiously inshore. Against a wonderful sunset of purple and green the sombre land loomed dark. After the rattle of the anchor chains had ceased to echo from the cliffs, an eerie stillness settled down, and we felt that we might indeed be the "first, who ever burst, into that silent sea."

Fig. 31

The "Noma" and a Flightless Cormorant at Tagus Cove, Albemarle

Fig. 32

FLIGHTLESS CORMORANT ON HER NEST

Observe the short, useless wings
This bird lived to a good old age in the New York Aquarium

Fig. 33

GALÁPAGOS PENGUIN

Captured in Tagus Cove, close to the Flightless Cormorant

RAINBOW CHASING

Quotations from the *Ancient Mariner* seemed constantly to occur to us among these islands. Perhaps the presence of the albatross, seen on the first day, influenced us; certainly "water, water, everywhere, nor any drop to drink" was so appropriate as to be positively painful.

With the help of a scaling ladder we managed to land on a flat rock at the mouth of the ravine and had a half hour of daylight before the sun sank. This meant only a short walk, and an unsuccessful search for a fresh-water pond. In a shallow cave near the shore, many old dates were carved, one of 1836. An ant-hill in volcanic dust yielded six ants. The most conspicuous things were white long-petalled flowers, two to each plant, with half a dozen elongated leaves growing in the dust, into which their roots penetrated a bare half inch.

As we landed, pelicans were fishing in the dusk, but the greatest thrill came when we saw our first flightless cormorant swimming fearlessly near the boat. It watched us and now and then dived deeply, once passing far down under the boat. Just before dark as we were rowing back to the yacht, a bird swam across our bows, which looked very much like a shearwater, although one of us thought it was a penguin. This decided us to remain at least long enough on the morrow to learn more of these remarkable birds.

We returned after dark in an unenthusiastic frame of mind as far as fresh water was concerned. The chart said there was fresh water at Tagus Cove but by this time our attitude toward the chart statements was like that of most people toward the weather bureau's prophecies. There was nothing left to do but to go to Chatham Island, the only inhabited one of the Galápagos. Here there is a small penal colony, and here we knew we could get water, for the inhabitants must live, and besides we had read of schooners that in recent years had stopped there for refreshment.

GALAPAGOS

In order to reach Chatham on time, the Captain announced that we should have to leave Tagus Cove at ten o'clock the next morning. So we girded up our cameras, guns and collecting paraphernalia, and turned in early, ready for a few precious morning hours of the hardest, quickest work we could possibly achieve.

In this equatorial, twilightless world it was as dark as midnight at five o'clock when we came on deck into the cold sweet air. It was too chilly for comfort. A scant half moon hung in the zenith, but it was eclipsed by the brilliancy of Scorpio which was almost twisted about it. The Southern Cross lay flat on its side near the horizon, Gemini was low in the west, and Venus the magnificent rested like some wonderful unearthly beacon, exactly on the summit of the mountain east of Tagus.

On all sides of us were the high, mysterious cliffs, dark and silent as they had been when they looked down upon the ships of the old buccaneers, of the whalers, of occasional ships of war of various nations, even upon Lord Byron in charge of the dead bodies of the King and Queen of the Sandwich Islands lying in state.

Not a sound of night bird, or distant bark of dog in this desolate land; only motionless heaps of cinders, piles of dust undisturbed by any breath of air and mean foliage hanging limp and quiet as the rocks themselves. Only once was the sight and sound of this oceanic isolation broken, when one of the most brilliant meteors I have ever seen shot across the sky, with a blazing glare of light and a low, sinister hiss, vanishing beyond the distant crater rim as if it had plunged within—a worthy sanctuary for this inorganic wanderer from outer space.

While waiting for breakfast we collected the half hundred moths which had flown on board during the night, and loaded the boats ready for instant disembarking. The water in our shut-in bay was smooth and black and it merged indistinguishably with the shadows of the surrounding cliffs, so that the *Noma* seemed to float at the bottom of a narrow well.

After hastily snatched cups of coffee we divided into three parties, one going the rounds of the cove to see what the waters and overhanging cliffs had to offer, another went up to the head of the bay and inland along the ravine. I chose to attempt to land on the northern side farther out toward the open water, and if possible to strike upward to the colonies of birds along the edge of the cliffs. As it turned out this arrangement was as perfect as if we had known beforehand what awaited us.

I will let Ruth Rose tell first the story of the landing party:

We had separated into three parties in order to cover as much ground as possible, and across the cove we could see tiny figures scrambling on the cliff's face as they performed Herculean feats in getting themselves and their battery of assorted cameras up to the almost inaccessible spots where cormorants and boobies had chosen to build their nests. Another boatload cruised about the bay fishing and keeping a sharp lookout for penguins.

The surf had subsided during the night, so that we found it comparatively easy to land at the mouth of the water-worn ravine, where in more propitious rainy seasons our longed-for fresh water evidently poured down to waste in the sea. At the side of this gully was a little cave where visitors of former years to this inhospitable spot had carved the names of their ships and the dates of their arrival.

We advanced up the dry watercourse for a few hundred feet, noticing wistfully the marks of torrential rain that at some time had, and at some time would again, run down this rocky channel. It was a case of "jam yesterday and jam tomorrow, but *never* jam today." The breeding season for the black finches seemed at its height, for in every other bush we found their deep, round, grassy nests, a few containing young, but the majority with two or three little speckled eggs. Their owners regarded us with slight concern, though they were not so fearless as those of James. Indeed, at no time were the finches so amazingly tame as the mockingbirds and doves. Turning to the left, we scrambled up the side of the gully, dislodging chunks of lava and parched friable dirt. When we reached the top, we looked down another slope, equally steep and much longer, to a perfect little crater lake cupped in the hills. It was one of the most beautiful things we saw in the Galápagos, not only for the perfection of its setting, but also for the wild hope it roused in our parched souls.

The rim of the sun was just above the horizon, and high over our heads dragonflies and butterflies hovered and swooped in full sunlight; we were not yet touched by the direct rays, and far below us the little lake shone silvery in the clear twilight preceding dawn. The perfect moment was brief; in the space of a few seconds the sun shot up to shine on all alike, and we, breathless with excitement, were sliding and tumbling down toward a supply of water ample for a hundred yachts and only separated from our anchorage by a narrow ridge. The last thirty yards of the descent was a mass of frightful thorns, but we ripped through them almost regardlessly, and in company with a small avalanche started by our tumultuous descent, we literally fell into the lake.

It was as salt as the sea.

There ensued an interval during which we picked out thorns, and swallowed our disappointment; then we examined the lake from a scientific, instead of a thirsty, viewpoint. The water was beautifully clear and as far round as we went was a narrow ledge of worn rock just under the surface, masquerading in a white crust of salt, as coral. We waded along almost waist-deep, starting a duck from the solid bank of thorny bushes. Once I stepped off on what looked like hard white sand, but I began to sink so rapidly that I hastily struggled out again with the help of out-stretched, rescuing hands. The opposite shore looked as evenly planted with low trees as though an orchard had been set out on its precipitous slopes, and of course it looked infinitely more attractive than the place where we were. As the Spanish proverb says, "Distant fields are always green." Lacking time and a boat, however, we were reluctantly compelled to give up exploring and resume collecting.

Although the Galápagos are classed as tropical, being directly on the equator, collecting insects there was a microscopic proceeding. It was a meagre field compared to the abundance of a tropical jungle, where strangely formed and coloured life can be taken in any casual scoop of a net through dense undergrowth. Here, on these barren islands, collecting was an affair of grubbing search under heavy stones and of close examination of each parched leaf. The majority of the resulting specimens were so small as to require a lens to determine their classification, even in a broad general way, and the few large, brilliant forms of insect life were soon acquired in such numbers as to make further collecting needless. The one exception were the dragon-flies, which we decided were the only wary creatures of the Galápagos. At one time and another each of us had our experience of leaping wildly in the air after these elusive beings, which never seemed to alight but eternally danced overhead just out of reach of our desperate efforts. Sadly unlike the dragon-flies, we always did alight, unmistakably, among the broken lava rocks.

The thing that made the infinitesimal insects interesting was the fact that little of this sort of collecting had been done in the archipelago, so that many of our painfully-acquired specimens were hitherto unknown as occurring in the Galá pagos, or were even new to science. In an hour's collecting at the head of Tagus Cove, we secured twenty forms new to the islands, and ten wholly unknown.

When a long blast from the *Noma's* whistle warned us that time, coal and water were fleeting, we put an end to our collecting, retraced our steps down the gorge, and waited for a boat to take us off. There, within five minutes, we secured the two rare creatures that we had come to Tagus Cove to see. On a shoulder of rock above the ravine sat a large, dull-coloured bird; it turned its head and a spark of clear greenish blue glinted from its eye and a long, goose-like neck stretched out. Only the sight of a great auk could have been more thrilling, for here was a flightless cormorant, a bird probably doomed in a few years to an extinction as complete as the great auk's. Found only in this small area of the archipelago with no adverse conditions nor dangerous enemies to combat, its numbers nevertheless seem to be rapidly decreasing.

Hardly had we begun to breathe again after this excitement, when a very small sea-lion pup was seen swimming under water just below our perch. It shot up to the surface, shook itself vigorously, and turned into a little Galápagos penguin, paddling along now like a fat, high-shouldered duck. This was almost too much joy; we scrambled hurriedly over impossible declivities, risking our legs and necks in our delighted efforts to keep pace with the comic little bird as it skirted the shore. And so ended the efforts of party number one.

As for my own party, with Bill Merriam and John Tee-Van I set off for the seabird cliffs near the outer mouth of the cove.

Skirting the cliffs we discovered an inlet with the slope at the head less steep than on each side, but the end proved so narrow that the slight rollers surged up and down a height of about eight feet and made landing unthinkable. Near the entrance we found a half-concealed water-worn gully, with a narrow ledge of lava just above waterline. Leaping out on this we passed out cameras, guns and tackle on the upward lift of the boat. Then began a ghastly climb, up and down, up and down, over slopes just under the sliding point of loose clinkers. The surface was composed of my old friends, flat slabs of lava, like manhole covers, balanced and resting on a thick layer of volcanic dust. We never dared climb in line, for time after time, a careless step would upend one of these rocks and it would go careening down hill like a runaway cart wheel, starting sub-avalanches at every touch. With a forty pound moving picture camera in one hand, a 3-barrel shotgun in the other, and a game-bag which was diabolically clever in getting in the wrong places, my feelings were far from science when my foothold gave way. At such a moment I would sink flat and spread eagle as much as possible, chewing and eating dust which had taken part in the birth of Albemarle. Slowly I would move down hill, with a movement as sickening as that of a circular earthquake. Rock after rock would hurtle to the bottom and splash into the black, sharkful water. More than once when I seemed actually to be gathering momentum, my eye caught sight of the *Noma* riding so peacefully at anchor, and I would have given much to be on board of her. Within a few seconds time I mused, apparently for hours, on the insanity which impelled me to tempt fate thus,—when a fraction of a degree's greater pull of gravity would precipitate myself, camera, gun, clinkers and dust into the depths of Tagus. Then my foot would catch in the precarious roots of some small plant, which in my present plight seemed as a mighty oak.

I even welcomed the painful assistance of a cactus,—anything rather than that feeling of utter helplessness, when the whole earth seemed sliding downward with you. The moment I achieved some kind of a grip, my mind went ahead to the boobies and, possibly, cormorants awaiting me and I would slowly and painfully arise, and from a serpent's progression, attain that of a quadruped, and on hands and knees creep to the nearest point, for another attack upon the slope above. My helpers could do nothing to help, nor could I offer assistance when they were in trouble. The greatest kindness was to keep as far apart as possible. In the cold chill of early morning, we were panting and soaked with perspiration when we reached the summit of the succession of knife-topped ridges back of the colonies of birds.

From now on, our way led gently down hill, with no more under foot trouble than would be offered by a ploughed field of jagged stone. Dodging the thick growth of thorn brush, we encountered a veritable chevaux-de-frise of cobwebs. Here were the same zigzag-backed spiders as elsewhere, but of the largest size and with webs as thick and strong as elastic cord. With arms full of apparatus, and every pore dripping, lame and sore from our frightful climb, it was no added pleasure to have hundreds of sticky weblines across eyes and ears and face, with the spiders themselves crawling everywhere from cap to knees. I sometimes stopped, seized a single great strand, lifted it and snapped it back of my head without even nearly breaking it.

A few of the giant, brilliantly coloured grasshoppers flew up before us and were now and then entangled, but everywhere were scores of close-wrapped mummies of big sphinx moths which had been caught during the preceding night. In the early morning dawn songs arose in all directions from the throats of little black finches. The mockingbirds which came to look at us were silent for the most part, many of them accompanied by full-grown young.

I will never forget the scene when once we could drop our loads, stand erect, and let our pounding pulses gradually quiet down. Landward, one's eye followed the low slope, leaping the dozen gorges up and down which we had toiled, and watched it slowly steepen until it culminated high against the sky on the crater rim of Tagus mountain, eight hundred feet above us. It was covered with lava and volcanic ash, with meadows between, of the equally terrible bur-grass sheltering the myriads of hooked seeds. Here and there were groups of small trees, with the scanty foliage greyed with dust and yellowed with the beginning of the pitiless drought which was closing down for the coming eight months. To the west lay Narborough, with its symmetrical slopes standing out like bronzed steel in the dawn light.

Northward, the two great volcanic peaks which we had encircled yesterday, showed clearly, the daily cloak of clouds having not yet formed.

Immediately behind was the marvellous even-sided cove, with its steep slopes sliding sheer into the water. Beneath, on the rim of the cliff, giant breakers dashed foam over and around rugged piles of lava, while one mighty, flat-topped cube was completely covered with huge marine iguanas, all sprawled out in the sun.

We finally reached the rookeries, which we found to be a pure culture of blue-footed boobies. To our intense disappointment we saw no trace of cormorants from water-level to cliff top. The nest of the boobies were on the very edge of the cliff and when I first approached, several dozen were gathered together, many full-sized brown birds with quivering wings importuning their parents for food. When I was a few yards away they dropped off the cliff and floated gently outward, some of them, banking up wind, turning and drifting back. When a few feet from my face they put on brakes and hung in mid-air, craning and twisting their necks to get a good view of me with their great yellow eyes. The remaining birds were all sitting on eggs and refused to leave. When we wanted photographs and moving pictures of their eggs I had to push the bird back. Even the males, easily distinguishable by the small size of the pupil, were quite fearless, and faced me unflinchingly as I walked up to them. Solan goose is a good name for them, for in the general clamour of a rookery, the dominant sound is the strident, goose-like trumpeting of the females. The egg or eggs are laid simply in a hollow in the ground worn by the birds' bodies, or in a crevice of the lava. I selected five sets of eggs, three of one, and two of two eggs each, and this was about the general proportion.

After taking hundreds of feet of film, and dozens of plate photographs we picked up several birds and wrapped them in game bags, preparatory to carrying them back to the *Noma*, and at this moment a boat with the members of party number two appeared at the foot of the cliff and yelled up something about penguins. Even after the vision of the night before we hardly expected to find penguins here, and this thrilling news made us pack up in all haste and return. None of us will ever forget the trip back. Bill Merriam had the great moving picture tripod and the bag of live boobies, and if the accidents and unexpected and impossible things which happened to him on the back trail, could have been filmed, it would have eclipsed the utmost of Hollywood's effects. The three legs of the tripod appeared alive and the birds were miraculous in their ability to get a wing free at a critical moment. Once there were three flapping wings and I almost feared that Sindbad's experience with the roc would be repeated with Bill. How we managed to get safely down that nightmare place I cannot explain, but at last we fell into a life-boat, an inextricable mass of cameras, men, guns, tripods, boobies and bags.

We went full speed straight across the cove to a series of caves just in time to see Jim Curtis ladle out several penguins. They were all in the dark at the back of the cave and obligingly came out one by one and stood still until the net closed over them. When taken out in the boat they made no attempt to bite or escape, accepting their new environment and friends with perfect equanimity. I examined the cave and from the number of feathers and abundance of sign, I am certain these birds were bred there. We captured two full-grown birds of the year and an adult female. These are the smallest penguins in the world, and by far the most northern of their kind, almost all others being Antarctic. On our return trip, far north of Tagus, but still on the west coast of Albemarle and about six miles north of the equator, I was watching the shore with powerful glasses resting on the rail when I distinctly saw four penguins waddling over the rocks. They jumped in feet first, as they usually do, came up and swam around as long as we were in sight. This is the first time these birds have been seen north of the equator and while relatively not an important new fact, yet it is interesting as extending the distribution of the order *Sphenisciformes* into the northern hemisphere.

We had hardly recovered from the excitement of the capture of the penguins when word came from another exploring party that there were cormorants on their nests just across the cove. Sending to the *Noma* for more films and plates, we went across and found two magnificent flightless cormorants sitting on their nests. The two nests were fifteen feet apart on slight projections of lava rock on the steep slope, and about twenty feet above the water. Each bird had a single egg and nothing would induce her to desert it. Even when one of our party, climbing along further up hill, stumbled, and sent two lava slabs hurtling down very close to one of the birds, she only stood up and shifted her position slightly.

The nests were well-formed masses of débris brought from the water, chiefly dried seaweed, desiccated starfish, fish bones and other aquatic jetsam. The wings were very small, held ordinarily close to the body, but out sideways at a sharp angle when the bird was sunning itself. When I approached closely or waved my hat the bird rose high on her toes, opened and snapped her mandibles, uttering a regular cormorant croak, but louder and more resonant than that of our common species

After we had taken quantities of film and photographs from all possible distances and positions, Gilbert Broking and I approached. When close, I seized her as quickly as possible. I reached for the mandibles, but quick as a flash she dodged, leaped off her nest and when I caught her body, she raked my hand fore and aft with the cruel curved tip of her beak. Three times she went to the bone before I could secure her. Broking had in the meantime rescued the single egg. I carried her down to the boat and handed her over without further mishap to one of the party. I then captured the other bird, getting full measure of wounds from her as well. The quickness of the head and neck was in marked contrast with the slow gait of the birds on land. Their usual mode of progression is an awkward waddle, the whole body of the bird moving in rhythm with the short legs and great webbed feet. When they meet an obstruction, as a lava block, they bend down a little and leap upward with both feet at once, sometimes clearing six inches and often resorting to a series of penguin-like hops, wholly unlike the habit of any species of flying cormorant. When disturbed by me, and during the few seconds when I fumbled my grip and the bird was attempting to escape, it fell flat on its whole length and pushed ahead with its webbed feet, recalling the gait of certain Antarctic penguins, when going full speed over hard snow or ice. The eggs are elliptical in shape, of a pale bluish-green, much concealed by a whitish deposit of lime and about one and three-quarters by two and three-quarters inches in size. These are giants among cormorants, lack of flight having resulted in a decided increase of dimension and body weight. The larger of the pair which I caught alive weighed eight and a half pounds as against an average of six and a half pounds of those which I have collected in British Guiana. Compared with the Farallone cormorant the percentages are as follows: the Galápagos birds are one-third longer in total length, the culmen is one-third longer, the tail is 9% longer, and the wing is 40% shorter. Although the Galápagos bird appears to be the largest of living cormorants, yet the recently extinct Pallas cormorant of Bering Island, which disappeared about one hundred and eighty years ago, reached a weight of fourteen pounds. As an example of wing degeneration the bird I had captured was even more striking than the classic great auk, for while in the latter the wing formed 24% of the bird's total length, in the cormorant this relative measurement is only about 19½%. As we might expect, the legs and feet are large and of great strength, and the swimming webs are aided by an additional expanse of skin in the form of a broad, stiff flap extending down the tarsus. They swim much lower in the water than other cormorants, often with only the head and neck exposed, the back being completely submerged. When diving, the body is humped suddenly forward or actually thrust up, forward and down in a real forward dive. Long distances can be covered under water, the wings held loosely, but motionless, while the great totipalmate webs curl and uncurl with great power and a machine-like regularity.

The stomachs of these birds contained the following:

number one, three eels, of which two were *Scytalichthys miurus*, and one *Gymnothorax dovii;* number two, a good sized octopus; number three, one small fish, well digested.

Like most of the other birds which live among these black lava islands, the cormorants are dull in colour. The bill, pouch and feet are all dusky brown or black, while the plumage is drab brown, with sometimes a little greenish iridescence on the upper surface, set off by a few white, hair-like feather-filaments on the head and neck. The eye is the only exception to this colour scheme, being of a clear, glittering Italian blue.

The inconspicuousness of these birds, large as they are, is attested by the fact that they escaped the attention of Darwin, and all expeditions down to twenty-five years ago. They are confined to parts of the coast of Narborough and the adjoining western side of Albemarle, and are very near the danger line of extermination. One of the last expeditions to the Galápagos killed twenty-six of the birds, and they have been thoroughly measured and examined, so it is to be hoped that very few additional ones will be needed. The two females which I captured alive, are in perfect health today after almost a year in the Zoological Park and Aquarium, and so aggressive are they that the keepers have to be on their guard against attack. I secured the two eggs and the nests complete. Being, by this time, completely laden down with specimens and photographs and every container overflowing, there was nothing to do but to remove my shirt and wrap the nests carefully up in it.

From a landscape which superficially seemed utterly barren of life, we had extracted the greatest possible profit to be gained from photographing, both stills and moving pictures, paintings in water colours and oils, specimens of all kinds, plants, flowers, minerals, insects, fish, lizards, birds' nests and eggs, and notes scrawled on every imaginable scrap of paper.

RAINBOW CHASING

At 10.30 A.M. we drew up the anchor and steamed from the cove, after as strenuous a five hours as most of us have ever spent.

The next eight pages are from the journal of Ruth Rose:

We arranged comfortable accommodations for the two live cormorants aboard the *Noma*, one of which promptly laid us an egg like a well-trained fowl, and we transported them to New York without mishap. The three penguins that were captured made irresistible pets, and were so tame that it was almost impossible to keep them at arm's length long enough to take their photographs. Lifted from their box, they waddled busily about the deck, crowding each other in their earnest attempts to get under foot and be stepped on, and sometimes managing to hop over the high threshold into the smoking-room where they carefully inspected the phonograph. With foolish little stumpy wings outspread, they hurried from one person to another, looking up into each face in an appealing, anxious way which meant that fish were ardently desired.

Benjamin, our pet sea-lion pup, was fond of the phonograph also. When allowed to roam, he always turned in that direction and after much flopping, hoisted himself over the sill and fell with a thud into the smoking-room. On one occasion he managed, by super-sea-lion efforts, to climb to a leather-covered seat and from that to the top of the phonograph, where he was discovered lying cozily in and on someone's new straw hat.

We left Tagus Cove before noon to retrace our voyage of the preceding day. Steaming round the northern point of Albemarle, we turned southeast, establishing what is probably a record in equator-crossings, having done so four times in about twenty-six hours. We worked feverishly all day on the many new specimens acquired, and it was a weary lot of scientists and assistants who slept that night while the *Noma* slowly approached Chatham.

The slowness had two motives; first, to reach our anchorage in Wreck Bay by daylight, and second, to save coal, for the water crisis was now further complicated by a coal shortage. It was like Paris in wartime, when each week brought its fresh "crise." We had now attained the perfection of the vicious circle; we must have fresh water in order to go on burning coal, but if we went on burning coal we should not have enough coal to go on looking for fresh water. The brain reeled.

The *Noma*, of course, had condensers, but on a cruise of this length it was not possible to carry enough coal to run both condensers and boilers.

Soon after sunrise Wreck Bay opened before us. The ominous sound of the name made the captain doubly cautious, and visual evidence justified his attitude. We were confronted by the masts of a submerged vessel that had come to grief on hidden rocks, nothing below the crow's nest being visible above water. There is a treacherous reef just south of Wreck Bay; in heavy weather it would be plainly seen, but in such a calm as prevailed when we were there, there was only an occasional piling-up of smooth water a little higher than the gentle heaving of the whole ocean.

On consulting the pilot-book the preceding day, we had been somewhat nonplussed to find the following information: "There is no pier at Wreck Bay. A pipeline brings fresh water to the end of the pier."

Thus we were allowed to believe anything we liked, and soured by experience of charts, we refused to believe anything at all. This was just as well, for we found on arriving that there was no pipeline, no fresh water, and only the remnant of a very shaky pier.

Wreck Bay, principal and only port of the Galápagos, was the thriving metropolis that one would expect to find. It consisted of two buildings, one of which was the only lighthouse in the archipelago, visibility four miles, according to the chart. That is not bad for a gasoline lamp on a pole. Five miles back in the hills is the settlement of Progreso, worked with convict labour, where sugar cane, coffee and fruit are cultivated.

The arrival of a yacht in such a spot was naturally the signal for everyone who could find the means to come aboard. The first up the companion-ladder was a portly, middle-aged Ecuadorian, dressed in some striped material that nearly resembled bed-ticking. He informed us that he was Governor of the island, Captain of the Port, and Keeper of the Lighthouse. Perhaps he was also the crew of the captain's gig. He consulted his watch with an important flourish at least once every five minutes, and since it was impossible to believe that anyone would have a pressing engagement in the Galápagos, we concluded that he was merely displaying his jewelry. The watch was certainly worth showing. It was perfectly square, of the finest brass, about four inches across and two inches thick, heavily sculptured with trees, rivers and castles.

Next came a sunken-cheeked ancient who announced as he stepped on deck, "I'm Johnson of London." In ten minutes we were in possession of the main facts of his long life. For fifty years he had lived among the islands, on Chatham for the most part, and he had not heard his native tongue for so long that it had become more foreign to him than Spanish. His opinion of those who had been his neighbours for a half-century was almost unprintable, and though we knew from experiment that the protean Governor understood no word of English, we felt excessively nervous as old Johnson made very personal remarks about him in a loud, cracked voice.

The subject of fresh water was broached, with some difficulty, as Johnson was very deaf, and it was necessary to piece out conversation with the Governor from a few words of Spanish and a great many gestures. For a time it looked as though all our troubles were over; they both assured us that there was plenty of water to be had for the asking. Then it appeared that the water was to be brought down in "tankas" from Progreso over five miles of the worst trail in the world, and we had mental visions of rivalling the length of Johnson's stay in this spot, if we waited for forty tons to be collected in any such manner.

Our visitors were invited to have a drink, of a liquid less precious to us than water, and we then discovered that the Governor had an English vocabulary after all. He was letter-perfect in the word "whiskey." As for Johnson of London, all speech forsook him for a while and then he murmured, "God, I ain't seen any whiskey for six years."

When they saw that real ice was a part of the treat, they beamed with satisfaction, and after being established on the forward deck with a clinking glass and a cigar apiece, the Governor overwhelmed us with a flood of rapid Spanish, while the English vagrant became confidential on the subject of buried treasure.

On second thought, it is unjustifiable to call him a vagrant. Anyone who has stayed anywhere, especially the Galápagos, for fifty years, can hardly be said to be consumed with wanderlust.

When he was a young man, he said, he had come out to build a road on Chatham. He was to receive a thousand dollars for the job, but so far the Ecuadorian government still owed him four hundred and he was waiting for it. Patience on a monument is too nervous a simile to be applied to his case.

He knows the very spot where the buccaneers buried their plunder, but to us he located it no more definitely than to say it was on Albemarle, which is nearly a hundred miles long.

"Thirty years ago," he said, "I built me a scow to go look for that treasure. But I couldn't get nobody to go with me."

He paused to reflect and added, "That is, I could have got somebody but—" and he drew a finger across his throat, rolling a significant eye at us. A penal colony is probably not the ideal place to find companions for a treasure hunt. He remembered Ralph Stock of the *Dream Ship* which touched here a few years ago, and after a few drinks recalled vividly Darwin's visit in 1835, and Admiral Porter's in 1813.

Presently, these two departed, promising gifts of turtles. Our next visitors were not long in arriving. They had seen the yacht from Progreso and hurried down the trail as fast as their horses could bring them. One was the young patron of the settlement, a slim, young Ecuadorian. His father started the plantation forty years ago, and was murdered in an uprising of the convicts in 1907.

The other men were three Americans who had arrived at Wreck Bay six weeks before as passengers on a small sloop from Guayaquil. They questioned us eagerly about Indefatigable, and said that their purpose in coming to the Galápagos was to look into the agricultural possibilities of that island. We who had so recently and painfully explored its devastated shores listened to them speechlessly. When they heard that we intended to return to Conway Bay they explained that they were waiting at Progreso in the hope of getting some ship to take them to Indefatigable, and they now proposed that we should do so. The story sounded so mysterious that we said vaguely perhaps it could be done, and put off giving a direct answer to the amazing request that we should help three men commit suicide by transporting them to a desert island, to which they did not intend to take either food or water, and leaving them there until a problematical boat should arrive. We told them at once that we could not possibly spare them any supplies from our diminishing store, but they waved away that objection with the cheerful remark that they intended to live off the country. So far as food went, it would no doubt be easy to sustain life on birds and crabs but we tried to convince them of the utter aridity of the island as far as we knew it. Their confidence was quite unshakable, however, and although we painted a word picture of Indefatigable which made Purgatory look like the Elysian Fields, they insisted that there must be fresh water there somewhere and that of course they could find it.

Their purpose in going there seemed strange, to say the least, and we made puzzled faces at each other behind our callers' backs. To put the finishing touch on the mystery, one of the men introduced himself as Mr. Nemo.

Some of us withdrew to discuss it privately and were unanimous in our dislike of being responsible for a proceeding that so strongly resembled marooning. When the refusal was gently but firmly conveyed to our compatriots, it transpired that they had been so sure of a favorable reply that they had brought all their belongings aboard, prepared to stay. They agreed that they could understand our point of view, however, and our fear of being blamed if they came to grief. We have often wondered since if they reached Indefatigable, and what they thought of its agricultural resources.

Señor Cobos, the young Ecuadorian, had in the meantime been discoursing in fluent French, which was a relief after our Spanish efforts with the Governor. He had been educated in Paris, which was hardly calculated to make him blissfully happy during the six years he had spent superintending the Chatham plantations. "*Ce n'est pas gai*," as he said with a sigh.

He comforted our anxiety about water by informing us that just behind the beach was a *petite rivière* where we could get an ample supply, not clean enough for drinking but good enough for the boilers. We brushed aside the question of our own thirst; that was only future thirst anyway, but the boilers' need was more immediate. A party went ashore and found the *petite rivière* to be a brackish lagoon, the water from which would have ruined our precious boilers in a day. We could only suppose that Señor Cobos shrank from paining us by a disappointing report.

Johnson of London lived in a shack behind the lighthouse with his mestizo wife. He came out to talk with the shore party and when the situation was explained anew to him, he promised us water from another source. Everyone promised us water. This time it was to be found on the south side of Chatham, in Freshwater Bay. Johnson would go with us to show the place where, he said with a wide, free gesture, "I'll show you tons of fresh water, right on the coast."

Hope does indeed spring eternal. Hope alternately sprang and collapsed in our breasts during our rainbow chasing. We got under way as speedily as possible, taking Johnson with us, a proceeding which savored a little of seizing a hostage, as I am sure we had reached the stage where if he had not fulfilled his promise, we would have thrown him to the sea-lions. It is my opinion that the old man, having been in the Galápagos for fifty years, thought it time for a little change and a day on a yacht with plenty of whiskey and cigars, appealed to him. However, he was as good as his word. He showed us tons of fresh water.

We steamed past a shore line of black cliffs, backed by rolling hills that bore a faint resemblance to poor pasture land. In one place we saw through the binoculars a dozen cattle grazing on the high slopes, though the fodder must have been of the scantiest. Early in the afternoon we reached Freshwater Bay, and there at last we saw what we had almost come to believe was non-existent anywhere in the world—wild fresh water. Two waterfalls, one quite small, the other of good size, plunged over the edge of the cliffs into the sea. Niagara or Kaieteur never looked so wonderful. We all gathered on the forward deck and cheered. But after our preliminary thrill, the question of how to get any of it on board made our faces wan and grey. Johnson was quite right. There were tons of fresh water on the very coast, but breakers were dashing all along the shore, and dashing with particular violence below the cascades. It was impossible to hold a small boat in the surf, even if there had been any hose long enough to reach from the top of the cliff, where the stream was, to such a boat. There was no anchorage for the *Noma*, so she steamed slowly back and forth consuming precious coal, while we leaned on the rail gazing thirstily at the vast quantities of water running to waste in the ocean. It was living the part of Tantalus. We wished we had been in a position to share Vancouver's sang-froid, who, writing of the Galápagos in 1795 said airily:

As for fresh water, that great need of navigators, some reported that these islands had large brooks and even rivers, while others thought that they had it only in small quantities, or even that they wholly lacked it. This point is of small importance because of the proximity of the Cocos Islands, where springs that never dry up seem to water each part of the island, and offer to vessels all the water of which they can have need.

As the Cocos Islands were over four hundred miles away, we found nothing of immediate assurance in this suggestion.

Two ship's boats put off to investigate the water falls, and through the glasses we could see them jockeying outside the terrific surf breaking at the base of the cliffs. Scanning the coast closely I saw an opening in the steep, overhanging shore, where a mass of light vegetation looked suspiciously fresh and green. With Curtis and McKay I put off in the four-oared gig, and we rowed straight in. The approach proved to be so deep that there were no breakers, only great rollers which remained unbroken until they washed up on to a beach of rounded pebbles. I went overboard and allowed myself to be stranded on this strange shore, clinging prone to the shifting stones as the wave rushed back, and scrambling out of reach of the next. When I had surmounted the eight-foot windrow of cobbles, I saw at my feet a large, quiet pool into which a stream was flowing through banks of solid green, the outlet apparently being through the beach itself. I tasted it and did a Robinson Crusoe dance of joy.

We planned a scheme which seemed feasible, and sent off to the *Noma* for a hundred and fifty feet of rubber deck hose and the big life-boat. I swam ashore with one end of this, the other remaining in the anchored boat. It lay on the water like a great, black, undulating sea snake. Over the windrow we lugged the end of the hose, and when close to the pool, I fitted it with a large tin funnel and held the end as high as possible. The others formed a bucket brigade and began to pour the precious fluid into the hose and in five minutes the sailor in the life boat yelled that it was flowing out steadily. Relieving one another we actually managed to fill the boat as deeply as we dared, with about four tons of fresh water.

During one of my resting spells, I went into a cave at one side of the open beach and found it a wonderful place and full of life. The entrance was low, so low in fact that every high wave actually filled it, but just inside, it enlarged into a lofty cavern. At the entrance noddies were nesting and I found a fork-tailed gull's egg, although I saw no birds of this species. Seizing a favorable moment, I rushed in, waist deep and cleared the low arch before the next roller filled it. As I entered, two seals swam past, brushing against me. The air was so foul inside that I could not stay long, but I saw hosts of crabs and other sea creatures, and I smelled the unmistakable odor of bats, but saw none. The light was most remarkable. When the surface of the water was low, a fairly clear, indirect light illumined the cave, but when the mouth was water-filled, the light changed to a faint, weird mottling of green and white which played and flickered with incredible rapidity over the walls and ceiling, making the lofty ceiling more distinct than the more direct lighting had done. Huge mollusks crawled about the dripping walls, and invisible drops fell with musical splashes during the brief intervals when the sea was still.

I let an outrushing wave wash me along, and brought up with my face close to a brooding noddy tern on her single spotted egg. When I returned to the water party I saw two monarch butterflies flying slowly about, and we stopped work for a moment, but lacking a net, found it impossible to capture them. There was not the remotest chance of being mistaken in the species, and therefore *Danais plexippus* (or *Danais archippus* as I knew it when a boy) is definitely added to the fauna of the Galápagos Archipelago. On several islands I had seen a milkweed in blossom, so this wide-ranging butterfly should have no difficulty in establishing itself.

Triumphantly we towed the deep-logged life-boat back to the *Noma*, only to meet dire prophecies on the part of the Captain. We could not stay where we were all night as there was no anchorage on this side of the island, and it was madness to try to enter Wreck Bay after dark past those wicked reefs. Yet Johnson of London had to be returned. So as our present mode of steaming back and forth was consuming coal at a prohibitive rate, we must then and there give up rainbow chasing, if we hoped to reach Panama at all.

The greatest disappointment was not being able to return to shore and capture some of the fish which I had seen in the fresh-water stream. They must be very interesting, whether

FIG. 34
JOHNSON OF LONDON AND HIS ECUADORIAN WIFE

FIG. 35
THE CHATHAM LIGHTHOUSE AT WRECK BAY

FIG. 36
MEMBERS OF THE EXPEDITION IN DARWIN BAY, TOWER ISLAND
Broking, Cooper, Rose, McKay, Wheeler, Merriam, Mitchell, Hoffman, Beebe

FIG. 37

GIANT TORTOISE CLIMBING A HILL OF LAVA

This was the only specimen we saw, capturing it on Duncan
Testudo ephippium Günther

long residents of Chatham's fresh water, or recently adapted immigrants from the sea.

At dusk we reëntered Wreck Bay and dropped anchor once more. The lighthouse was bravely sending forth its small beam into a fresh-waterless world when Johnson was set ashore, with an armful of magazines ranging from *Vanity Fair* to the *Atlantic Monthly*. As he went down the companion ladder his valedictory speech floated up to us:

"Well, I've lived near eighty years and I've read about it, but I never seen a woman in pants before."

CHAPTER VIII

THE LAST RAID

By Ruth Rose

THE water and coal crisis aboard the *Noma* had reached the critical point where something must be done and that without delay. The situation was complicated by the facts that if we burned coal in order to cruise about and search for water, we should not have enough fuel left to reach Panama, even if we found the longed-for fluid; if we did not burn coal in such a quest, we should expire of thirst before we reached Panama anyway. Our talents as well-diggers being nil, and having neglected to provide ourselves with a witch-hazel rod, we reluctantly decided that our Galápagos explorations were over, unless we wished to add to the interesting relics of bleached bones which are so usual a feature of Galápagos landscapes.

Our time being thus painfully limited, we made the most of the few hours that remained to us in these desert islands so tantalizingly full of interest and of unexplored mysteries. The plan was to make what in burglarious circles is referred to as a quick clean-up. Indeed we had moments of feeling like burglars, and remorseful ones at that, when invading a spot that had been undisturbed by trampling humans for many generations of the small, busy lives here working out their destinies.

THE LAST RAID

To read of the old whaling days in this archipelago is to see a vision of shouting men, wild in the freedom of a day ashore after long months of seafaring, striking one of these isolated Edens like a pestilence and rushing aboard ship again with dozens of creatures which, aside from the giant tortoises, served no purpose except the amusement of the moment and which in a few hours were tossed overside, limp bundles of feathers or sprawling scaly limbs. Among the marauders, however, there must surely have been now and then one who slipped away from his noisy shipmates to watch quietly the prehistoric parade of black lizards, or to win the easy confidence of a mockingbird.

We were forced to combine the rôles of collectors and nature-lovers, for while our impulse was to leave such primitive perfection exactly as we found it, we had a purpose to fulfill and a commission to execute, and an overdose of sentiment would have made us fail in both. Like omnia Gallia, we were divided into three parts, so that our depredations could be as widespread as possible in the course of one short afternoon. One section was detailed to get sea-lions at the rocky amphitheatre known as Guy Fawkes, another was to search the islet of Eden Rock where Amblyrhynchus was most numerous, and capture as many as might be, and the third division was to raid the vicinity of Conway Bay for fish, and for the accessories of vegetation, rock and co-existing animal life which should later show in a habitat group the home life of *Amblyrhynchus cristatus*.

This third division numbered five,—two Artists, the Lawyer, the Doctor and myself. We had half a day in which to get a representative collection, and we planned feverishly to make every moment count. We left the ship with a vast deal of paraphernalia, as usual, for the naturalist collector must be prepared for every eventuality, and his baggage must include instruments for the capture of minute insects, as well as containers for the wonders of the deep, in case a tidal wave should be so kind as to deposit them at his feet.

Camp Harrison, divested of its tents, looked quite forlorn, but our faithful red-billed oystercatchers waddled hopefully to meet us. They had soon learned the significance of seines, for they had never been disappointed in their expectations of crumbs or rather scales from the scientists' hauling. They had been the first inhabitants to greet us when we landed and their friendliness had never gone unrewarded.

Our first duty was, like that of the wage-slave, to the hungry mouths at home. We had splay-footed, green-eyed cormorants and irresistibly comic penguins on board the yacht, and they must be provided with food on the voyage to Panama; also we hoped for success to the sea-lion hunters, and all this meant that we must fish and fish and fish again. But it was blazing midday and dead low tide, and the Lawyer shook his head gloomily as he unrolled his favorite seine. To paraphrase the Mock Turtle, he was Curator of Dredging and Diving and Seining in Coils, and into this fishy occupation he put the earnestness and enthusiasm with which he would attack a doubtful legal proceeding back in mythical New York. His ominous forebodings were fulfilled and it would be hard to tell whether his or the oystercatchers' disappointment was greater when the first two hauls yielded exactly half a fish each.

Having so many things to do, we let the cove lie fallow for a time, while we turned our attention to crabs, which were an important item on our shopping list, for these brilliant crustaceans are inseparable adjuncts to any landscape containing Amblyrhynchus. Their scarlet backs and bright blue under-surfaces spangle the black rocks of the shore in innumerable swarms, supplying color that is quite lacking in their

scaly neighbors and their burned-out background. Among
them whisk tiny crabs, like little evil gnomes, coal-black as
the lava itself and inconceivably rapid of motion. Their
speed must be what keeps the race extant, for crabs are
notorious cannibals, and these little chaps must be preyed
upon to an extent that encourages swiftness. Their large
relatives are the comedians of these black, surf-drenched
coasts, and helpless laughter at their antics often made doubly
difficult the pursuit of them over slippery jagged rocks.

One of these big scarlet-backed creatures sidled slowly
toward me, lifting first one and then the other of his huge
front claws to his mouth, conveying some invisible nourishment. His stately gait, his color, his popping eyes and his
gestures made him look like an apoplectic old General saluting
ambidextrously as he reviewed his troops. I waited till he was
close, and then clapped a butterfly net over him. There was
a wild scurry, an ineffectual clutch on my part and he was out,
jumping from rock to rock in a series of monkey-like skips, and
twinkling his eight legs derisively. I tore after him, round a
point of lava where he had vanished; the next moment he
rushed back in my direction, and the Artist and I collided
heavily as she tried to claim him for her prey. Our nets
struck at the same spot, where the crab no longer was, met
and tangled with a crash, and we both slid gently into the
water. It was only knee-deep and we expected to be neck-
deep before the day was over. When we waded out to
return to the chase, all the crabs in our vicinity were placidly
going on with their peculiar grazing. Across the intervening
point we saw the Lawyer intently bounding from crag to
crag with his quarry scuttling before him, while a war-whoop
from the Doctor announced a catch. The other Artist had
withdrawn himself from the big game hunting and had set to
work to finish a portrait of the *Noma* lying at anchor, with

Eden for her background. Few yachts have had so favorable
a setting.

So far as crabs were concerned, we were all mighty hunters.
In half an hour we had rounded up a sufficiently large herd of
them and had insinuated their countless legs into a series of
bamboo bird-cages. A crab has only eight legs, but by the
time a large one has been persuaded through a small door,
even a zoologist would be tempted to classify him with the
centipede. As for the bird-cages it was apparently a point of
honor on this expedition never to put anything to the use for
which it was intended, beginning with the *Noma* herself,
designed to cruise serenely through smooth waters with pretty
ladies drinking tea on her immaculate decks. She has seen
hard service in the Bay of Biscay since her regatta days,
but I doubt if anything the war did to her could have
jarred her sensibilities as our conversion of her must have
done.

The library, for instance, was made a storeroom for fish-
nets, tents and birdskins; incidentally it was also my bedroom.
The daintily appointed sun-parlor was turned overnight into
a crowded laboratory, lined with unfinished boards for tables,
on which dissection did its messiest. The after deck was a
menagerie, the forward deck was cluttered with barrels of
vegetables and cases of bottled water, as well as five-gallon
milk cans, such as line the platforms of country stations at
dawn. But these cans were for the preservation of specimens
and contained things that would curdle the milk they were
intended to hold. It seemed to me that the last word had been
said when the Scientist insisted fiercely that I keep a baby
seal in my bath-tub. The motto "It's all for Science," with
which he was accustomed to console us for any exigency
failed for once to awe me into acquiescence.

To return to our crabs. They were stowed away in the

cages, and the Lawyer took another judicial look at the sea. The tide was still too low, so we set to work to gather background for the lizards that we hoped were being collected on Eden's rocky shore. The business of "getting atmosphere" is usually a vague one, not easy to define, but in this case most of the required atmosphere was all too palpable. Take cactus, for instance. Try to cut down a ten-foot cactus and land it, whole and undismembered, on the jagged broken rocks among which it grows. Then lift it up tenderly, lift it with care, and bear it a hundred yards to the beach. Then sit down to nurse your wounds and extract the spines and wonder how it is to be conveyed into a small boat and from there to a ship's deck.

After some discussion, we arranged a complicated rig of rope passing round the cactus and over several limbs of nearby bushes, which, the Lawyer assured us, would brace the cactus from every angle. Then he explained the angle at which he intended it should fall, and the Artist and I were told off to stand with a gunnysack swung between us like a hammock, catch the cactus exactly amidships as it tottered down and lower it gently to the ground. The theory was perfect. A few strokes of a hatchet cut deeply into the pithy trunk and then as the crucial moment approached, a brush of wings across our heads made us look up. A beautiful hawk alighted on a branch not four feet from us. Quite fearless and very curious, his bright yellow talons gripped his swaying perch, and his yellow eyes gleamed as he cocked his head to inspect our extraordinary presence and procedure. We all gaped in admiration at his glossy dark back and wings and at his pale fawn breast, mottled with dark streaks. The faithless hand that held the bracing rope relaxed.

"Look out!" shrieked the Artist, at the very moment that the ten-foot cactus fell across my thinly clad back.

In the side-shows of the circus are freaks called Human Pincushions. I instantly qualified for such a position.

The Artist, well-meaning and nobly self-forgetful, darted to me and with bare hands brushed briskly at my writhing form. She only brushed once. The situation remained unchanged, except that now we could be billed as the Twin Pincushions. Also our beautiful specimen cactus had broken across my shoulders and now lay in several pieces among the rocks. The hawk remained quite calm and watched our antics, while a few thousand spines were hand-picked from my person, leaving only a few dozen that eventually worked out or in, I am not sure which.

After infinite labor and lacerations, several whole cacti were obtained and the Lawyer, astride a crevasse in the lava, relayed them along and directed their manœuvring as far as the beach. There we took counsel and coldly determined to wash our hands of them. Figuratively speaking, of course. Not soap and water, but tweezers, were what we needed. The unanimous verdict was that the lazy brutes who had done nothing all day but chase lizards and stalk sea-lions could themselves get these interesting botanical specimens into the boats and aboard the *Noma*.

Then we filled gunnysacks with lumps of the pitted black lava, that is to Amblyrhynchus his four walls and roof, another sack with fine white sand and two more with the endless variety of shells that windrowed the beach.

There is a great deal to be said on the subject of those shells. In vast quantities they strew the beaches, long barrows of them marking high-water line. Like second-hand bookstores, the only way to get past them is to ignore them. Once stop to look and you are lost. Wonderful shapes and colors are here for the finding, and the longer you stay to search, the finer become your appreciations, until you are exclaiming over and treasuring shell specks hardly larger than a pinhead. Perfect specimens are here,—a tiny translucent Japanese fan of scarlet, a cowry all mottled purple and cream; but to an amateur conchologist may prove even more fascinating those water-worn and rock-chafed shells, that are like faces on which adventure and hardship have carved characters so different from their original aspects that those can now be hardly guessed.

It was a piratical scene. Five wild-looking individuals dragging heavy sacks across a desert island beach, surely suggest unburied treasure and the pity was that there were no uninitiated observers to be so excitingly misled.

Fishing time at the big cove had not yet arrived, but we decided to try our luck at some of the tiny bays that indented the northern shore. The final disgrace to a member of this expedition would have been to make a change of clothing for the purpose of going in the water. Bathing-suits were worn, to be sure, but they were as likely to be the costumes for climbing a volcano as to be used for aquatic pursuits. Shoes were always worn into the water as a protection against coral or lava or kindred jagged things. One simply walked into the water fully dressed and later the clothes dried on the person. Equatorial sun and trade winds take no time to dry the wettest garments.

So the Lawyer and I unfurled the thirty-foot seine and stepped into the pleasantly cool water. A big pelican sat smugly on a bit of lava nearby and watched us with that double-chinned expression of disapproval peculiar to pelicans. He probably regarded seining as unsportsmanlike, particularly as he had never profited by it. As we waded in, the other Artist, of the sterner sex, came running up to help us. He paused to move his watch from trousers to shirt-pocket, and as he went deeper and deeper, I saw him over my shoulder take his watch from his shirt and put it in his teeth; just then the Lawyer stumbled over a submerged rock and became even as the rock. As he scrambled up, the Artist tucked his watch in his hat-band and spread his hands in resignation. It was a perfect piece of pantomime.

With the water lapping gently in our ears, we turned shoreward, holding the ends of the seine far down and slanting forward. We lacked the red sashes and earrings traditionally associated with this pursuit, but the situation was not without its picturesqueness. Not everyone is privileged to share a seine with an ex-Assistant Secretary of the Treasury. When we made the final rush up on dry sand and opened the bag, we had a good, though not spectacular, haul. At least we saw nothing to get excited about at first, as we put the fish, all of species that had grown familiar to us, in a basket. Then we came upon a little flounder, the first we had seen in these waters, and so flat was he, and so marvellously protected by his color and pattern, that we had very nearly thrown him aside as a bit of sand. Of course we had nothing in which to keep him alive. No matter how much impedimenta is carried on a collecting trip, the moment always arrives when some imperative lack is discovered. With the gesture of one who risks sunstroke in a good cause, the Lawyer snatched off his hat and filled it with water. It was all for Science. It was a leaky aquarium but it served to keep the precious fish alive across the few yards that separated us from our equipment.

Some one started to rush away with the flounder, when a shriek of excitement rang out. In the seine lay a worn silver dollar and for a moment wild thoughts of buccaneers and sunken galleons raced through our minds. Then the silver piece proved to have a mouth and eyes, scales and fins, and long streamers that flowed back far beyond its tail. It was a moonfish, and to find it and the flounder in the same net was to see a meeting of extremes beyond imagination. The moonfish, a mere tissue of a creature, moving upright through the water like a round razor-blade, could have lectured the flounder on the incomprehensible theme of height, while the flounder, flapping his way along the sandy bottom, explained to the moonfish the mysterious Third Dimension of width.

Two more flounders and five moonfish were taken in successive hauls, and for a time a sort of ambulance service, or First Aid to Fishes, was established, as we galloped madly from the seine to the landing-beach with captives struggling in the Lawyer's dripping hat. We had nothing in which to keep them alive except a five-gallon milk-can, which it was manifestly impossible to move across the intervening rough country.

High tide came at last, and back at the big beach we began to get the netfuls that had made us scornful of anything less than a whole school of fish at each haul. Again and again the seine was drawn in practically the same spot and the supply seemed as inexhaustible as the widow's cruse. Each time the net came up on the sand, the bag was a mass of lashing bodies, from two-inch minnows to twelve-inch mullets,—a mass that separated into such diversities as narrow trumpetfish beautifully striped in shades of blue, and black and white puffers that chattered in staccato as they blew themselves up into absurd round balls, positively inviting a puncturing pin.

Our oystercatchers, gulping the titbits that we occasionally spared them, were joined by some beautiful grey lava gulls, and they all travelled to and fro excitedly, following us up like barnyard fowls rather than wild creatures. A shark's fin tacked slowly from side to side in the cove, as hopeful of scraps and leavings as any dog, and two small sharks swimming along the bottom scurried out of our way. A round black spot under water, that might have been a rock or a patch of seaweed, proved suddenly active as we waded closer, and drifted slowly like a cloud shadow along the bottom. Here was something much more to be respected than sharks, and our efforts to capture it were attended by a distinct shrinking sensation in those portions of us that were under water. No matter how quietly we tried to surround it with the net, that shadow-like patch, which seemed to move so aimlessly, nevertheless slid always just out of reach, and then suddenly it was no longer there at all. For the next few minutes, I don't think anyone's mind was entirely concentrated on food fish and I am quite willing to admit that I wasted valuable time in pausing to peer earnestly, nose touching the water, at perfectly harmless rocks and weed, because they were submerged and dark and roundish, and so might have been the straying sting-ray.

Across the bay came the lizard hunters, rowing hard from mile-distant Eden. In tow they had a curious craft, rigged that morning,—a small skiff, with a ridgepole laid on uprights at bow and stern, serving as support for chicken-wire that was nailed tight along the gunwales. In this vessel there now scrambled a seething mass of black and scaly forms, and as she grounded on the beach we cheered for R. B. Amblyrhynchus. If the S. S. *Cedric* and the S. Y. *Noma* are correct terms, who is to find fault with R. B. *Amblyrhynchus?*

The news from Eden was briefly told; four determined naturalists armed with nooses on fish poles had spent a busy afternoon in swinging amazed captives through the air, and had profited to the extent of forty large marine lizards. It was a rather thrilling idea to think how few were the people who had ever seen them, and that we were privileged to take back the first living specimens to the Zoological Park and the Natural History Museum.

A series of explosive sounds from the direction of Guy Fawkes made the name seem for the first time appropriate, but it was only the motorboat in a fit of temperament. The sea-lions were arriving, and incidentally the hunters of the beasts. They were the Surgeon and our two Tired Business Men, sunburned to a color that Wall Street never saw, and quoting sea-lions as active but falling. To be exact, two had fallen,—a big bull and a cow, and the chase was described with zest.

Their object had been to secure a family for a group in the American Museum of Natural History. The party had ascended the slope on the outermost of the Guy Fawkes Islets where were sleeping or sunning themselves a herd of sixty or seventy sea-lions, bulls, cows and young. The Chief Hunter squeezed between two rocks and walked toward the north shore of the island, when he came suddenly upon a large sea-lion near the water. As it was in a favorable position he shot and killed it instantly, when to his surprise a pup rushed from behind its parent and dived into the water. Almost in the same second a monstrous tiger shark rose, swung swiftly around, seized the young sea-lion and sank. It was so near the surface and our Hunter that almost half of the mottled body swept out above the surface, and the slap of its tail drenched him. Eighteen or twenty feet was a conservative estimate of its length, and indeed these sharks have been killed over thirty feet long. This was a vivid explanation of the reason for the pup seals keeping close to the lava ledges.

The bulls were all up on the hillside some eighty feet above the water, so when one was selected, it was necessary to prod and urge it downward, until it reached a point close to the water, from which it could be rolled into the boat.

Several large, marine iguanas were caught, and the Hunter gave a vivid account of the Chief Fisherman capturing one by sheer speed, making as wonderful a pick up as was ever cursed out by a football coach; there was no temptation to fall on the lizard on these lava rocks.

The great limp bodies of the sea-lions sprawled in the boat, where they had been tumbled after enormous exertions and many experiments in the use of levers and fulcrums. The thing, however, that sent us all splashing into the water to the side of the boat, was the sight of little, sleek, dark heads weaving to and fro above the gunwales. Four baby sea-lions had been captured alive by the simple method of walking over and picking them up. Each one was a goodly armful, and there was something appealingly puppy-like in their big brown eyes and wrinkled noses, while, as we discovered afterward, to see one amble down the deck was to see the original of the walk that made Charlie famous.

Here was the second museum group complete, and the equatorial sun, observing his year-round six-o'clock precision, was far down in the sky. We made one last desperate sortie for final loot. The Scientist suddenly remembered that no Galápagos group could be true to life without at least a dozen of the giant locusts that catapulted from every bush at our approach, so he careered away among the scrub, pursuing locusts till they became exhausted and sank to earth, when he promptly fell on them. The Photographer saw a particularly brilliant specimen of the little scarlet-cheeked lizards that rushed across the sand, so he added another pair of flying legs to those that already adorned the landscape. The Artist and I saw various bits of characteristic vegetation that we had neglected to collect, and forthwith fell to cutting trees like a pair of demented beavers. In the distance the Surgeon was crawling over the lava, stalking some invisible treasure. Everywhere were scurrying forms and clouds of sand, and through it all the Lawyer and the Doctor doggedly seined on.

The male Artist appeared, with a beautiful hawk clasped

in both hands. He had thrown a stick at it and knocked it over, and he was feeling rather sheepish at such an unsportsmanlike proceeding. And after an apologetic explanation of his method, he added, "I could have got two at once if I could have made the stick go round a corner."

After that he was known as the Boomerang Artist.

The *Noma*, silhouetted against the blue bulk of a distant volcano, had a heavy stream of smoke issuing from her stacks and we imagined her tugging at her anchor, or whatever substitute a ship has for pawing the ground. It was not necessary to imagine what the captain was doing; we knew he was walking up and down impatiently, for unless we got under way by sunset, we stood a fair chance of being towed into Panama, which is, I suppose, the final sea-going ignominy. So with a last look at our surroundings, in an attempt to impress every detail on our minds, we scrambled aboard.

Behind the motorboat we towed the life-boat, loaded down with cactus, rock, sand and various things, our care of which would have been the final proof of insanity to the casual observer. Behind the life-boat bobbed the *R. B. Amblyrhynchus* with its unique passenger list. As for us, we perched and draped ourselves among and over cameras, nets, sea-lions, anchors, and other odds and ends.

With the sky behind Eden a blaze of green and gold, the *Noma* moved slowly out of Conway Bay, continent bound, and we took what we then supposed to be our last look at our desert islands. As a Victorian novelist might have said, we little dreamed that in ten short days our eyes would once more gaze on these forbidding shores.

CHAPTER IX

GIANT TORTOISES

THE first reference in history to the giant tortoises of the Galápagos is found in a letter from the discoverer of the archipelago. The Bishop of Panama wrote a description of giant tortoises capable of carrying men upon their backs which he found on the islands to which he was carried by an accident of wind and current.

Ambrose Cowley and William Dampier were here on the same ship in 1684, and they both made brief references to the tortoises from a strictly utilitarian point of view. Cowley wrote: "Here being great plenty of provisions, as Fish, Sea and Land Tortoises, some of which weighed at least 200 pound weight, which are excellent good Food."

Dampier devotes more space to the first detailed description of these creatures:

"The Spaniards when they first discovered these islands, found multitudes of guanoes, and land turtle or tortoise, and named them the Gallipagos islands. I do believe there is no place in the world that is so plentifully stored with these animals. The guanoes here, are as fat and large, as any that I ever saw; they are so tame, that a man may knock down twenty in an hour's time with a club. The land-turtle are here so numerous, that five or six hundred men might subsist on them alone for several months, without any other sort of provision: They are extraordinary large and fat, and so sweet, that no pullet eats more pleasantly. One of the largest of these creatures will weigh one hundred and fifty or two hundred we'ght, and some of them are two feet, or two feet six inches over the callapee or belly. I never saw any but at this place, that will weigh above thirty pounds weight. I have heard that at the isle of St. Lawrence or Madagascar, and at the English forest, an island near it, called also Don Mascarenha or Bourbon, and now possessed by the French; there are very large ones, but whether so big, fat and sweet as these, I know not. There are three or four sorts of these creatures in the West Indies. One is called by the Spaniards, hecatee. . . . These tortoise in the Gallipagos are more like the hecatee, except that, as I said before, they are much bigger; and they have very long small necks and little heads."

Dampier also quotes what was told him by Captain Davis who made a second voyage to the Galápagos in 1685:

"Captain Davis came hither again a second time, and then he went to other islands on the west side of these. There he found such plenty of land turtle, that he and his men eat nothing else for three months that he staid there. They were so fat that he saved sixty jars of oil out of those that he spent. This oil served instead of butter to eat with dough-boys or dumplins; in his return out of these seas."

Woodes Rogers in 1708 described the tortoises as follows:

"Some of the largest of the land-turtles are about 100 pounds weight, and those of the sea upwards of 400. The land turtles laid eggs on our deck. Our men brought some from the shore about the bigness of a goose egg, white, with a large big shell, exactly round. The creatures are the ugliest in Nature, the shell not unlike the top of an old hackney coach, as black as jet; and so is the outside skin, but shriveled and very rough. The legs and necks are very long, and about the bigness of a man's wrist; and they have clubfeet, as big as one's fist, shaped much like those of an elephant, with five thick nails on the fore-foot and but four behind, and the head little, and visage small like snakes, and look very old and bleak. When at first surprised they shrink their neck, head and legs under their shell. Two of our men, with Lieutenant Stratton and the trumpeter of the *Duchess*, affirm they saw vast large ones of this sort, about four feet high. They mounted two men on the back of one of them, which, with its usual slow pace, carried them and never regarded the weight. They supposed this could not weigh less than 700 pounds. I do not affect giving relations of strange creatures so frequently done by others; but when an uncommon creature falls in my way, I cannot omit it. The Spaniards tell us, they know of none elsewhere in these seas, but they are common in Brazil. . . . I saw no sort of beast, but there are guanos in abundance, and land-turtles almost on every island. It is strange how the latter got here, because they cannot come of themselves and none of that sort are found on the main."

From Captain Colnett's account of his visit to the archipelago in 1793:

"An officer and party, whom I sent to travel inland (this was on Chatham), saw many spots, which had very lately contained fresh water, and about which, the land tortoise appeared to be pining in great numbers. Several of them were seen within land, as well as on the sea-coast, which if they had been in flesh, would have weighed three hundred weight, but which were now scarcely one third of their full size.

". . . I found the turtles, however, far superior to any I had before tasted. Their food, as well as that of the land tortoise consists principally of the bark and leaves of trees, particularly of the mangrove, which makes them very fat; though, in rainy seasons, when vegetation is more general, their food may be of a more promiscuous nature. . . . The land tortoise was poor at this season but made excellent broth. Their eggs are as large and their shell as hard as those of a goose and form a perfect globe. Their nests are thrown up in a circular form and never contain more than three eggs which are heated by the sun, an hole being so contrived as to admit its rays through its daily course. The shell is perfectly smooth and when highly polished, receives a beautiful and brilliant black."

On James, he said, "the woods abound with tortoises, doves and guanas, and the lagoons with teal." And he waxes enthusiastic over the delicacy of tortoise meat:

"But all the luxuries of the sea yielded to that which the island afforded us in the land tortoise, which in whatever way it was dressed, was considered by all of us as the most delicious food we had ever tasted. The fat of these animals when melted down, was equal to fresh butter; those which weighed from thirty to forty pounds, were the best, and yielded two quarts of fat; some of the largest, when standing on their feet, measured near a yard from the lower part of the neck. As they advance in age their shell becomes proportionatcly thin, and I have seen them in such a state, that a pebble would shatter them. I salted several of the middle size, with some of the eggs, which are quite round, and as big as those of a goose, and brought them to England."

In 1813 Captain Porter gave the most accurate and detailed description of the different species of Galápagos tortoise; he first landed at Hood, and of these tortoises he said:

"Here wood is to be obtained, and land tortoises in great numbers, which are highly esteemed for their excellence, and are remarkable for their size, weighing from three to four hundred-weight each. Vessels on whaling voyages among these islands generally take on board from two to three hundred of these animals, and stow them in the hold, where, strange as it may appear, they have been known to live for a year, without food or water, and, when killed at the expiration of that time, found greatly improved in fatness and flavour."

Captain Porter, after the capture of three British ships wrote:

"We found on board of them, also, wherewith to furnish our crew with several delicious meals. They had been in at James' Island, and had supplied themselves abundantly with those extraordinary animals the tortoises of the Gallipagos, which properly deserve the name of the elephant tortoise. Many of them were of a size to weigh upwards of three hundred weight; and nothing, perhaps, can be more disagreeable or clumsy than they are in their external appearance. Their motion resembles strongly that of the elephant; their steps slow, regular and heavy; they carry their body about a foot from the ground, and their legs and feet bear no slight resemblance to the animal to which I have likened them; their neck is from eighteen inches to two feet in length, and very slender; their head is proportioned to it, and strongly resembles that of a serpent. But, hideous and disgusting as is their appearance, no animal can possibly afford a more wholesome, luscious and delicate food than they do; the finest green turtle is no more to be compared to them in point of excellence, than the coarsest beef is to the finest veal; and after once tasting the Gallipagos tortoises, every other animal food fell greatly in our estimation. These animals are so fat as to require neither butter nor lard to cook them, and this fat does not possess that cloying quality, common to that of most other animals. When tried out, it furnishes an oil superior in taste to that of the olive. The meat of this animal is the easiest of digestion, and a quantity of it, exceeding that of any other food, can be eaten without experiencing the slightest inconvenience. But what seems the most extraordinary in this animal, is the length of time that it can exist without food; for I have been well assured, that they have been piled away among the casks in the hold of a ship, where they have been kept eighteen months, and when killed at the expiration of that time, were found to have suffered no diminution in fatness or excellence. They carry with them a constant supply of water, in a bag at the root of the neck, which contains about two gallons; and on tasting that found in those we killed on board, it proved perfectly fresh and sweet. They are very restless when exposed to the light and heat of the sun, but will lie in the dark from one year's end to the other without moving. In the day-time, they appear remarkably quick-sighted and timid, drawing their head into their shell on the slightest motion of any object; but they are entirely destitute of hearing, as the loudest noise, even the firing of a gun, does not seem to alarm them in the slightest degree, and at night, or in the dark, they appear perfectly blind. After our tasting the flesh of those animals, we regretted that numbers of them had been thrown overboard by the crews of the vessels before their capture, to clear them for action. A few days afterwards, at daylight in the morning, we were so fortunate as to find ourselves surrounded by about fifty of them, which were picked up and brought on board, as they had been lying in the same place where they had been thrown over, incapable of any exertion in that element, except that of stretching out their long necks."

Captain Porter sent the chaplain of the *Essex* to survey a large island which they were unable to recognize from any charts in their possession. It was undoubtedly the one that is now known as Indefatigable, but Mr. Adams, the chaplain, named it Porter Island. Porter says:

"He stated that it abounded in wood, and that land tortoises and green turtle were in the greatest abundance, the former generally of an enormous size, one of which measured five feet and a half long, four feet and a half wide, and three feet thick, and others were found by some of the seamen of a larger size."

On his second visit to the islands, Captain Porter has this to say of the tortoises of James:

". . . we began to lay in our stock of tortoises, the grand object for which every vessel anchors at the Gallipagos Islands. Four boats were despatched every morning for this purpose, and returned at night, bringing with them from twenty to thirty each, averaging about sixty pounds. In four days we had as many on board as would weigh about fourteen tons, which was as much as we could conveniently stow."

The account which Charles Darwin wrote when he visited the Galápagos in the *Beagle* is well worth recounting. He was only twenty-six years old at the time, and yet the keenness of observation is well illustrated in this narrative:

"I will first describe the habits of the tortoise (*Testudo nigra*, formerly called Indica), which has been so frequently alluded to. These animals are found, I believe, on all the islands of the archipelago; certainly on the greater number. They frequent in preference the high damp parts, but they likewise live in the lower and arid districts. I have already shown, from the numbers which have been caught in a single day, how very numerous they must be. Some grow to an immense size: Mr. Lawson, an Englishman, and vice-governor of the colony, told us that he had seen several so large, that it required six or eight men to lift them from the ground; and that some had afforded as much as two hundred pounds of meat. The old males are the largest, the females rarely growing to so great a size; the male can readily be distinguished from the female by the greater length of its tail. The tortoises which live on those islands where there is no water, or in the lower and arid parts of the others, feed chiefly on the succulent cactus. Those which frequent the higher and damp regions, eat the leaves of various trees, a kind of berry (called guayavita) which is acid and austere, and likewise a pale green filamentous lichen (*Usnera plicata*), that hangs from the boughs of the trees.

"The tortoise is very fond of water, drinking large quantities, and wallowing in the mud. The larger islands alone possess springs, and these are always situated toward the central parts, and at a considerable height. The tortoises, therefore, which frequent the lower districts, when thirsty, are obliged to travel from a long distance. Hence broad and well-beaten paths branch off in every direction from the wells down to the sea-coast; and the Spaniards by following them up first discovered the watering-places. When I landed at Chatham Island, I could not imagine what animal travelled so methodically along well-chosen tracks. Near the springs it was a curious spectacle to behold many of these huge creatures, one set eagerly travelling onwards with out-stretched necks, and another set returning, after having drunk their fill. When the tortoise arrives at the spring, quite regardless of any spectator, he buries his head in the water above his eyes, and greedily swallows great mouthfuls, at the rate of about ten in a minute. The inhabitants say each animal stays three or four days in the neighbourhood of the water and then returns to the lower country; but they differed respecting the frequency of these visits. The animal probably regulates them according to the nature of the food on which it has lived. It is, however, certain, that tortoises can subsist even on those islands, where there is no other water than what falls during a few rainy days in the year.

"The tortoises, when purposely moving towards any point travel by day and night, and arrive at their journey's end much sooner than would be expected. The inhabitants, from observing marked individuals, consider that they travel a distance of about eight miles in two or three days. One large tortoise, which I watched, walked at the rate of sixty yards in ten minutes, that is 360 yards in the hour, or four miles a day, allowing a little time for it to eat on the road. During the breeding season, when the male and female are together, the male utters a hoarse roar or bellowing, which, it is said, can be heard at the distance of more than a hundred yards. The female never uses her voice, and the male only at these times; so that when the people hear this noise, they know that the two are together. They were at this time (October) laying their eggs. The female, where the soil is sandy, deposits them together, and covers them up with sand; but where the ground is rocky she drops them indiscriminately in any hole. Mr. Bynoe found seven placed in a fissure. The egg is white and spherical; one which I measured was seven inches and three-eights in circumference, and therefore larger than a hen's egg. The young tortoises, as soon as they are hatched, fall a prey in great numbers to the carrion-feeding buzzard. The old ones seem generally to die from accidents, as from falling down precipices: at least several of the inhabitants told me, that they never found one dead without some evident cause.

"The inhabitants believe that these animals are absolutely deaf; certainly they do not overhear a person walking close behind them. I was always amused when overtaking one of these great monsters, as it was quietly pacing along, to see how suddenly, the instant I passed, it would draw in its head and legs, and uttering a deep hiss fall to the ground with a heavy sound as if struck dead. I frequently got on their backs, and then giving a few raps on the hinder part of their shells, they would rise up and walk away;—but I found it very difficult to keep my balance. The flesh of this animal is largely employed, both fresh and salted; and a beautifully clear oil is prepared from the fat. When a tortoise is caught, the man makes a slit in the skin near its tail, so as to see inside its body, whether the fat under the dorsal plate is thick. If it is not, the animal is liberated and it is said to recover soon from this strange operation. In order to secure the tortoise, it is not sufficient to turn them like turtle, for they are often able to get on their legs again."

My friend Mr. R. H. Beck has probably had more experience with Galápagos tortoises than all other men together. He has made many voyages with the express purpose of collecting them, and in 1905 he wrote a very interesting account of his adventures in collecting these remarkable chelonians. This was published in the seventh annual report of our New York Zoological Society, and as this has only a local circulation, I have reprinted parts of it:

GIANT TORTOISES

"In recent years several expeditions have visited the Galápagos Archipelago, and over one hundred and fifty tortoises, dead and alive, have been taken away. The last lot contained over fifty specimens, and was brought to San Francisco in August, 1902, in the schooner Mary Sachs, and taken across to London for the Honorable Walter Rothschild.

"The largest tortoise in that shipment was obtained near the top of a large crater, eighteen miles from the shore, and twelve men were required to carry it the last six miles of the distance. So far as known this specimen is the largest ever taken from the Galápagos Islands, and is larger than any other of the dozens seen there by the collectors.

"It is only within the last few years that the home of these very large tortoises has been invaded by man, but the rapidity with which they are being killed, and the reason for their destruction, leaves us but little hope that they will survive any longer than did the American bison after the hide hunters began their work of extermination.

"A description of the south side of Albemarle Island, where the largest tortoises live, will give one a fair conception of the vegetation and general character of the other mountains and islands occupied by tortoises.

"The first 1,000 feet above sea level, which slopes up gradually, is composed of rough lava, in the cracks of which trees and bushes take root and grow during the three or four months of the rainy season. In the next 1,500 feet vegetation is more abundant, and the vines and bushes form a serious impediment to travel. On the trees, ferns and orchids grow in large numbers.

"From about 2,500 feet upward the forest ceases, and long, rank grass and brake-ferns form the principal growth. At this height, during the summer season, a heavy fog hangs over the mountain almost continuously, and here a majority of the tortoises spend their time from May until January. On the mountain particularly described, on Albemarle Island, are thousands of cattle, descended from a few placed there years ago.

"Years ago a gentleman from Guayaquil settled here with some labourers for the purpose of killing the cattle for their hides; and upon finding it would take three or four years to do this, he established a ranch at the upper edge of the forest, where the cattle and tortoises were most abundant. We spent two weeks at this ranch, collecting and studying the tortoises. On our walk up to the ranch, ten miles by trail from the shore settlement, we counted over thirty tortoises in the last three miles, and it was quite evident that at this place were more tortoises than we had ever expected to see in their native state.

"At the time of my visit, the majority of the tortoises were in the open glades and sunny parks in the upper edge of the forest. In every such place along the trail, and near other trails traversed near the ranch, they could be seen feeding, walking about, or quietly sleeping with their heads against the base of some bush or tree where they had dug a form in which to lie. The form which a tortoise occupies is similar in shape to that of the common hare in California; but instead of facing outward, as does the hare, the tortoise always faces inward.

"The tortoise seemed to have no regular time for feeding, being at all hours of the day eating or walking about. During the middle of the day, if the sun is shining, they keep in the shade of the trees, but if it is cloudy many spend the time wandering back and forth on the trails. We were told by the natives that in the summer the tortoises go up to the top of the mountain; and this statement confirmed my observations of similar habits of other species in the archipelago.

"We found that the tortoise trails extend up and down the mountain side for miles, one of the objective points at the lower part of the range being a rocky basin where water collects during rains. By centuries of constant use these rocks have been worn so smooth that it is almost impossible to walk over them after a rain, while they are wet. Once we noticed four tortoises slaking their thirst at a rocky pool near the trail, but during our stay at the ranch the rainfall was so great that every little hollow in the ground held water, and a tortoise could get a drink anywhere.

"One afternoon, while standing under a tree during a heavy downpour, I was surprised to see a big tortoise come slowly down the hill through the wet grass, walk into a rapidly-forming pool of water, take a long drink, and then lie down in the pool. When he settled down the depth of the water was only two inches, but in a few minutes it had increased to eight inches, and he seemed entirely content, until his attention was attracted to a female tortoise, which also came to the pool to drink. That attraction was the stronger, so he left the water and set out to make her acquaintance.

"After the rain had ceased, I went down the trail some distance and saw another tortoise lying in a hollow filled with water. He remained there all night apparently, for on our return the next morning he was still in it. These two observations rather tended to disprove my theory regarding one of the causes of the annual migration which effects nearly all the species of Galápagos tortoises. I had formed the opinion that the migration was partly due to the slightly colder weather and heavy rains high up on the mountains during the winter season, but it would seem from the actions here cited that these causes have but little to do with it after all. With this species (*Testudo vicina*), it might be the mating instinct that causes them to wander down three or four miles from their summer home.

"Love affairs were in full progress during our stay, and the amorous exclamations of the males could be heard at a distance exceeding three hundred yards, even in the thick forest. Being told by the natives that the largest tortoises were on the top of the mountain we took a couple of burros and went up to secure the largest specimen obtainable. It was necessary to skin this animal, for the reason that even two burros could not carry such a specimen alive.

"In a little valley, ten miles from the ranch, and but a short distance from the mountain's top, we found a number of big fellows, considerably larger than any seen at the lower altitude around the ranch. This valley was undoubtedly the home of the patriarchs of the mountain, and a better spot for their development could not be found. There were several deep ponds of water, and judging by the number of cattle present, the grass was of the sweetest. The absence of female tortoises at this height was very noticeable, not over half a dozen being seen out of probably seventy-five individuals observed. Whether the wild dogs, which are so numerous and ferocious here, have eaten the females, or whether, as is not unlikely, they had gone lower down on the mountain, we were not able to determine. We saw a number of skeletons of tortoises that the dogs had killed, and noticed that, as a rule, the females being the smallest were the ones to suffer. However, we saw a couple of males over three feet long, showing that when hunger is keen enough even the large ones are killed. The presence of these oldest and largest males at this elevation during the rainy season would indicate that they travelled but little, and after noting their surroundings we could see no reason why they should ever journey more than a thousand feet away from the water-holes. A constant abundance of grass, scattering trees, plenty of bushes to lie under when the sun shines too fiercely, and water within easy reach, surely should constitute conditions conducive to long life and perfect contentment.

"An interesting scene was several times witnessed here, when one, and sometimes two, of the little birds (*Geospiza fuliginosa*), so common on the islands, would fly up on the back, head, or neck of a tortoise to pick off the minute grass seeds that lodged in the folds of the skin, in the corners of the mouth, and in the nose. We noticed it on several occasions, and the tortoises never evinced the least displeasure in the proceedings of their small visitors. Once, while watching a bird picking seeds from the nose, and even reaching over to pick them from the mouth of a large tortoise, the thought occurred to me that the story of the little bird that enters the mouth of the crocodile for flies is now not nearly so improbable as when first read years ago. Once I saw one of the brilliant red fly-catchers ride along on the back of a giant tortoise for some distance, flying off now and then to catch an insect, but immediately returning to its odd perch, apparently enjoying the ride as much as we enjoyed watching it.

"The actions of the tortoises living in the hollows and small valleys along the mountain-top were very similar to those of the cattle that occupied the same range. Walking cautiously over a rise we would see perhaps three or four at a water-hole, drinking, and dispersed in the open valley would be others busily nibbling at the short grass. During the heat of the day many would be seen lying in the shallow pools of water that the heavy rains had formed, or under the bushes near by them. One hot day I saw two large tortoises and two young bulls lying side by side under a small tree. Nearby were other cattle, and another large tortoise was headed for the tree, having just left a water-hole a few rods away.

"After seeing on this mountain dozens of tortoises of good size, one wonders where the small ones are; but after spending a few days a-foot and seeing the many wild dogs in that region —descendants of those left years ago by sailing vessels—we can only wonder that so many of the large ones remain. From the time that the egg is laid until the tortoise is a foot long, the wild dogs are a constant menace, and it is doubtful if more than one out of ten thousand escapes. We certainly saw none, and the natives told us that the dogs ate them as fast as they were hatched.

"In November we found several nests in the lower edge of the forest. Of these, two had been rifled, and the broken egg-shells were what first attracted our attention to them. All the eggs found on that date (November 12th) were perfectly fresh, and we saw two or three newly dug holes with tortoises but a few feet from them. Most of the nests found were in well-travelled cattle or tortoise trails. They were so placed that the sun shone on them but a few hours each day; when it did it was very hot. Ordinarily it was very difficult to recognize the site of a nest, the very slight elevation in the trail, or slightly fresher-looking earth being our sole guide. Several times we imagined that we had discovered nests, and prodded about with our sticks, and dug with our hands until finally we realized that we had misinterpreted the signs.

"On finding our first nests in the trail the old adage, 'Don't put all your eggs into one basket,' was forcibly brought to mind. This is a rule that is followed by the tortoise, for within a radius of fifteen feet four nests were found, each containing eight to seventeen eggs. The holes were about fifteen inches in depth, and nearly a foot in diameter. The eggs were placed in layers of three to six, the first layer being on the soft soil on the bottom, separated from the next by an inch or so of dirt, and the second layer separated from the third in the same manner. The dirt surrounding the eggs was loose, but the top of the hole was covered to a depth of three or four inches with a very hard crust that had probably been formed by the tortoise lying on it and working from side to side in the same manner that we frequently noticed them working down a form to lie in.

"Judging by the size and number of the eggs found in several of the tortoises that we dissected, it would seem that one or two nests are finished at a given period, and a week or two later the remainder of the eggs are laid. From ten to twenty eggs were ready for extrusion together, while twenty or thirty more were from one-half to two-thirds the normal size. A peculiar fact regarding the tortoises that inhabit this mountain is the scarred appearance of the shells of those living near the top as compared with those living near the base, or, for the matter of that, any of the other species in the archipelago. The young tortoises near the top are very smooth, but with hardly an exception the old ones show irregular spots on their shells that thus far have not been satisfactorily accounted for. It seems hardly possible that they could have been made by a shower of lava, and yet, within a mile of the spot where the old patriarchs of this species are found, there are a number of living volcanoes which might, a hundred years ago, have burst forth in explosions of sufficient violence to hurl lava for miles.

"The roughness alluded to cannot be due, as has been suggested by some persons, to the tortoises having fallen and rolled over rocks, for were this the case, the tortoises living on Duncan Island, where the ground is much rougher and rocks more plentiful, should be scarred worse than any, but no marks upon them indicate any such experiences. While at the ranch, where nearly fifty men were at work, we were amazed at the reckless and heartless manner in which some of the natives destroyed the tortoises. The proprietor informed us that only the males were killed, but we noticed that the working people made little distinction in the sexes when killing for food. Some evenings, two or three men coming in from different directions would each carry in his hand a small piece of tortoise meat, and a pound or so of fat with which to cook it. Of each tortoise killed not over five pounds of meat would be taken, the remainder being left for the wild dogs that swarmed about.

"One Saturday evening I had occasion to go down the trail a mile or so, after some of the natives had departed for the shore settlement, where all the women and children lived. I found a large tortoise, three feet six inches long and hundreds of years old, which had been cut open with a machete, but apparently not more than three pounds of meat had been taken from it. A little farther on lay a dead female, from which nothing had been taken save a string of eggs and a very little meat.

"At the rate of destruction now in progress it will require but a few years to clear this entire mountain of tortoises, and when we see the methods pursued by the proprietor in getting tortoise oil for shipment to the mainland, we know that the large tortoises can last but a few months after the work of the oil hunter begins in earnest.

"To show what has already been done by oil hunters, I took two photographs at the water-hole, where lay the largest number of tortoise skeletons. There were about one hundred and fifty skeletons at this pool, and a half mile away, in another depression, were about one hundred more. While there were more skeletons at these two places than we saw elsewhere, frequently ten or fifteen were observed in other basins where the tortoises had gone for water.

"The outfit of the oil hunter is very simple, consisting merely of a can or pot in which to try out the oil, and three or four burros for carrying the five- or ten-gallon kegs in which it is transported to the settlement. After making a camp near a water-hole, and killing the tortoises there, the collector brings up a burro, throws a couple of sacks over the pack-saddle, and starts out to look for more tortoises, killing them wherever found. A few strokes of the machete separates the plastron from the body, and ten minutes' work will clear the fat from the sides. The fat is then thrown into the sack, and the outfit moves on.

"When the burro is well laden, man and beast travel back to camp, where the oil is tried out. Each large tortoise yields from one to three gallons of oil. The small ones are seldom killed, because they have but little fat. By daily visits to the few water-holes during the driest season, in the course of the month the hunters get practically all the tortoises that live on the upper part of the mountain.

"When we first stepped ashore at the settlement we saw a number of casks lying on the beach, and learned on inquiry that they contained eight hundred gallons of tortoise oil. In a large boat, under a nearby shed, were four hundred gallons more. While we were there the boat sailing between the island and Guayaquil left for that port with those casks and a cargo of hides. The value of the oil in Guayaquil was about $9.00 (American) per hundred pounds. While the tortoises are as plentiful as we saw them, this price yields a fair profit to the hunters, but two more raids such as that shown in the photograph will clear that mountain of all the fair-sized tortoises upon it, and then the oil business is ended.

"The photograph of seven tortoises at a pond was taken ten miles from the ranch at 3,500 feet elevation, where the hunters had not yet been, but soon these will be discovered and go the way of the rest. Those that the hunters overlook will be killed by the wild dogs as soon as the cattle are decimated. Between men and dogs the creatures that live on the ground must soon disappear. On the northern end of the island the land iguanas have been exterminated, although they were much more numerous and better able to escape than are the tortoises. Near Tagus Cove, the best harbour in the northern part of the island, situated at the foot of another large mountain of volcanic origin, there were formerly, if the number of trails can be taken as a criterion, hundreds of tortoises. - Today it is a hard matter to find one, for it appears that this species (*Testudo microphes*) has been used as food for whalers more than any other.

"Quite a large valley extends along the southern base of the mountain, near Tagus Cove, and here the tortoises were in the habit of coming every rainy season when the grass was young and tender. Several smooth rocks of irregular basin shape, where water collects in rainy weather, indicate by their worn appearance the visits of innumerable tortoises in days gone by. On the small sandstone ridges which terminate at the valley can be seen trails two feet deep, and just the width of a tortoise. It must have required centuries of travel to wear such trails as these. Higher up on the mountain, among the thick trees and bushes, the trails often extended around the mountain side from one basin or grass flat to another; but here, instead of quenching their thirst at water-holes, as do those on South Albemarle, the tortoises usually have to be satisfied with what moisture they can get from the cactus leaves, as do the species on Abingdon, Duncan, and Indefatigable, at least to a large extent.

"¡It seems remarkable that soft-tongued tortoises should be able to eat the sharp-spined cactus leaves, but that they do so, and greatly relish them, is proven on several islands by the way the cactus leaves and blossoms disappear from under the trees. On the north end of Albemarle, where still another species is found, we noticed several small cacti, the growing leaves of which had been partly eaten by tortoises.

"On Duncan Island, a few miles from Albemarle, lives *Testudo ephippium*, the species with which I first became acquainted, and the remembrance of my first sight of a Galápagos tortoise in his native haunts will never be forgotten.

"After climbing at least 1,500 feet up the mountain, we came to an extinct crater, filled with a growth of bushes and trees. The floor of the crater was several hundred feet below us, and the steep sides, covered with loose rocks and thorny bushes, made the descent very difficult. After reaching the bottom, the members of our party separated, and each of us went looking about independently to find a tortoise. We had seen a number of wallows where tortoises had been lying in the mud, and each member of the party was on the alert to find the first living specimen.

"I wandered through narrow lanes, over little grassy meadows, and sometimes went under bushes on hands and knees to avoid their thorns. The high walls of the crater towered aloft all around me, and the intense stillness of the place was broken only by the drone of a cricket. I easily imagined we were back in a bygone age, and then, as a large tortoise, with neck outstretched, ponderously appeared from behind a thick bush, I felt that it would not be surprising to see a pterodactyl come flying over the rim of the crater, or a megalosaurus rise out of the bushes near by.

" We found several tortoises in that crater, and after considerab'e search, discovered a steep trail leading up to the mountaintop. Going up this trail the next day, a large tortcise was found placidly devouring a fallen cactus limb. On a hillside farther on, the trails from one cactus tree to another were as plain and as well worn as cattle trails are on a well-stocked range in California.

"We soon learned that the easiest way to get about through the thick brush was to use the tortoise trails, even to the extent of crawling part of the time. This island (Duncan), being much lower than Albemarle, has less fog and rain, and there are times when the tortoises get no water for months together. We found, however, that they knew every good-sized cactus tree on the hillside and in the valley where they lived, for there were no leaves nor limbs lying about under the trees, as was the case in other places.

"We camped for a week on that mountain-top, and captured altogether nearly thirty live tortoises, which were later on sent to Europe. We were much chagrined, however, at finding no very small specimens, but soon came to the conclusion that the large rats, of recent introduction, and now common everywhere on the island, eat the young as soon as they are hatched.

"There are still a few tortoises on Duncan Island, and probably will be for some years to come, unless the natives should elect to visit it and hunt them with dogs, in which event they would be quickly exterminated. They live in a space of less than five square miles, and I doubt if fifty still remain.

" In 1901 the natives of Chatham Island, where there is a large sugar and coffee plantation, sometimes visited other islands to procure iguanas for food, their supply of cattle having been exhausted. It may happen at any time that a few

FIG. 38

FRONT VIEW OF GIANT TORTOISE

FIG. 39

OUR GIANT TORTOISE SWIMMING TOWARD THE "NOMA"

FIG. 40

GIANT TORTOISE IN THE NEW YORK ZOOLOGICAL PARK

FIG. 41

SHELLS OF HUNDREDS OF GIANT TORTOISES IN ALBEMARLE

Killed for their oil by natives of Ecuador

expeditions will stop at Duncan Island long enough to clear the tortoises of that spot from their native home.

"So far as known, *Testudo abingdoni*, of Abingdon Island, is practically extinct. We secured two specimens in 1901, but on another trip, after thoroughly hunting over the ground where tortoises were formerly common, only a single fresh trail was discovered, where a lone tortoise had passed a few months before. The part of the island where this species lived is fairly easy to travel over, and therefore it was not a difficult matter for the hunters with dogs to make a clean sweep. While there are about forty square miles of surface on this island, not over seven or eight are suitable for tortoises, and for some of the other species the proportion of suitable ground is still less.

"A very few years will probably see the extinction of two or three of the present living species, and while a few specimens of the others may linger for a much longer time, they, too, are bound to disappear under the attacks of their enemies. The ease with which these long-lived reptiles may be kept in captivity, and the great interest displayed by the public in watching the ponderous movements of a 500-pound tortoise, hundreds of years old, make them mostly inmates of a Zoological Park."

The last important scientific expedition to the islands was that from the California Academy of Science. Over two hundred and fifty specimens of tortoises were collected, and fifteen species were recognized. Van Denburgh sums up the status of these wonderful creatures seventeen years ago, as follows. The generic name of all is *Testudo:*

Island	*Name*	*Status in 1906*
1. Abingdon	*T. abingdoni*	Rare
2. James	*T. darwini*	Rare
3. Jervis	*T. wallacei*	Very rare
4. Duncan	*T. ephippium*	Fairly abundant
5. Indefatigable	*T. porteri*	Not rare
6. Barrington	*T. sp.*	Extinct
7. Chatham	*T. chathamensis*	Nearly extinct
8. Hood	*T. hoodensis*	Very rare

Island	Name	Status in 1906
9. Charles	T. elephantopus	Extinct
10. Narborough	T. phantastica	Very rare
11. Villamil, Albemarle	T. güntheri	Abundant
12. Iguana Cove, Albemarle	T. vicina	Numerous
13. Tagus Cove, Albemarle	T. microphyes	Fairly numerous
14. Bank's Bay, Albemarle	T. becki	Fairly numerous
15. Cowley Mt., Albemarle	T. sp.	Rare

While in the extremely limited time at my disposal I could make no extended search into the interiors of the islands, yet we always kept on the watch for signs of tortoise, and except for a few scattered bones at Tagus Cove, found nothing. Finally we decided to make a short trip with the express purpose of finding a tortoise, and four of the party took the motor lifeboat and went from Seymour Bay to Duncan, a distance of twenty-four miles. They landed on a rocky shore protected by a small island, and climbed the walls of the northern crater. The rim was about nine hundred feet high and the inner slope quite as steep as the outer. The crater was a large one; the bottom about three-quarters of a mile across. Two of the party remained on the rim and the others made the descent. Robert McKay was the fortunate discoverer of the first and only Galápagos tortoise which we were to see. He and Broking made their way to the bottom of the great crater, and found it so exhausting and fearfully hot that they rested for a while in the scanty shade of a clump of cactus. They then began to search carefully for traces of the great reptiles, keeping vocally in touch with each other and with the other members of the party high up on the volcano's rim, for the acoustic properties of this Galápagos amphitheatre were perfect.

Much of the bottom consisted of dried mud, and in it were numerous tortoise holes or forms and the tracks of their feet. The holes, McKay says, looked exactly like shell-holes made by soixante-quinze shells. About half-way across, while crawling on hands and knees through thorn bush, he came suddenly upon the tortoise, announcing his find with a yell which reached all over Duncan.

A pole and ropes had been provided and the tortoise was lashed to it and slowly carried up over the terrible lava and through the thick growth of cactus and other thorny plants. It was even more difficult making the descent outside the volcano, especially as the sun had gone down and in the dark it was impossible to avoid striking against thorns and falling over loose clinker slabs. It was late when the boat was finally reached, and the course set for home in the moonlight. The party reached the *Noma* at eleven-thirty that night, with a tremendous respect for Beck who had transported numbers of much larger tortoises from many miles inland to the coast of the islands.

Our specimen was *Testudo ephippium*, and was twenty-two inches long and weighed forty-two pounds. We got a splendid series of moving pictures of it under various conditions.

I was interested to see its ability to climb up extremely steep and rough lava slopes, which no ordinary tortoise could possibly surmount. In spite of frequent slipping it kept obstinately ahead in any direction it had once chosen. The edge of the shell was deeply scalloped and bent alternately downward, and several times on a steep slant I saw it slip helplessly back,—saved from a long tumble by a shell rest on an outjutting rock, giving time for its elephantine feet with their remarkably long and sharp claws to reach about for a foothold. Once it voluntarily backed a few feet and went around a precipitous area.

Over rather soft mud, the great flat soles left almost no trace, but the claws sunk in deeply, making curious, isolated, vertical holes. When first put down on level ground, it reared its great prehistoric head and neck high aloft, looked slowly about and chose the precise direction it wished, when it lowered its head and set off resolutely. When suddenly alarmed, the head was drawn in with a sharp hiss and the two great armoured wrists folded together in front.

When first set down near the coast of Indefatigable, it turned at right angles, forcing me to one side, and set out stolidly for the interior of the island. The second time it was placed facing the sea, and it described a half circle at once. In fact no amount of reorienting, turning, or twisting had any effect. It was wholly out of sight and hearing of the sea, and yet never failed instantly to head straight for the heart of the island.

Although it was relatively a moderate sized individual, yet we had the satisfaction of seeing at last this representative of the great hosts of which Dampier and Porter wrote. It seemed resigned to fate and after we had handled it a few times it ceased even to hiss or withdraw its head completely under the shell.

Mr. Van Denburgh, who studied the tortoises brought back to California in 1902, came to the following conclusions:

"The various races of tortoises of the Galápagos Islands differ from one another chiefly in shape. There are no real differences in structure, such as are found in the lizards and snakes of the archipelago. The relative values which should be attached to these differences in shape are extremely difficult to estimate. Therefore the tortoises do not throw much light upon the history and development of the archipelago. Some points, however, are of considerable interest.

"Tortoises either live, or are known to have lived, upon Abingdon, Chatham, Hood, Charles, Barrington, Indefatigable, James, Jervis, Duncan, Narborough, and Albemarle islands. The last named island supports several races. The tortoises of Barrington are practically unknown. Each of the other nine islands had its own peculiar race of tortoise, and on none of these nine islands has evidence of more than one race been found. This lends particular interest to the fact that several races of tortoises occur on Albemarle Island.

"Although these tortoises can live for at least several days floating on the surface of the ocean, they are absolutely helpless in the water. They are unable to swim, and can only float and drift at the mercy of the winds and currents. When they drift on island shores, they are usually so battered and injured on the rocks that they only live a few days thereafter. The fact that each Island except Albemarle, has one and only one race of tortoise, is evidence that interchange between the islands has not occurred, for such interchange would result either in preventing differentiation or in the presence of more than one race on an island.

"If the transportation of tortoises from one island to another does not occur, there is little reason to believe that tortoises, at some time in the past, have drifted over the vastly greater distance from some continent, and have reached each of the eleven islands on which they have been found. Nor do we know whence they have come. The evidence offered by these tortoises, therefore, seems to be against the view that these are oceanic islands, which have been independently thrust above the surface of the water, and have received such animals as have drifted to them. We must rather adopt the view that the islands are but the remains of a larger land-mass which formerly occupied this region, and was inhabited by tortoises, probably of but one race; that the gradual partial submersion of this land separated its higher portions into various islands; and that the resulting isolations of the tortoises upon these islands has permitted their differentiation into distinct races or species."

I am in perfect agreement with Van Denburgh in regard to the formation of the Galápagos by subsidence, and I go even farther and can see no explanation of the origin of the flora and fauna except through a former, direct connection with the mainland along the Central America-Panama line. But he is mistaken when he says that these tortoises are "absolutely helpless in the water," for we tossed the Duncan individual overboard and took hundreds of feet of moving picture film of it as it floated two miles from land. Not only did it float upright with no attempt at balancing, but considerable of the anterior part of the shell was exposed, making it possible for the head to be lowered under water, held easily clear of the surface or raised high above it.

But the most surprising thing was the ease and excellence of its swimming ability. The reptile would swim toward the row-boat which I occupied, and finding it too high, would turn and swim over to the *Noma*, stretching its head high along the waterline. Then it steered its way to the companionway. This was with, across, and against the very appreciable current at will. I could see the throat vibrate in breathing, without any detectable lowering or elevation of the body. So for a time at least these creatures have perfect control over themselves in the water.

A week later this tortoise died without warning. The body was somewhat wasted, but hardly enough to account for so sudden a termination of life in these slow living, slow dying animals, although the stamina of the last few remaining survivors may be very weak. The lungs and small intestine were heavily congested and I am inclined to think that it was a quantity of salt water which the creature may have swallowed which caused death. This would negative any possibility of the tortoises being able to make their way over wide expanses of water, either from the mainland or from island to island, in spite of their unusual swimming ability.

Our Duncan specimen had been feeding on green and dried grass and leaves and a few Bursera berries.

GIANT TORTOISES

In the New York Zoological Park there is at present a Galápagos tortoise which is thirty-eight and a half inches in length and weighs two hundred and sixty-eight pounds. Compared with our specimen, this shows an increase of about 60% in length, and more than 700% in weight.

These tortoises are said to be the oldest living beings in the world, four and five hundred years being claimed for the largest.

We know nothing of the ancestors of these great tortoises. A fact worth recording, however, is that giant tortoises closely resembling these have been found fossil in Cuba.

CHAPTER X

GAME FISHING IN THE GALÁPAGOS

By Robert G. McKay

THE possibility of deep sea fishing from the sportsman's point of view was anticipated by the lay members of the party to some extent, although no real idea could be gained from former accounts just what to expect nor for what to prepare. Certainly no one foresaw the sporting paradise that was waiting for rod and reel.

Having done a small amount of Tuna fishing at Catalina, and some tarpon and sailfish fishing in Florida, our choice of tackle was dictated by these experiences and was found to be lacking in variety, quality and quantity. We found that the greatest difficulty in sea fishing was to adapt your tackle to the size fish you were going to hook. We actually succeeded in landing no really large fish for the simple reason that when we hooked any of the monsters we deemed ourselves lucky in saving the rod.

Our first experience was the most usual one in tropical waters. At Key West someone thought of shark,—over went one of the big hooks in a piece of meat and in a very few moments an eight foot harbour shark was being dissected by the scientists. She was carrying three unborn young. These were to all appearances perfect fish and undoubtedly this fact is what gives rise to the generally publicly accepted theory that the shark swallows her young when they are threatened with danger. The fact is, if a shark swallows her young, or any one else's young, they are carried through the normal digestive apparatus and never quite appear the same again.

At Colon the fishermen began to have a real foretaste of what was coming. The builders of the Panama Canal and the present operatives, the latter largely made up from the former, form a proud organization. From gate tender to the boss of all you will find a worthy pride in having been associated with that gigantic undertaking, and an esprit de corps that is only too rare among government workers. They form a group of out-door, red-blooded, pioneer stock that one finds wherever great constructive battles with nature take place. These men are natural sportsmen. As an outlet for this instinct they have formed a fishing club known as the Tarpon Club. Their club house stands on the high bank of the Chagres River, half a mile below Gatun Dam. Gatun Lake has two outlets, one through the locks, the other over the Gatun Dam which controls the flow in the old Chagres River. Gatun Dam is a two hundred foot high, six hundred foot wide block of concrete, regulating the waters of Gatun Lake through nine enormous gates; from the foot of the actual dam extending down the river five hundred yards is a smooth, gently sloping concrete apron completely flooring the space between the high concrete walls on each side that here form the river bank. Running evenly and smoothly over this apron, its depth regulated by the number of gates open at the top of the Dam, comes the sweet water of Gatun Lake. At the lower end of the apron which is cut off sharply, this water boils off into the natural rock of the river bed and goes tearing through as fine a bit of rapids and pools as one could ask for. Fish of all species climb towards this barrier for two reasons—because the fresh water from the lake cleans them of salt water parasites, and because of a constant flow of food coming down with the rich waters of the lake—food that tempts by its change after the monotony of the diet afforded by the sea. The club house stands on the shore high above where the rapids begin and here takes place the most unique fishing in the world. Tarpon, snooker, yellow-tail, jack and countless other varieties of fish come up the river and continually storm this rush of water. The fisherman, shoes armed with spikes against slipping on the grass-grown concrete, stands or wades along this boiling edge three or more feet from the lip, a half mile sweep of knee deep water everlastingly pushing him towards the edge and certain destruction. Every step must be calculated, every spike must be made to grip. To make him a little more careful he has only to look a hundred yards downstream where a small sand bar makes out from the shore, lying on which are some black logs which from time to time slip back into the water without a ripple, each crocodile showing two little bumps that travel just out of water.

A ten foot rod with plenty of play, a heavy two handed butt, at least two hundred yards of No. 15 cuttyhunk and the regulation tarpon reel, leader and lure is the ideal tackle for this Chagres tarpon. Allow the bait to run out through the rapids and with luck, you have a fish before it sinks in the first pool. He can then be played from the apron and landed at either shore end, or the fisherman can leave the apron from either end and play his fish, moving down the bank of the river. We saw no gaff at the Club. The common practice is to land, estimate and put back with an appreciative "I thank you."

This Chagres tarpon does not run as large as his Florida cousin, one hundred pounds is a good one, but he has all his natural fighting qualities, and the combination of this, the precarious footing and the fast water makes most other fishing seem tame. Snook and yellow-tail, delicious to eat, can be caught at the same spot. Lighter tackle, smaller hooks and a proportionate amount of sport can be had. If a tarpon takes your snook lure, which is very probable, you can only pray that a slack line and the next jump will solve your problem.

We were told that larger tarpon can be caught by trolling up the Chagres river from its mouth, seven miles below the dam. We did not have the time to take the all-day trip from Colon, and anyway were too much fascinated to leave the rushing waters of the apron. The members of the Tarpon Club spared neither their equipment nor themselves to make our visit pleasant, and they most certainly succeeded.

On our return to Colon we tried trolling for tarpon up the old French canal. Four of us caught five splendid fish in one short morning, besides enjoying a beautiful launch ride through what now appears to be a natural tropical river. At the head of this canal is a lake some hundred acres in area. This lake was full of tarpon and there was never a moment when we could not see fish breaking and rolling over in their characteristically lazy manner.

We heard rumours of wonderful fishing to be had at the Cocos Islands on the Pacific side. When we arrived at Panama these rumours hardened down, and though we were going miles beyond those islands it gave us some idea of what was in store for us at the Galápagos.

Four days later, crowding the rail of the bridge of the *Noma* we watched our Islands grow from hazes to outlines, from outlines to green mountainous realities. As we came abreast of the Seymours with James standing up off our starboard bow, a school of porpoise were playing under our forefoot, the silent fins of several sharks were lazily pushing along the surface, we looked down on a large yellow-white seal steaming along at great speed, and better than that from the fisherman's point of view was a monster sailfish jumping clear of the water in the peculiar travelling manner of the species.

Schools of smaller fish, mackerel, bonita, albacore, sardines and others made their characteristic "signalling" spots on the smooth sea, while flying fish broke away from our bow wash in long glistening curving flight to tumble back into their other element. Sea birds came to wonder at us, their eyes actually inquisitive; became satisfied in a moment that we were useless and returned to their taking toll of the sea.

Anchoring at Conway Bay, fishing for the moment had to give way to our keen desire to see what was on land. Returning to the *Noma* later that afternoon, excited and delighted at what we had found, we were met by the crew with requests for fishing tackle, hand lines and hooks. They had caught many varieties of small fish but had lost all their light tackle to shark and unknown monsters. Looking over the side into the clear water we could see hundreds of small fish and deeper below them the lazily moving forms of sharks. Luckily we were able to get together enough hand lines and hooks to fill the demand, and as a result the forty odd members of the crew were able to add to our variety of specimens and at the same time to their mess.

Several of us started out the next morning with the heaviest tackle we possessed, our object being merely to troll around and across the bay. We immediately began to catch fish. Golden spotted Spanish mackerel and black and greyish-black groupers gave us a busy morning. The fish would strike at anything, squid, spoon or piece of white canvas. None of these fish in fairly shallow bay water ran over twelve pounds. The same fish on lighter tackle would have provided some real sport. Some may scorn the idea of a Spanish mackerel being a game fish, but hook a twelve pounder on a bone fish tackle and you will find he can fight quite as well as the far famed bone fish itself. Several savage and unseen strikes carried away leaders or broke new hooks. What these were we have no means of knowing; we *do* know that they were too heavy for our frail tackle. A seven foot shark took all of an hour to subdue, even with the help of many shots from a Lugar and several charges from a shotgun at short range. Many of our catches were torn or cut in two by sharks. An experienced fisherman will know when a shark is after his hooked fish, for up the long line comes an instinctive feeling that his catch is more terrified of something than he is of the gnawing hook. When this happens it is better to give out slack and take a chance of losing him, allowing your fish to escape his natural enemy by his own efforts. The shark, when he sees his intended victim able to take care of himself, will give up the attempt, and if your fish is well hooked you can start over again.

Fish savagery is always a striking feature of sea-fishing. Large fish are wary of attacking other large fish, but the moment either one seems to be in trouble or incapacitated he immediately becomes a victim. The attack seems more savage than the kill of the jungle, and the smell of blood arouses much the same instinct among fish as it does among the jungle carnivora. The struggle for existence,—for food,—that takes place in the black depths of the sea is more fierce than that on land. No land carnivora of which we know will normally kill and eat its own young,—none of the sea carnivora when hungry allows family ties to interfere with appetite. It is merely a question of size and speed. Individual schools of the fish that habitually travel in schools,—namely, sardines, bonita, tuna, mackerel and albacore,—are made up of fish of the same size; no fish two inches shorter is safe among his two-inch longer friends. It is also lucky for our food fish supply that sharks are comparatively clumsy and do not breed with the same prolificacy.

Most of the larger fish—from five to thirty pounds—were caught with a tarpon-rod and tackle. Due to the inexperience of some of the men who fished, in trying to land their fish too quickly some three dozen spoons and leads were lost.

When we landed at Tower Island, where large fish were as plentiful as minnows, we only had one spoon left, which was safely guarded. It was suggested that we use white canvas, cut in strips 1' x 6" long. We had some difficulty at first with this kind of tackle on account of the water birds who would follow in large numbers, diving down and catching it in their bills before the fish got a chance at it. It took some care to pull it out without hooking them, but by adding a little weight to the line to keep it slightly submerged, this was eliminated. The fish seemed to go crazy about the canvas, and no sooner had we thrown it back into the water and before it had gotten two feet below the surface we caught another. In less than an hour, while the boat drifted, we had the bottom of the large boat covered—some four hundred pounds.

Having lost all our swivels we were compelled to manufacture some. This was accomplished by dipping one end of a piece of steel wire into hot solder. This gave us a wire with almost a perfectly round ball at one end. By taking two twenty-two calibre empty shells and piercing the butt ends with a hole large enough to allow the wire to pass through and inserting the piece of wire into each empty shell with soldered ball, pressing the two shells together to about one-half of the shell length, one lapping over the other and carefully soldering the joints to make them secure, loops were made at each end of the wire, for fastening the line, making a fairly strong swivel.

During our days at the islands we caught hundreds of fish.

The food fish were consumed on the yacht, while rare specimens went into alcohol. Albacore, tuna, mackerel, bonito,
jack and grouper formed the main body of our catch. We
worked hard to find the sword-fish family. We saw them,—
gigantic ones,—but in the short time at our command we were
not fortunate enough to hook one. Enormous mantas or
giant rays would rise from the water, seem to flap their wings
in the air and crash back with a glorious splash. Whether this
leap is made to shake off parasites, to come down upon a
school of fish for food, or for fun, we do not know. In any
case it is an awe-inspiring sight and makes one wonder what is
still hidden in the ocean that no man has ever seen. We would
not even venture to guess how many hundreds of pounds
some of these fish weighed. Certainly great sport could be
had with a powerful boat and the right harpoon and line.

Sea turtle abound. On a launch trip to Duncan Island
from Seymour Bay we passed through a school of hundreds
floating lazily with the current and covering over a mile of
sea surface. Nearly all the steep high beaches that we found
formed nesting places for these turtles, the beach at James
Bay being particularly adapted for this purpose.

There is a profusion of bait and the bait net can be cast
from any of the dozens of small sand beaches on the north side
of Indefatigable. We used live and artificial bait and could
see very little difference in the result. In our attempts to
catch sword- or sail-fish we used live bait and the same tactics
as when fishing for sail-fish in Florida. Our not getting any
strikes from these fish is very understandable to anyone
who has tried this fishing; it takes time and luck. We had
neither in this case.

The best fishing grounds that we discovered,—although
there are many others,—are on the north side of Indefatigable
from Eden to the Seymours, the channel between Indefatigable

and Seymour, and in and around Darwin Bay, Tower Island. The deep flow of the Humboldt stream through the thoroughfares formed by the islands is alive with all the big game fishes.

To sum it up from the sportsman's angle, Galápagos waters contain all the large game fishes of Florida, with the exception of the tarpon, plus the game fishes of the Pacific.

I am of the opinion that we can include the sail-fish among the fishes seen, but cannot be absolutely certain.

The fish have not been fished on these grounds and are abundant, have plenty of food, are consequently in excellent fighting condition and run to great size. The fishing grounds are not marked by charts and guides as in Florida and this adds a pleasant factor,—here you are exploring and fishing at the same time. If you find a bank, current or channel, where your great game fish abounds, you may chart it, name it, fish it, and it is yours.

Fig. 42

Tarpon Caught at Colon

Fig. 43

Hauling the Seine off Indefatigable

The net is full of Mullets

Fig. 44

The High Veldt-like Interior of South Seymour, with a Herd of Wild Goats in the Distance

CHAPTER XI

O
N our return to the Galápagos, after recoaling and
watering in Panama, we anchored in Seymour Bay,
off the northeast coast of Indefatigable. Far to
the west we could just make out Eden and Guy Fawkes about
twenty miles away. Daphne Major was only four miles to the
northwest, and South Seymour lay two miles due east. This
last was the third islet appendage of the mother Indefatigable
which we looked forward to exploring. As usual we were
fortunate in finding a landing place, a great cove or bay, a
half mile across, with black lava reefs reaching far out around
it, high basaltic cliffs along one side, and the bend of the
crescent a wide beach of brownish white sand, shell-studded,
and altogether lovely in the sunlight. Just back of the beach
we found our first fresh water—a pool far down among lava
boulders, and a quarter of a mile to the south was a series of
deep pools with only a slight tinge of brackishness. The bottom was green, but the water quite clear, and, for a ship-
wrecked man, drinkable. One of our first discoveries in the
pools was wholly unexpected and added the entire order
Anostraca of the leaf-footed crustaceans to the Galápagos
fauna. Our hunter in chief, William Merriam, the most
enthusiastic and ingenious worker who ever outwitted fish,
flesh or fowl, called to me from one of the pools that he had

found some crazy fish which swam on their backs; hardly
believing the thought which came into my mind, I rushed
over and saw several *Branchipus* swimming slowly around.
The last time I had seen one of these was in a pool of melted
snow in early April many years ago in the Orange Mountains
of New Jersey. When a boy I knew exactly what puddles
would produce these most wonderful fairy shrimps, and
within a day or two of the schedule each year they would
appear. After a few weeks, the puddle and the last shrimp
dried up, vanishing until the next year. Even the northern
ones were beautiful, iridescent in the early spring sunshine,
with pink feathery appendages, but these, in this equatorial
pool, were exquisite. As they swam about among the maze
of tangled algæ near the bottom, they were almost transparent,
pale turquoise blue, with a film of green internal structure,
orange-red tail and jet-black eyes. These strangely beautiful
creatures, so delicate that they survived removal to the yacht
only for an hour or two, probably appeared for a few weeks
during the rainy season, the eggs drying up the rest of the year.
These Seymour shrimps were *Dendrocephalus cervicornis*, and
all that I saw were males, with most elaborately modified
antennæ. This species has been recorded from Argentina
in salt water, never from these islands, nor from what was
practically fresh water.

In the same pools were three species of water or whirligig
beetles, one of which, of medium size, and etched horizontally,
was undescribed. Their active, voracious larvæ moved about
beneath, feeding on anything which they encountered, including one another.

Scattered here and there over the rough lava gullies and
slopes were good-sized bushy trees of *Cordia lutea*, some
fifteen feet high. These were a mass of large, yellow, trumpet-
shaped blossoms, and were alive with a very few species of

insects. The commonest was the yellow butterfly *Callidryas eubele*. Twenty which I captured from a single tree were divisible into three sharply demarcated types,—first, normal yellow males and dark-banded females; second, pale brownish white individuals, quite lacking all green or yellow tinge, this appearance seeming to be caused by a very even, regular thinning or actual bleaching of the scales, with very little apparent wear and tear of the wings; third, a considerable proportion of small-sized butterflies. The interesting thing was that in all the specimens secured, there were no inter-mediates; the normal, large butterflies averaging fifty-nine millimetres in extent of the wings, and the small ones forty-nine.

A few day-flying moths were on the flowers, especially the small painted micro-lepidoptera, *Atteva hysginiella*, which crawled over the blossoms and leaves in hundreds. There were in addition numerous brown and also green lace-winged flies, *Chrysopa*, both of which were unnamed species new to science. One other insect was abundant, a small, elongated, brown beetle, a new species of *Asclera*, while a few minute Chalcids flew about, one of which I caught. This represented the high tide of abundance of Galápagos insect life, and was characterized, as always, by its tameness and silence. No net was needed to catch these butterflies and lace-wings; they could be picked from the leaves or flowers with the fingers. Once as I sat quietly, I caught a cricket's chirp, as I had on Eden. This was a faint, repeated note, lasting a full minute, most remarkably noticeable in this land of silent reptiles and few bird voices and utter absence of that humming undertone which fills our northern meadows. I beat the bushes and swept with the net for a half hour at a time, but usually got only a few spiders. To find an insect under a stone was an event. On a flat, dried-up pool bed back of the beach I caught my first wasp, a tiny red-bodied *Larid*, running about on the damp soil, which proved to be a new species.

In a hollow I found the perfect skeleton of a goat with very beautiful, long, curved horns, and close by there were eight skeletons close together, all of full-grown animals.

Great *Epeira* spiders hung their webs from bush to bush, in such numbers that after a half mile of walking the front of my shirt was a network of grey silk. These spiders were the most abundant of all the fifty-odd species recorded from the archipelago. Many were an inch in length, with wide-spreading, pale orange legs, dark body, a wide, scalloped, cream-coloured marking down the back, and others on the sides. So strong were the webs that no insect was safe and I often saw the largest sphinx moths and the giant grasshoppers hanging like rolled-up, silken mummies. Once a young *Geospiza* finch flew into a web, and hung for a few seconds fighting to free itself. I went toward it, when it fell to the ground and with difficulty flew off.

Among other spiders was one which we called the Union Jack, *Lathrodectes apicalis*, a species found only in the Galá-pagos. This lived among the lava rocks, in crevices, and spun a little sheet of web before its door. It was strikingly coloured; the legs and cephalothorax dull leaden brown, the abdomen black with a yellow band around the anterior border. Three great slashes of scarlet extended from below over the posterior half, converging toward the back. It was closely related to the red and black spider of our own States, which is everywhere feared as a creature whose bite is deadly.

A remarkable thing about Galápagos life was the presence of so many groups, many of which were represented by only one or two forms. For example, there is only a single praying mantis or devil's coach horse, and here, on Seymour, I found two specimens, dead leaf brown, and quite wingless. One was near a small mass of foam-covered eggs, a blunt-topped, elongate bundle, glued to a blade of grass. It was freshly made, with the sheltering froth hardly dry. Two weeks later, on my table in the yacht's laboratory, the egg case burst forth into twenty-one small devil's ponies, measuring only 6.5 mm. or about a quarter of an inch in length. They were quite unlike their parent in colour, having the body brown, but the terminal joints of the forelegs, all of the posterior pairs, antennæ and eyes, boldly banded with black and white.

On a bit of sandy meadow I came across several ant-lion pits, from one of which I rescued an ant, which had come to grief like Morrowbie Jukes. If the truth be told, however, I substituted only scientific immortality for sudden neuropteric death, for I consigned him to the tender mercies of Professor Wheeler who was myrmecologizing on the opposite slope.

Laden down with gun, nets, jars and botanical specimens, I returned from the fairy shrimp pools to the spot where we had piled our luggage at the summit of the beach, when I caught the movement of some great creature among the paint boxes. I approached from a hollow on the landward side, and silhouetted against the sky there appeared a veritable dragon, nosing about the lunch baskets. In a moment I recognized it as *Conolophus subcristatus,* the other giant iguana of the Galápagos.

Every minute that I remained in the islands made them seem stranger, and farther removed from the humdrum modern life of the mainland. Conan Doyle or H. G. Wells could write many chapters of straight description of the Galápagos Islands and their inhabitants and have it fit in perfectly with a tale of ancient geological epochs, or indeed of another planet. Volcanic activity had only just ceased. In 1825 great eruptions took place and within the present decade smoke and fire were seen coming from some of the craters. Today, the accidents of subsidence have resulted in an Age of Reptiles. From shore line to mountain top this class is in control, dividing the land between them, and the only weakness of this simile is the fact that the tiniest lizards alone are carnivorous—the large ones being vegetarians, so there is no exciting survival of the fittest by direct attack and defense. The only indigenous land mammals are a few mice and a bat, so there is no competition in that class.

I walked slowly and quietly, and the big reptile clawed the luggage. Remembering the harmlessness of the marine Amblyrhynchus, I dropped my bundles, crept up and seized it by the tail. After a tussle I secured a better grip and held it securely. Hardly had I straightened up to examine my prize when I saw in the distance, up and down the beach, the other members of the party, each vainly trying to conceal something dragging behind him as he struggled along. It turned out that not one had failed to secure a specimen of these giant lizards, and there were five separate and individual chagrins, as the realization came that no one could claim the first land iguana.

We trussed up our captives and sat down to gloat, when two families of mockingbirds came pell mell, shrieking their joy. How they knew the Conolophus were helpless, I do not know, or it may be that they were enacting a usual habit, accepted reciprocally by the saurians. The birds alighted on heads, bodies and tails and began to pick off the ticks from the skin of the reptiles, pushing their heads far under the armpits, and between the toes. I wanted the ticks myself for quite a different purpose, so with forceps and vials I anticipated the mockers, greatly to their disgust. They proved to be a new species of tick, since named *Amblyomma williamsi*.

I have no idea what it was that prevented us from handling these lizards as carelessly as we had the sea iguanas, but early in our experience with them two Conolophus were accidently brought together, when they seized one another ferociously and bit unmercifully. Later one got a good grip upon a shoe and actually chewed through the leather. This warned us and we treated them with infinite respect thereafter. Indeed they proved to be as wholly unlike their cousins in temperament as in colour. There seems to be no doubt that these two Galápagos lizards are quite closely related, Conolophus having continued more along the conventional iguana line of terrestrial life, while the black offshoot took to the shore and the water, and found life possible and pleasant among lava reefs and sea-weed. Amblyrhynchus became decolourized, losing all the bright greens and pinks of the South American iguanas, slipping over its scales the greys of the lava.

The iguanas of the South American jungle can occasionally be irritated into biting, but they are rather adepts in delivering telling blows with whip-like swings of their long tails. Both the Galápagos species have changed in this respect, Ambly having no enemies and no defense, and Conolophus with its shortened tail trusting to the great power of its jaws, and its small but sharp teeth. It must be said in their defence that never but once did I ever see wild ones attack one of their own kind, and never each other or their aquatic relatives after capture, when turned loose in the same enclosure. But in the excitement of actual capture or when being carried by the tail, they bit any and everything which came within reach. When we were getting specimens of ticks from Amblyrhynchus we would look the whole reptile over, even prying its mouth open to examine the curious tri-lobed teeth. But with Conolophus this was an entirely different operation, and to avoid being bitten one of us was compelled to give his whole attention to a grip on the neck and body while another pulled off the required ticks. Their great, broad bodies and rounded tails, emphasized the slimmer stream-lines of the water iguanas. In colour the land-iguanas were quite the gayest inhabitants of the islands.

I found that those which we found on the lowlands back of the beach were only outliers, strays from the main colony. My next excursion took me north toward the rugged basalt cliffs, leading to the higher part of Seymour. I found an easy sandy trail zigzagging upward, and halfway to the top I looked over a barrier of thorn scrub and saw a monster slowly crawling about a clearing and plucking something. I found that he had no fear whatever of me, and walking to within ten feet of him, I saw that he was feeding on the yellow buttercup-like blossoms of a species of prostrate plant, *Tribulus*. The leaves were thick, very tough and extremely hairy, and were not touched. But wherever a flower appeared, the ungainly lizard crawled toward it and plucked it with a quick, sidewise jerk of its head. The slight grip of the rootlets in the soft sand often resulted in the whole plant coming loose, Conolophus then raised first one, then the other front foot, and held the plant down while the head stretched far upward, pulling off the refractory blossom. Professor Wheeler and I followed suit and ate some of the blossoms, finding them very agreeable, with an early spring aroma of a taste. So deliberate and tame was this iguana that when I called for the moving picture camera, Mr. Tee-Van had no trouble in approaching closely and getting many yards of excellent film of a hungry Conolophus.

Part way to the top of the plateau we saw two very large, bearded billy goats fighting like mad, pushing each other back and forth on a narrow ledge of the cliff. Their tall, lyre-shaped, spiral horns could not do much damage, as neither could back far enough off to get sufficient impetus for a dangerous charge. Close by, a small white kid was watching the combat and bleating mournfully. Around a corner of the

Fig. 45

GIANT LAND IGUANA

Conolophus subcristatus (Gray)

At midday every cactus and small bush has one of these great Iguanas asleep in the narrow line of shade

FIG. 46
HEAD OF THE GIANT LAND IGUANA
These creatures recall ancient extinct reptiles in their great size and majestic mien

cliff appeared Harry Hoffman, our indefatigable Landscape Artist, and at the sight of this apparition bearing easel, paint boxes, seat and cameras, the goats ceased their quarrel, and bounded up the steep side of the lava wall, followed by the youngster.

From the summit of a huge boulder at the top, we had a splendid panorama, extending from distant Daphne, and the *Noma* two miles out, to the long low profile of Indefatigable, all but the coastlines wrapped in a swirling mass of dark purple rain clouds. Below was the great sandy cove, outlining the zone of the fresh-water pools with its tangled vegetation; sweet-scented *Bursera*, yellow-flowered *Cordia*, the strange plant *Parkinsonia*, with its recurved spines and down-curved, weeping willow branches, and the dark, all-spined *Discaria*. Behind me was a complete surprise—a scene wholly different from any I had encountered on other islands. It was South African veldt and nothing else; a great stretch sparsely clothed with low, rounded *Bursera* trees and cactus, grassy between, with occasional out-cropping lava. I thought to myself, it would be perfect if only an antelope were here, and at a second glance there were the antelope! Seven forms were bounding swiftly along, tails up, stopping now and then to glance at me, and then on again in single file. My glasses spoiled it all, and showed of course the forms of wild goats, but even then the illusion was hard to overcome, the wide-spreading horns, at least head on, looking exactly like the horns of Grant's gazelle.

We found the animals very abundant and exceedingly diverse as to temper. The first flock was as wild as any well-hunted African game, yet later I walked within a few yards of some on their way to the pools. One of our party very sensibly took to a low tree when several files of sixty or seventy animals came past. Instead of being frightened at his demonstrations, they shook their heads threateningly and refused to stop. A dozen old billies attacking at once with their sharp-pointed horns would prove no mean antagonists.

Later on we shot one or two, and I was surprised to find them in as perfect condition of both flesh and pelt as I have ever seen wild game. Although the island is covered in many places with the abominable bur-grass, so that in walking twenty yards many of the thorny seeds are entangled in one's clothing, yet the hair of the goats was sleek and smooth, and even the well-developed beards and fetlocks were combed and burless.

I was not able to understand the numerous skeletons of goats which were scattered about. If the deaths were caused by thirst, the kids and younger individuals would probably have been the first to succumb, but all these were in their prime, with magnificent horns. With an abundance of cactus and succulent herbs there is slight chance of these hardy animals succumbing to the many months of drought. Twice after the first day I saw goats fighting with unbelievable fierceness and it may be that they actually kill their opponents now and then.

I had walked only a few yards over this Seymour savanna, when I realized it was the headquarters for all the Conolophus of the island. Every cactus, every small isolated bush of *Cordia* or *Acacia* or *Bursera* sheltered a lizard, and all big ones. Throughout all of our exploration of this colony I saw not a single lizard under twenty-four inches and most were three feet and even more in length. I sat down, and counted fourteen, all prehistoric, ancient-looking, all sprawled out in spots of greatest shade. In the case of the cactus, they were stretched along the line of stem shadow, while under the bushes they were often curled up in some small dark shaded area. Against the dull green of the sparse vegetation and the brown sandy soil they fairly glowed with colour. The great heads were rugged with rough, bright golden scales, which became green on the mosaic-like labials, and chrome yellow on the under side of the chin. The folds of the neck skin were whitish grey, with the hind neck and fore legs yellow. The body and tail were divided into large irregular areas of terra-cotta, black and yellow, the spines along the back taking to themselves the colour of their respective bases. It seemed to be a time of general moult and every lizard was in rags, with thin tissues of scaly skin blowing out from body and legs. As the sun rose higher and shifted its longitude in the sky, the Conolophus drew closer to the bases of the cactus and shifted sideways.

When disturbed by our close approach they made off slowly for their burrows, sometimes at a considerable distance. These showed as rounded holes, in sandy clay, leading obliquely down, under or around lava boulders, for an extreme depth of six feet. There was no pile of earth outside. When suddenly rushed the iguanas became panic-stricken and tore off in any direction. Once started, however, it was not easy to turn them, and they chose to barge straight ahead, dodging only at the last moment. When hard pressed, they frequently sought sanctuary at the heart of a thorn bush, and when once in such a place, refused even to be poked out. It was an easy matter to run them down within a distance of fifty yards. When safe behind their thorn shelter, they gave vent to their anger in regular lizard style of nodding. On the edge of the veldt country, they would occasionally dash on to a pile of lava, and, poking head and body as far as possible down a crevice, remain quiet, complacently ignorant of the fact that hind quarters and tail were in full view. When first lifted by the tail they struggled vigorously, and made several alarming but ineffective attempts to bend up and bite one's hand, before resigning themselves to fate.

In the stomachs of two which I examined were enormous

quantities of the leaves of several plants, especially *Cordia* and *Maytena*, and many flowers of the former. One had swallowed whole five cactus fruits and another three. I spent considerable time lying quietly and watching these splendid creatures, and when one's point of view was only a few inches from the ground, so that the head of the lizard cut a distant mountain, the creature might easily have been thirty instead of three feet in length.

Twice I saw Conolophus getting the cactus fruit which they seemed to love. To my surprise it was a deliberate attempt—the nearest to an intelligent action which I have ever observed in any reptile. Two were close together at the base of an *Opuntia*, and one stretched up and struck the spiny base, slowly but repeatedly with one foot. At first I thought it was making an attempt to climb, but I soon saw that no such effort was intended. Nothing happened at first, but finally two fruits fell at once. The second lizard rushed up and gulped down both, spines and all without a second's delay or mastication. The first leaped and seized its fellow by a foot, chewed for a moment, when the limb was jerked away, and the owner rushed off to a neighbouring cactus with a vacant shadow. Conolophus number one made no further attempt on the "sour grapes" but sank down, shut his eyes and went off into whatever dreams come to Galápagos iguanas,—perhaps of lands where spiny fruit fell into one's mouth and there was none to steal.

I saw circumstantial evidence of these lizards actually eating the pads of *Opuntia*, spines and all, but I could hardly believe this, until, in a number of cases, following the capture of those which I took on board the *Noma*, I found their droppings to be a mass of full-length spines. How it is possible for any creature to swallow such needle-length and steel-hard spines and not perish, I cannot conceive. The contrast of such food with the delicate golden petals of which they are so fond, was startling.

Conolophus is essentially herbivorous, but that it occasionally partakes of animal food I can attest, from finding one and three grasshoppers respectively in the stomachs of two individuals. This is very unusual, and I question the accuracy of the statement made by the mate of a vessel, that they were seen drinking the blood of a slain goat.

That these lizards are agents in the dissemination of plants was everywhere evident. Wherever old droppings of *Conolophus* lay on beach or veldt, there a colony of new-sprouted seedlings was visible, growing strongly. Not only did these lizards feed on the ground but they also climbed to the top of the bushes and low trees, and remained sometimes an hour cropping the flowers and foliage. Charles Darwin in his Journal, written when Naturalist to the *Beagle*, says, "To obtain the acacia-leaves, they crawl up the low, stunted trees and it is not uncommon to see one or a pair quietly browsing, while seated on a branch several feet from the ground." Just back of the Seymour beach I saw a Conolophus twenty feet up a *Maytena* tree, clambering awkwardly over the flat, dense mass of foliage. When I came nearer he gave a leap, missed his footing and crashed headlong to the ground, landing on his side and bouncing from the force of the impact. Unhurt, he sprang up and ran full tilt for the nearest pile of lava.

The favourite resting pose when awake was with tail, hind legs and flanks close to the ground, but fore body raised on almost straight fore legs, and head well up,—a very statuesque dignified pose. The rite of challenge was a series of slow nods. When asleep, they usually were sprawled out flat, often with legs flung back, soles up. When sound asleep I have stalked and caught them, but a careless step would awaken them. The pineal eye is slightly functional, and when sleeping in the sunlight, my hand, waved between them and the sun, would awaken them.

When awake and watching me approach, the first sign of nervousness was a slow movement of the tail tip, almost catlike; then a lashing of it from side to side, crocodile fashion, although as I have said, the tail seemed never to be used in defence at close quarters. Next came a single, long, upward nod of the head and neck, followed by six or eight in quick succession, bringing the head well up. Then, if the supposed danger increased, the mouth would be opened wide, and closed several times, the head remaining still, or moving threateningly in every direction. No sound except a sharp hiss was ever uttered. Finally the Conolophus moved slowly away, or if in extremity, this changed instantly to frantic flight.

When walking slowly, the lizard dragged its body, and the tail, as in the small Tropidurus, was held up in a gradually ascending curve. The faster the pace, the more twisting became the movement of the body, the legs reaching out with greater and greater effort. The gait is actually a rapid, alternate stepping, with violent lateral wriggling of the body and tail. The tail is held horizontally, and the body is clear of the ground for about half of each stride. At a moderate pace, the creature actually brought to mind a great rhinoceros on the trot, although its steeply arched back and general appearance made me think of *Stegosaurus* or some other giant reptile of the past.

So obstinately did they hold to the direction first chosen, that when another of the party was chasing one, I have known it to rush straight at me when I happened to be in its path, and to dash between my legs. Occasionally an instance of remarkable stupidity was evident. Several times when we were busy collecting insects, or preparing for lunch, a giant Conolophus would appear from somewhere, rush through our midst, and come to a sudden stop in the shade of an absurdly small bush, flopping down, and gasping and wheezing like a fat old gentleman out of breath. Here he would remain throughout all our subsequent activities, apparently assured that his huge yellow, red, black and white form was wholly concealed by the slight shadow of a few leaves.

Conolophus was once abundant on a number of the Galápagos Islands,— James, Indefatigable, Albemarle, Narborough and South Seymour. But it has been exterminated on the first two or probably on the third as well. Only on Narborough and South Seymour does it still hold its own. The destruction of this splendid saurian is probably due to occasional raids of human beings who have killed them for food, and especially to dogs and pigs which have become established on several islands, and doubtless destroy the eggs and young. This would account for their extermination in the three large islands.

I have eaten many scores of Mexican and South American iguanas and found them delicious, and I can readily believe some of the old voyagers who dwell on the toothsomeness of Conolophus. But every one we collected was too precious to sample. Unlike Amblyrhynchus, they accepted captivity in an equable spirit at once. We fed them at first on cactus fruit and then began to draw on our cold-storage cabbage, lettuce and bananas, and in a very short time every one was eating greedily, taking food from our fingers. At the present date of writing, seven months after capture, only one death, and that due to a mechanical injury, has occurred among the eighteen specimens collected and brought back alive to the New York Zoological Park.

The Amblyrhynchus were confined to the shore and I never saw Conolophus within a hundred feet of their marine relatives, but the bright, active little Tropidurus ran about everywhere, frequently over the very bodies of the iguanas.

Once I saw a scarlet-faced female rush up on the broad back of a Conolopus, whereupon both the giant and the dwarf went through their comical nodding rites, solemnly, as if sealing with approval some important saurian pact.

Although the plants on the upland veldt of Seymour were rather sparse and low, a few species of insects were fairly abundant. Large black bees flew about, making the most of the meagre supply of nectar. These were females, and twice only did we see the yellow bodies of the males, which avoided our utmost efforts at capture. Diminutive replicas of our northern white-lined sphinx moth were common, flying about and feeding in the full glare and heat of the noon-day sun. They measured three-fourths as large as the specimens from New York. This shows the dwarfing effect of the islands, as does the red fritillery butterfly *Agraulis*, many of which had three-fifths the spread of wing of typical northern individuals.

Perhaps the most interesting discovery of the day was a bee to which clung numerous, small, elongate triungulid larvæ, over a hundred clustered about the wing bases of the big insect. No member of the *Meloidæ* or oil beetle family is known from the Galápagos, and yet these were very clear evidence of the occurrence of at least one species.

If these triungulids are really the larvæ of an oil-beetle, and their development is like that of our northern forms, then their history is an exceedingly interesting one. They hatch in the shape of a primitive six-legged insect, and climb at once to a flower. Here they await the visit of a bee, attach themselves to it, and are carried to the nest. They devour the egg of the bee, and feed to repletion on the store of honey, passing through a series of three most remarkable stages of larval development, each time becoming more helpless and grub-like, until finally they change to a pupa, which becomes a soft-winged beetle.

Another happening within the hour revealed the catholic character of the intellectual riches of Galápagos. We climbed down to the shore beyond the basaltic promontory, and in a sheltered cove, well up beyond ordinary tides, were six bleached skeletons. Many of the smaller bones were missing but the huge skulls and most of the vertebræ and curved ribs were there. To my scanty knowledge they represented only the remains of a school of small whales, but we collected the lightest of the skulls and its curious network of cervical vertebræ, all coalesced into a maze of a single bone. As it proved, this latter was the most diagnostic bone of the entire skeleton, and without possibility of doubt, identified the animals as false killer whales, *Pseudorca crassidens*. Extensive search of the literature has revealed nothing but ignorance in regard to the life and activities of these creatures. The first one known was described from sub-fossil remains from a fen in Lincolnshire, England, and named by Sir Richard Owen. Long afterwards, small herds of living animals were located on the coast of Denmark, and another year specimens from near Tasmania were found to be identical, although almost exactly on the opposite side of the world, eleven thousand miles away. Another false killer whale has been taken in Davis Straits, west of Greenland, and is now in the National Museum in Washington. My skull and cervical vertebra, salvaged from the remains of the six specimens on the shore of South Seymour in the Galápagos, seem to complete the records of this sea mammal. This mysterious creature, so rare, yet so widely distributed all over the globe, is said to be wholly black, and about fourteen feet in length.

The days we spent on Seymour were perfect, clear and never hot wherever the trade wind had a chance to reach. About ten o'clock the central mountains of Indefatigable would disappear in their daily cloak of rain, and three hours later the clouds began to drift in our direction over the lower lava slopes. They rushed on, gathered, piled up almost overhead, with the perpendicular lines of rain hiding the very beach of Indefatigable a short distance away. Even at this climax the whole storm might be dissipated without a drop reaching us, or it might empty its whole contents in our immediate vicinity, while the *Noma* was dry, in hot sunshine. So local were the rains here that a person might die of thirst on an islet half a mile from the mainland, or be drowned in any lava hollow a few yards inland. It was tantalizing to see the daily downpour a mile or two away, when a few hours of rain on our stretched canvas would have prolonged our stay in the archipelago for days.

FIG. 47

A GIANT LAND IGUANA CAPTURED BY THE HISTORIAN OF THE EXPEDITION

FIG. 48

FRONT VIEW OF HEAD OF ONE OF THE VICIOUS GIANT LAND IGUANAS

FIG. 49

MODEL OF DAPHNE MAJOR LOOKING TOWARD THE WEST

The island is three-fourths of a mile in diameter

FIG. 50

CRATER OF DAPHNE MAJOR FROM THE EDGE OF THE RIM LOOKING TOWARD THE WEST

CHAPTER XII

DAPHNE—A CRATER NURSERY

OF all the islands which appeared and passed before us when we first raised the archipelago and made our way along the northern coast of Indefatigable, Daphne delighted us most. I think we were almost disappointed at the unexpected size of the larger islands; they stretched away for mile after mile, with the central peaks dim with mist and distance. But here was an islet, or rather two, which we could grasp in their entirety. To distinguish them I dubbed them Daphne Major and Minor. Each consisted of a single crater, but Minor rose so abruptly from the sea, to a flat bushy top, that it was quite unclimbable.

Daphne Major was three quarters of a mile in diameter with an even, steep slope all around, rising to an almost perfect crater's rim. It was five miles north of Indefatigable and the same distance west of South Seymour.

On April twenty-third I first visited Daphne Major, leaving the *Noma* and going due north-west in the big life-boat, towing a four-oared gig. Trolling was exciting as usual and we caught a new species of mackerel and two groupers, while a three foot, twenty-six pound bonito or albacore with yellow fins, *Germo alalunga*, fought me for fifteen minutes. Besides this a valuable tackle went to some monster whose weight felt as if it were somewhere in the hundreds.

Today was shedding day in the world of cirripedia, and as far as the eye could see there were thousands of small greyish-white patches scattered on the surface of the water. I scooped up one of these bits of limp fluff and under a hand lens counted ninety-two ghostly, shed skins of barnacles.

Half-way to the island, yellow butterflies began to pass us and recalled that on three previous days and once on my first voyage to the archipelago I had observed a similar migration. It is a common but mysterious habit in mainland representatives of closely related forms, and, as I have described elsewhere, in British Guiana it often takes place on an extraordinary scale. There the usual direction was north-north-west; here it seemed invariably south-east. On Indefatigable I watched it two days in succession, the insects flying low over the water from the direction of James and Albemarle and continuing due south-east across cactus and craters. Again on the backbone of Seymour I saw many keeping on their course straight out to sea in the teeth of the trades, headed for Chatham, and finally at Freshwater Bay brave little Sulphurs were fluttering past on their inexplicable, compass-true path, headed for the open sea and certain death. Yet all around were others flying from cotton blossoms to the blooms of the Cassia—each as yellow as their own wings—paying no heed to their travel-stricken fellows in mid-air. Soon perhaps they, in their turn, would follow. The Galápagos martin has abandoned all his migrations, the sulphur butterflies of these islands are still slaves to an instinct which seems to us unreasonable, useless, almost inimical.

Only once or twice have human beings ever landed on Daphne. Beck has visited it once and the California Expedition spent parts of three days there. That it has a very respectable isolation age is shown by its gecko—a small, vacuum-toed lizard—which is distinct from that living on

Indefatigable, five miles away, differing both in scales, tubercles and length of the head.

Owing to the greater attraction of the surrounding larger islands, this little crater in the midst of the blue water has been almost neglected. Nineteen species of birds and three reptiles have been recorded from it, but no plants, grasshoppers, ants, butterflies, moths, spiders or snakes.

The most astonishing thing about the various islands of the Galápagos is their superficial similarity and their actual diversity. As we approached Daphne, we saw the same rugged lava surface, the identical scattered cactus, the low, half-dried shrubs. And yet when we had explored the island we agreed that it was like nothing we had ever seen elsewhere. Our first glimpse into the crater and the surprise it held, was one of the memorable events of the trip.

First we had to find a landing place. The light south-east April trades were blowing, sufficient to keep a white line of surf dashing up along the shore. As we went around the island there were three places which looked promising, but when we steered close inshore the cliffs developed an inward slope which only a fly or a gecko could overcome. On the east side a flattened, rocky ledge showed just below the level of the water and on it were two big sea-lions on their backs, with fore flippers held straight up out of the water. Now and then a big roller came in and lifted their great bodies and bumped them down again. But not until we came close and poked them with an oar did the creatures right themselves and with a gasp of astonishment dive into the depths.

Sharks were everywhere: they took our bait and tackle, and almost every grouper hooked was bitten in two before we could haul him in. And yet down through the clear water we could see groupers and sharks swimming about in fairly close association with no hint of alarm or pursuit. But the moment the prick of the hook sent the fish whirling about in a panic, the sharks gathered and attacked it. It may have been the few drops of blood from the gills, but I think in this instance it was a sense of something wrong, an instant reaction to the abnormal which drew them. After we had landed three or four more or less mutilated groupers and mackerel near the submerged ledge, a dozen sharks, four to eight feet long, milled around and around the boat.

Back to the south side we went, and there, where the crater's slope came evenly down, a projecting craggy rock offered foothold. On a rough day this would have been impossible, but today we landed without much trouble, leaping out when the boat lifted to a level with the top of the lava ledge.

Along the lower part of the slope was a scattering of cactus and bushy portulaca, which sent strong roots deep into the crevices and afforded excellent hand support. The slope was not quite steep enough to hurtle down shale to the bottom, but a disturbed clinker would roll and slide many yards before it stopped. I headed for the top at once, and reached the rim with only three interruptions, all of which were of interest. The rest of my party scattered along the slope or climbed obliquely.

One of the first cactus which I passed had a trio of sturdy ground finches, *Geospiza fortis*, an adult pair feeding a full-plumaged young bird. Ultimately all three were secured, but ten minutes of watching revealed most remarkable instances of vagaries of habits which insular life had induced. The first reaction of the birds at my approach was curiosity. The female flew out from the spiny pads and alighted at my very feet, chirping and turning her head from side to side. If her avian brain permitted relative comparisons I could have been to her only a sea-lion, harmless but strangely upright and deformed. The youngster tried to follow, but his mastery of the air was too uncertain and he swung back to an unsteady landing. As I squatted quietly, both parents began feeding their offspring, bringing seeds—a strange diet for a young bird—and twice, small striped caterpillars. Then without warning the male picked up a bit of dead grass and flew with it to a bulky nest fixed in the crotch of a young Bursera bush close to the cactus. The female flew up at once, took the grass and began weaving it into the rather frowsy structure. Then followed several feeding trips and again both parents visited the nest, the female entering and remaining out of sight until I took her out by hand. She pecked rather irritably as I lifted her out, but showed no fear and remained on the perch upon which I placed her, preening her ruffled feathers. A single fresh egg was in the nest.

When I examined the birds later, I found that the male was far from being in full breeding plumage, although in complete breeding condition, and the female had a fully-formed egg about to be laid, and another yolk ready for early development. In size of beak and wing the male was identical with the corresponding bird from the islands of Abingdon and Bindloe, fifty miles to the north of Daphne, which has been considered a distinct species by Ridgway and a subspecies by Rothschild under the name *fratercula*.

Here in one family party of these little finches I had evidence of the mating of birds of considerable difference in size of body, wing and beak, the male breeding in partly immature plumage, the female assiduously feeding a young bird, not more than two weeks out of the nest, and at the same time fixing up the same or another nest which already contained an egg of a new laying. Assuredly a mad country for birds and butterflies!

The comparative dimensions of the immature and the adult male birds were rather significant. The youngster was seventy per cent in weight and eighty-seven per cent in length that of the parent.   The beak and tail, eighty-six per cent and eighty-seven per cent, had lagged, apparently being of less vital importance to the young bird at this stage of its life than the wing, ninety-one per cent, and especially the leg and foot which had reached ninety-seven and ninety-eight per cent respectively.   The male was essentially a black bird, the female was characterized by brownish stripes, but the fledgling was the most negatively colored bird I have ever seen. Brownish and greyish white are all we can call him,—he was the essence of neutrality, of dull, patternless shades and tones. The nest, as in all these Galápagos finches, was an untidy mass of grass and bits of leaf and wild cotton, rounded and roofed over, with the entrance, as seems invariably the rule, on the north-west side, away from the winds and rains.   The egg was white, finely and rather faintly spotted with purplish brown, and with a few more distinct lines and dots near the larger end.

A third of the way up the slope I caught sight of a bit of white under a shelf of lava and turned aside to investigate. Well sheltered from discovery except from the angle where I had been, I found one of the most beautiful of all birds—the red-billed tropicbird, with one of the most unpleasant of all voices.   Although classed in the same family with the snake-birds, man-o'-wars, cormorants and pelicans, these well named tropicbirds stand apart.   In flight they progress by a curious quick beating of the wings, wholly unlike the soaring of the frigatebirds or the slow flapping of gulls.   The plumage is of purest white, finely barred above with black, the beak scarlet, and from the tail, two long slender feathers reach out far behind.   Few birds are more attractive on the wing.   But now as I peered beneath the ledge of rock I saw the Mr. Hyde phase of this bird—every feather on edge, eyes flashing, red, wide open beak showing its teeth-like serrations. No Galápagos native this, but a bird of the open water along both oceans, stopping here for the breeding facilities, but bringing with it all the fear and anger against a threatening intruder, which its ancestral experience had implanted. As I leaned down, two small Tropidurus lizards ran out of the crevice, showing no fear of the irate bird. Sheltered on all sides, armed with sharp, saw-toothed beak, the Daphne tropicbirds have absolutely nothing to fear, and yet within a few feet of the crevice lay two mummified young birds and one adult, while in a narrow niche at one side of the nesting ledge was a cold egg, as beautiful in pattern and color as the nycthemerus mosaic of the birds' feathers. The ground colour was a pale pinkish white, with everywhere a multitude of fine specks and dots of rich purple. The egg was easily explained,—the shuffling, awkward gait of the old birds must have given it a flick which rolled it into the narrow corner. But the dead birds were a puzzle, and the only explanation was family jars—tragedies of jealousy.

Pushing the irate mother tropicbird to one side, I revealed the cause of all her emotion—still another phase in the life of this sea-fowl, a formless ball of smoke-coloured fuzz, inanimate as a tuft of dead moss until I touched it, when it snapped into the action of a badly made German toy, jerking the diminutive beak backward and forward, and uttering sounds which could be improved upon by the bellows within any mechanical doll. Its wings were wholly buried in the long fluffy down, the feet were barely visible and quite unable to support the weight of the little bird, and only when the head was stretched far out was the bluish grey face exposed, and the small, clear blue eyes.

Five months after finding the home of the tropicbirds, I was in my New York studio talking to Mr. Beck who visited Daphne in 1901. Among some photographs which he took at that time was one of the identical nesting crevice from which I took the young bird and the cold egg. So for twenty-two years, successive generations of these birds had been using the same site. The breeding season on Daphne is prolonged, at least from April to late November.

Here and there on the steep hillside were half-grown olive-footed boobies, showing conspicuously in their snow white down against the dark lava and green vegetation. They squalled at me as I went by and their parents soared past so close that the wind from their wings struck strongly on my face. At one spot where there was a strong up-draught, a bird family sat upon it with motionless, wide-spread wings, hung in space ten feet above the ground, looking me over from head to foot, as an aviator would lean out of his seat and examine a plane passing below. Beck saw a few individuals of this species on Daphne in April, 1901, but did not record them as nesting. Gifford found eggs and young in November, 1906.

From the sharp, uppermost rim of Daphne's crater the view was magnificent. The sea was smooth as a mirror and of deepest ultra-marine, even to the very base of the island, for there were no shallows nearby,—the crater rose abruptly from great depths. Within the visible horizon circle starting at the north, I could see James, Jarvis, Albemarle, Duncan, Nameless, Eden, Guy Fawkes and the whole majestic slope of Indefatigable, then South and North Seymour, and near at hand the black crest, just clearing the surface, of Wheeler Rock, due east of where I was standing. Finally, north-east, Daphne Minor rose abruptly, unclimbable, crowned with a fillet of good-sized trees.

I was on the south side of the crater's rim, which stretched around the entire island, broken by an indentation only to the east. The opposite slopes, facing the trade-winds, were bare shale with a low, grey-green growth of weeds and low plants.

FIG. 51

COLONY OF BLUE-FOOTED BOOBIES ON THE FLOOR OF DAPHNE CRATER, LOOKING EAST TOWARD
THE DIP IN THE CRATER WALL

Fig. 52

FIVE SPECIES OF GROUND FINCHES

In order of reduction of the size of beak

SOOTY GROUND FINCH—*Geospiza fuliginosa* (Gould)
STURDY GROUND FINCH—*Geospiza fortis* (Gould)
CACTUS FINCH—*Geospiza scandens* (Gould)
CONICAL-BILLED GROUND FINCH—*Geospiza conirostris* (Ridgway)
GREAT-BILLED GROUND FINCH—*Geospiza maguirostris* (Gould)

Twenty feet beneath me, sheltered by the crater's slope, was a dense zone of the aromatic-scented, pinnated-leaved Bursera trees, six to fifteen feet high. These shut off my view and I was about to turn back and explore the outer slope more fully, when I heard a distant whistle signal, one of those which we have used in jungles for years, meaning "go straight ahead." I obeyed, with the necessary concessions to gravitation, and forced my way through the dense growth. Black finches were at the height of courtship and on all sides were exploding their little songs, wholly neglected by the small, dull-hued females, who were intent on quenching their curiosity by a close examination of the strange creature crashing through their domain. Most of them were sturdy ground finches, *Geospiza fortis*, and they never ceased their wheezy song, *kah-lee! kah-lee!* uttered during a slow fluttering flight. When perched they broke out into an excited *kah-kah-kah-kah!* Sometimes every male would let go at once, making a terrific din. There were a few sooty ground finches, *fuliginosa*, and a pair of the great-billed ones, *magnirostris*. I secured the male of the latter, a record new to this island, and found it was breeding, but in exceedingly worn plumage and in heavy moult. The bill was enormous (Fig. 52) with high-arched culmen, just short of an inch in length, and almost as great in depth. With a view of direct comparison and without moving from the spot, I shot one of the sturdy ground finches and a single sooty. Here were three birds almost identical in their black plumage, measuring respectively six, five and four inches in total length, but with bills so unlike that they almost ran the full gamut in size. A table will show the difference in relative percentages, better than description:

	magnirostris	*fortis*	*fuliginosa*
Total length	100%	83%	69%
Size of bill	100%	60%	47%

To look at the bills of these birds in the hand, we would conjecture wholly different diets. The small, delicate mandibles of *fuliginosa* would seem adapted to insect food, or at least small, rather soft seeds. At the other extreme, the huge beak of *magnirostris*, almost as large as the entire head, would be equal to the hardest of acorns. As soon as I returned to the *Noma* I examined the food of the three birds with the following results:

Great-billed ground finch, *magnirostris*—three ants, two small caterpillars, two soft Bursera berries and twelve small uncrushed seeds.

Sturdy ground finch, *fortis*—four ants, one Bursera berry and six small seeds.

Sooty ground finch, *fuliginosa*—one small caterpillar, two spiders, fourteen small seeds.

Whenever possible I repeated this experiment of securing several birds of different species of the genus *Geospiza* within a limited locality, and always I found the same results. Birds utterly dissimilar in relative proportions of mandibles were feeding upon identical food, and food which usually showed no signs of being crushed.

Professor Wheeler and I made a fairly representative collection of plants of the areas we visited and found no seeds which might be considered as adapted for the huge crushing mandibles of *magnirostris*. Either the mighty beak was developed for coping with some source of food which has now disappeared, or, as seems more probable, the lack of enemies, or an environment inimical to variation, has resulted in a diversity of bill which is not directly adaptive, but reflective of relaxed environmental control. This is made more reasonable by correlated centrifugal varitions in food habits, plumage and breeding.

The Bursera tree zone ended abruptly and I stepped on an

out-jutting lava ledge, with an astounding view spread out before me. Down, down, down went the crater on all sides, and at the bottom hundreds of feet below was the floor—an enormous, round, flat area, dazzling white and dotted with what at this distance looked like a multitude of flies. My glasses showed them to be a host of nesting birds—big brownish white boobies. Not until some of my party walked out on the white sand did I realize the great size of Daphne's crater. They looked like beetles weaving over the surface. Now and then a tiny white moth would swing over the crater's rim sweep out in great circles and finally volplane down to the floor—and I realized I had been watching not a moth, but a great booby with a stretch of wings of more than five feet.

I focussed my Graflex camera, and as I was about to take a picture, a large dark bird crossed the mirror and alighted near two white specks. Looking over my finder I saw a black Galápagos hawk just settling on her nest, upon a ledge, fifty feet below me.

By walking around to the east I found a fairly easy descent, on through the gap in the crater wall, down to the bottom. Here I found pandemonium in the shape of six hundred, nesting, blue-footed boobies, *Sula nebouxi*, a pure culture of this species, monopolizing the entire crater, with no other birds in sight except a scattering of ground finches feeding in the surrounding bushes and the hawks far up the slope. Not only this but they apparently possess this crater throughout the year and have done so for years unknown. Eggs, and newly hatched young, and full-sized birds of two or three months were everywhere and other observers have recorded similar conditions in July and November, so there is apparently no month when the crater is deserted. Beck twenty-two, and Gifford seventeen years ago found the same single species colony in this crater floor as I did today. I counted over four hundred nests, this being rather a misnomer, however, for the only nest was the small spot padded down firmly by the birds' feet. On each such spot were one or two eggs, or a corresponding number of young birds, or a sitting adult bird. The photographs show the general appearance of the birds and their home better than any detailed description. The only thing of which the photograph gives no hint is the brilliant, pale ultramarine blue of the legs and great feet and webs. As we looked about at the white sand, black lava slopes, and brown and white birds, these shining blue feet dominated all other colours.

There were more sets of two eggs than of one, and about the same relative proportion of young birds. Much has been written about the almost invariable death of the later hatched of the two nestlings, either from starvation or from actual infanticide on the part of the parents. In this colony at least, this Spartan habit appears to be inoperative. There were dozens upon dozens of pairs of young birds, almost always of considerable difference in size, but both well nourished and strong.

I saw both of the parents brooding with their great webs under the eggs. When I gently pushed a female booby back from her eggs in order to photograph them, she sat protesting for a short time, and then shuffled forward and with the greatest care shoved her huge blue feet beneath the eggs. This booby ranges up and down the mainland from Lower California to Chili; and there it often lays three eggs, while two is the usual complement. The Galápagos blue-footed boobies never, so far as I know, deposit three eggs, while one is quite as commonly a complete set as two. It is interesting that while these insular birds have still retained resentment of human intrusion, so lacking in most indigenous Galápagos species, yet they have appreciably altered in number of eggs. This is very reason-

Fig. 53

Roofed Nest of Sooty Ground Finch

Geospiza fuliginosa (Gould)

These nests are placed in cactus or small shrubs, always closed over, and the entrance is away from the direction of the prevailing wind and rain.

Fig. 54

Mockingbird at Entrance of its Covered Nest

FIG. 55

BLUE-FOOTED BOOBY

Sula nebouxi Milne-Edw.

Parent, Chick, and a Dead Bird on the floor of Daphne Crater

ably explained on the theory of supply and danger, when we recall that the Galápagos birds have few or no enemies, while the continental boobies suffer constantly from the attacks of gulls and vultures.

It was surprising on both days of my visit to Daphne to find the crater so cool. Although shut in completely on all sides, yet there was actually more breeze here than outside, the heated air from the white sand evidently ascending, while cool, fresh air siphoned in from the notch on the eastern rim.

The boobies, low down on the crater floor, felt the heat however, and both newly-hatched young and parents panted with open beaks. The day-old boobies squirmed about, naked and ugly, while older birds showed a slight growth of down. This increased with age and soon became dense and long, until they appeared like powder puffs. Then gradually, between the shoulders, brown feathers began to appear, and in the month-old birds, the juvenile plumage was almost complete. The birds which were still clad in body down, but with well advanced wing feathers were most comical (Fig. 57). Sprouting from the snow-white down were the enormous wings and splay feet, the wings so heavy that they fairly dragged on the ground as the youngsters waddled along, appearing more like useless deformed outgrowths than *anlagen* of the wonderful pinions of the full-grown booby.

Throughout all their babyhood the young birds stuck close to the hardened spot which was home. Only when their down was well on the way to disappearance did they begin to mingle with their fellows, squads of six or eight wandering about aimlessly, keeping a sharp lookout for manna in the shape of warm-storage fish from the crops of their parents, descending from the heaven of the unknown outer world.

The blue-footed boobies which roosted upon lofty cliffs, such as those at Tagus Cove, had a wholly different view of the world. As I walked along the edge of that high cliff, I saw a full-grown young bird leave its nest for the first time. It launched out bravely enough, but wavered like an airplane which has slackened speed to below forty miles. It volplaned faster and faster and finally, terrified by the waves, it attempted to brake suddenly, with the result that it splashed in upside down with a most unbooby-like flop, bobbed up to the surface again and paddled frantically to shore where it clambered up among a mass of great marine iguanas who had been quietly sunning themselves.

Here on this sunken crater floor, the young boobies during all the first months of their existence see only the dry lava and half-baked vegetation, and overhead the unbelievably blue sky. They probably have to remain a full three months or more before their wings are strong enough to carry them around and up, and if a booby has any imagination or ability to be surprised, with what emotion must he experience his first rise above the crater's rim, and see the ultramarine sea stretching away on all sides. Up to that time salt water has not existed for him, and the outside world has been represented only by passing clouds, the trips of his parents and the semi-digested fish which he has plucked from the capacious throats of his mother and father.

Here and there were dead mummified remains of boobies. Among these I observed no very young birds, and only five adults, while all the rest were nearly grown young. The partly ossified skull was unmistakable even where the plumage had fallen off and blown away, and I believe that the crisis of the entire life is the achieving of the crater's rim, after the wings have acquired sufficient strength and before the almost mechanical cessation of feeding instinct on the part of the parents. There is undoubtedly a trenchant survival of the fittest, at this brief temporal period, in value to the race far transcending the elimination of young birds by enemy gulls in mainland colonies.

Even the adult boobies had difficulty in rising. They were absurdly tame, but by rather badly maltreating one and pretending to attack it, I managed to force a bird to flight. It flopped and scuttled along, knocking over sitting birds and sending half-grown boobies headlong, before it got enough impetus by rapidly paddling feet and ground-striking flops to clear the sand. But it never left the crater, circling around and around, higher and higher, only to drop to within three feet of my head and again settle on its eggless home—a tiny patch of trodden surface.

The noise, especially in the more densely occupied northeastern section, was so terrific that no conversation was possible. A blare of brazen, raucous trumpet-like notes mingled with squeaks and shrill whistlings. I killed a trumpeter and a whistler and found that they were female and male respectively, and I also recorded that the eyes of the two were wholly unlike. In equal shadow the pupil of the male was small, the iris almost clear yellow; in the female the pupil was about once and a half as large again and the yellow iris was more or less mottled with brown. There was considerable variation, and in some birds the brown of the juvenile iris was dominant. When held in full sunlight, the pupil of the female contracted almost to that of her mate. It would be exceedingly interesting to know whether this was correlated in any way with nocturnal habits. Unlike certain petrels, boobies are decidedly birds of the day and it is improbable that such an explanation holds good. On my return I found that Dr. Murphy had recorded corresponding sexual differences in boobies breeding on islands off the Peruvian coast, showing that these distinctions have nothing to do with Galápagos isolation.

I have spoken of the scattered dead birds, among which was a single, very ancient pelican—sole intruder of this blue-footed boobies' domain, which had been stricken in flight, as it passed overhead, or died after alighting. Under these dried corpses was a most interesting assemblage of insects and other arthropods. There were several species of beetles, but by far the most abundant were small, shiny black ones belonging to the same group as our familiar meal-worm beetles, which feed on our flour and cereals. The Daphne beetle is called *Stomion lævigatum*. The genus is found only in the Galápagos, and, like so many other creatures of this archipelago, is most closely related to a northern group.

This species has only been collected by Darwin, and from an unrecorded island. These beetles are evidently scavengers and scurry away from the light by dozens when the body of a dead booby is raised, or a loose stone overturned. The most interesting fact about them is the loss of the power of flight. Not only are the wings themselves abortive, but the elytra or wing coverts are soldered together. Yet here again we have considerable variation, and while in most individuals the elytra tear into small pieces, in others the suture between the once paired sides gives easily and the deeply concave cases rise and separate, a mockery, as there are no membrane organs of flight beneath. Here in the depths of this crater, these beetles have found no need to rise and whirr away to other haunts. The supply of dead bodies affords all that life could wish. They strive only to keep out of sight during the day-light when hungry lizards and finches are about.

In the same places as the *Stomion* beetles were occasional weird Solpugids—spider-like creatures, which ran with extreme rapidity. They represent a very old and primitive group and one found only in warm countries. Some of the larger species kill small mammals, birds and lizards, but these were probably enemies only of insects.

Under the stones along the edge of the crater floor I found geckos in abundance. Eight forms of these little lizards have been distinguished in the archipelago, recorded from sixteen islands. The Daphne Island gecko has been given a name exactly as long as its body, *Phyllodactylus galapagensis daphnensis*, although a binomial would be more correct, for there is no chance for direct intergrading with the geckos of other islands. I found several, about three inches in length, under loose lava slabs, and like all the other inhabitants of these islands they were unusually tame. If the bit of slag was lifted gently the lizard did not move except to cock its eye upward. It was ashy grey above, with spots and markings of blackish brown, which varied greatly in different individuals. In some the dark markings formed irregular cross bands; in the majority, they were quite regularly alternate, the light grey forming rhomboidal interspaces. A black line extended from the snout through the eye back to the ear. Beneath, the lizard was greyish white, finely peppered with brown. The eye, as usual, was the most brilliant and beautiful character. The pupil was vertical like that of a cat, indicating nocturnal activity, and in the sunlight it was narrowed to a mere slit. The dark line across the head was reflected in the iris by a brown shading, and elsewhere was brilliant gold, shot through with fine, zigzag, black lightnings. Even in these little lizards I observed an insular upsetting of habits, for twice, at mid-day, but when the sun was obscured by clouds, I saw them creep out from their shelters and run about in the light. As one's hand approached, they waited motionless, but at the first touch they slipped around the stone like a streak of grey light.

Like their congeners in Guiana and elsewhere, their tails

were ready to drop off at the first tug. Two which I saw and could not catch, had lost their tails and every specimen which we secured was tailless by the time it was in alcohol. The first gecko I saw was under a bit of stone which partly sheltered the egg of a tropic-bird. With all the speed of which I was capable I grasped at it, but a fraction of a second after my hand reached the little creature, it had vanished, while its tail leaped and twisted and danced madly about on the lava. Carefully reversing the flat slab again, I espied the gecko, self-possessed, quiet, calmly watching for my next move, just as if one-half of himself were not twisting and curling near by, lost to him forever, dying its own death. Never will I become accustomed to the strange detached frenzy which inspires the dying of this lizardless tail —a frenzy of motion which must so often attract and hold the attention of an enemy while the owner escapes. Already the eight little muscles had drawn together, without the loss of even a single drop of blood, and *Phyllodactylus daphnensis* would soon sprout and develope a new tail, ready in its turn to foil any attacking scientist or hawk—I can imagine no other enemy on this isolated crater. Their food consisted of what small creatures they could conquer in their lowly wanderings; one had devoured the caterpillar of a small moth, and a lace-winged fly, *Chrysopa;* another had found two caterpillars, two jassid bugs and a spider, and a third had devoured three *Stomion* beetles and several small seeds.

On my way up from the bottom of the crater, I worked sideways along the slope, and half-way up came to the nest of the hawk, which I had discovered from above. The bird left when I was a few yards away, but stayed about while we photographed and examined the structure. It had been started on a large out-jutting lava rock, and had been added to year after year until it reached ten feet up along the slope of the hill, an upright mass of six or seven feet, in places five feet thick. The material consisted almost wholly of irregular twigs and branches of Bursera, a tall, leafless tree of which grew at the upper edge of the nest.

The latest addition to the nest was a foundation of four inches of sticks, and a two-inch lining of fine grass and dry leaves. I was able to distinguish the underlying five or six annual accretions, and judging by this decipherable strata, the entire mass represented not fewer than fifteen to eighteen nestings.

There were two eggs still warm from the brooding bird, and at the base of the nest was a third, evidently of some pre-ceeding year, for although unbroken, it was quite dried up. The eggs were plain white and very broad ovals in shape. Their weights and dimensions were as follows:

	Weight	Dimensions
First nest egg	59.1 grams	55.3 × 44.7 mm
Second nest egg	58.8 grams	55.5 × 43.8 mm
Third old egg	6.9 grams	54.5 × 43.2 mm

No sooner had we pushed off from this delectable island than the power of the surprises it had afforded redoubled. Twice we had steamed close to it, and even with the glasses had seen no evidence of the sea birds which made it their home. For a week we had worked within five miles and had observed only a stray booby now and then. The birds must go straight to their fishing grounds far to the northward, and singly, unlike the long files and flocks which put off from, and arrive at, Tower, morning and evening. No external hint was ever given of the hundreds of boobies in the heart of Daphne. May the little island long keep her secret from devastating mankind —a single boatload of which could exterminate the whole colony in a few hours.

CHAPTER XIII

W HENEVER, as a boy, I read a book of exploration and adventure, I often wished that the author would, for a chapter, neglect the startling, ignore the high lights, the crisis of some mighty effort, or the thrilling dangers which either pass me by or are too soon forgotten to transcribe. I wanted the tale of an ordinary day, the happenings on the twenty-sixth of a month, which might well be confused with those of the twenty-fifth or the twenty-seventh. Hence this chapter.

At a quarter before six, when the first light of the mechanically regular tropical dawn came through the port holes, I awoke and sat up. This was my fiftieth morning in Cabin Number Seven of the *Noma*, and it had become home to me. For a few minutes I looked around and mentally compared this with other rooms and berths I had occupied; the carved cubby-hole in a Chinese junk, where I could never quite straighten out, with its ever changing wall pattern of blattids and heteropters which carried no scientific appeal; and again another memory, of a torture-shaped room where, I never quite understood how, four of us humans slept, shaved and in great storms clung, without knowing a word of one another's language, without an unpleasant thought or an infringement upon each person's individuality. I still remember with pleasure

276

FIG. 56

BLUE-FOOTED BOOBY, PARENT AND YOUNG

FIG. 57

YOUNG BOOBY WITH FEATHERS IN FULL GROWTH, THE HEAVY BLOOD QUILLS DRAGGING
DOWN THE WINGS

FIG. 59

NEST OF HAWK ON THE SIDE OF DAPHNE CRATER

There were two fresh eggs at the top, beneath which the remains
of seventeen old nests were distinguishable

FIG. 58

OUR SURGEON TESTING THE TAMENESS OF A GALÁPAGOS HAWK

our method, on particularly vile weathered mornings, of passing the time. I would point and say "coat," and then, acting out the word, add "swinging coat." And three voices with three inconceivably different and comic accents would repeat the composite sound "swin-ging coa-at." Then I had to say the phrase, or whatever they thought I had meant, in three terrible tongues. Reoriented with the brief review of these and other memories, I looked around, and realized what a luxurious home was mine. I saw now with new eyes my little white cane headboard, mirror, bureau, closet, chest of drawers, couch and curtains of bright cretonne, soft carpet, six electric lights and fan, perfectly appointed bathroom. We were in an almost waterless country, and our stay depended on our supply, so as usual I used White Rock for tooth-brushing and shaving, a tumbler each, and bathed in clear salt water, which, now and then, the most delicate of tiny jelly-fish shared with me. I donned my regular tropical costume of tennis hat, woolen shirt, khaki shorts and sneakers. On the first deck the sailors were making that jolly noise of washing down, the shifting siss and rush of the hose mingling with the piston-like scrape and rub of brooms and holystones. I went up to the sun parlor which had now become our laboratory, with its ever familiar sight of myriads of vials, aquariums, stains, microscopes, books, and jars of fish, flesh, fowl and invertebrates.

A glance from the deck showed the wonderful panorama—southward the ascending green slopes of Indefatigable, with the highest crater, mysterious and quite unclimbed by man, wholly free from clouds; a few miles to the west Eden, our first love in the Galápagos and close at hand Daphne Islets, looking like the open mouth of a submerged bottle and its stopper. To port, stretched the long, low island of South Seymour, its white beaches sending an invitation which I would soon accept.

The water was like a mirror, a flexible one which bowed now and then almost imperceptibly as a low silent roll slipped gently beneath the keel and on toward shore. As I settled to work, Wireless looked in, handed me the daily typewritten news-sheet and asked if I had seen the last nine-foot shark which had been hooked late in the night.

I glanced at the sheet and read almost without interest that Berlin had denied something, Milwaukee boasted of I have forgotten what, the Giants were certain this year of . . . forty persons killed in a train wreck; bootleggers had a new scheme; and Saskatchewan—this I read twice, this held my attention:

"Saskatoon Saskatchewan . . . Carl Lynn world war veteran and one of the best known trappers and mushers in the North Country is believed to have lost his life in a death battle with a pack of timber wolves. Believed his body devoured by wolves after he had killed six."

So the world elsewhere was not all anæmic and inconsequential! Somewhere else in the world things were happening as in the days before Ruhr, prohibition, and Bolsheviks were coined. I ceased to resent the fake fire-place in the dining-saloon, with its transparent glass coal, and the wireless was not an anachronism in the Galápagos when it could bring a message such as the one from Saskatchewan.

With vial and forceps I went down over the side to investigate the shark. It was quite unhurt, and when we drew it up to the surface it almost demolished a small boat and myself. I went over its body and the great gills carefully for strange parasites. Weird beings live here, some of which take hold so firmly that they burrow down into the dermis, others skim swiftly over the rough shagreen surface, evading all efforts at capture. Some are recognizable as shrimp-like beings, others have lost almost all their organs except mouth, stomach and ovaries.

What a boon to the human race would be the tooth arrangement of a shark. Within the mouth is a deep groove with many rows of teeth lying flat, like shingles, upon one another. As the teeth in use on the edge of the jaw become worn or lost, others promptly rise up behind and take their place, and throughout the lifetime of the shark this supply never diminishes.

Soon after breakfast I went ashore. Long since I had learned to adapt myself to these islands in the matter of clothes,—in light tennis cap, bathing suit and thick soled canvas sneakers I could make my way slowly inland, along shore or into the water at will, without all the discomforts of seed-filled or soaked clothing. For a long time I sat on a new beach—a small personal beach of Seymour, where perhaps no one had ever sat before me. Here I fixed in my mind the surroundings of this islet so far out in the Pacific. Two miles away the yacht rode gracefully, almost eclipsing Eden, our first explored islet. The water before me was deep turquoise, darker blue along the horizon, paling imperceptibly into the clear emerald at my feet. The light fawn of the sand began at the alabaster line of gentle surf and stretched smoothly up to my seat. So far all was usual,—with less color it might have been a Long Island shore in mid-summer. But there are no shell-holes on Long Island beaches and I was sitting on the rim of one of the dozens scattered along in both directions. Not only in appearance, but in actual play on words these were veritable shell-holes, for each was the nest of a great sea-turtle, and at the bottom a litter of egg-shells from which the turtlets had made their escape. Here was the transition to a tropical island.

And the windrows of shells flung along high storm mark

were as beautiful and varied as in story-book islands. There were delicately lettered and shaded cowries, cassias voluted and stained with enamelled pigments; arca and wing-like valves of shell tissue, and fluted and cross-fluted super-clams. Limpets with scarlet circled keyholes became more brilliant as they decreased in size. Over and against all this, jet black heads of lava rock were now revealed, now covered by the restless waves.

The first vegetation was timid growths, which dared much in sprouting close to high tide, clinging with fearful rootlets to the surface of the sand; here the myriad grey-green rosettes of *Caldenia*, whose tiny leaves were rich wine colour beneath, all frosted with a heavy coat of hairs. Beyond were equally prostrate *Amaranthus* and *Euphorbias*. Almost the first plant to raise its stems and tiny white flowers above the sand was a graceful *Heliotrope*, while farther behind me waved a small, orange-flowered *Lantana*, cousin to one of the world's tramps, which I have seen on the opposite side of the globe, on sandy shores of Ceylon, in the mud of the Straits Settlements, in my Jersey lowlands and along the country roads of England.

I stopped at this moment of writing, for a few inches beyond my paper I saw a three-foot snake creeping down the beach between my feet. Slowly I prepared and braced myself, and before he could escape I had him. He was probably as tame as the rest of the reptilian life, but he was too rare to take any chances with. He was dark brown above with two pale brown bands down each side, and beneath, the scales were a delicate pink; a harmless Galápagan native, feeding, as I found later, chiefly on grasshoppers.

I had scarcely packed up the snake when, with sudden darts, a beautiful male Tropidurus lizard approached, nodding violently with head and elbows when he caught sight of me. But my motionlessness resolved me in his eyes into a harmless bit of scenery, and he came and snatched a tiny ant from my shoe. Then a very young mocker flew over from the nearest bush, and the lizard flicked from view. This new youngster alighted on the sand, and another rustled the dry leaves behind me. One of them called, *peeent!* and I answered, when instantly I was struck full in the face by the young bird, who, on unsteady wing, had tried to alight on my hat but fluttered too low. For a moment he lay, spread-eagled, in my lap, both of us too surprised to move. Then he dived away and was called and fed by a parent.

Pelicans flew over and once from the sky a booby arrowed into the water, sending up a fountain of spray like a depth bomb. The pelicans made a great fuss over any fish captured, tossing it up and getting it just right for swallowing, but the booby rose from the water with closed beak, and at once beat steadily off as if the dip had been merely for a bath. This past-master angler was able to strike, capture, and swallow almost simultaneously.

The scarcity of sea birds in this great bay was a cause of constant surprise. In front of me, only four miles away, was Daphne Major, where as I knew were hundreds of nesting boobies, besides numerous tropicbirds, terns and shearwaters, yet it was a rare thing to the east or southward to see more than two or three boobies a day. Every tenant of Daphne, as I have said, must find its food far to the northward, out in the open water, and, at least at this breeding season, commute to and fro with almost no wanderings.

A slight shift of wind brought an odour to my nostrils,—an ancient fishy smell which would probably send most people down wind away from the source. But to a scientist such a hint is laden with hope of new things, and while it is true that a rose by any other name, etc., the converse is not at all equally true. To announce that chemical action, induced by the passing of time and by appropriate environment, is causing a certain accumulation of shell-contained matter to give off the gas CO_2, is to do away with much of the opprobrium attached to the bald statement, "The egg is rotten!"

The odour which was wafted my way made me sanguine of solving the mystery surrounding the deaths of the goats, and then there were always the byproducts of burying beetles and strange diptera. I shook off the sand and started down beach, when an object like a mottled stick caught my eye, projecting from a hole of a ghost crab, a few feet above the present tide level. I found it was a very beautifully marked moray-like eel, quite unhurt but lying motionless on the sand. When touched it came to life with a vengeance, and true to the reputation of its clan, made no effort to escape, but tried its best to reach and bite me. It was just two feet in length, rather stout, with a head like an arrow and a very short tail. The colour was pale buff with several alternating series of large mummy-brown spots, becoming smaller and paler on the head. The eyes were remarkable, being upturned in their sockets, so the very small slit-like pupils had vertical vision, while on the outer, lower part of the eyeball were several small, scarlet, hieroglyphic-like markings. Judging by the position and direction of the eyes and the character of the pupils, I should say the eel was nocturnal and a haunter of holes, or at least lived on the bottom, flounder-like, half hidden in the sand.

After a brief but active fight I got it into a collecting bag. Long afterward, when I learned its name, I realized that a scientific term could be apt—*Scytalichthys miurus*—the short-tailed viper fish. Nothing is known about its life or habits, for only two other specimens have ever been secured, both from Cape San Lucas at the extremity of Lower California, eighteen hundred miles northwest of my beach. (Plate VII.)

Having bagged my game, I again sniffed and started up

wind. But as is so often the case, the side issue proved more important than the main object, for the scent led only to a pelican, hesitating between decay and desiccation. My by-product hopes were fulfilled, for as I turned it over, I saw a number of scurrying beetles. I collected about a dozen red-shouldered *Dermestes* closely related to our bacon or carpet beetles of the north, only here, lacking these human articles of diet, they have to turn to feathers and dried tendons. Darwin collected the same species on James Island. A tiny iridescent blue *Necrobia* or bone beetle was also captured.

Under a neighbouring stone I caught a pair of minute rove beetles, the male of which had a large horn on the thorax. It proved to be a new species. Also near by I found a group of half a dozen white toadstools, the only ones I saw in the whole archipelago.

The most abundant insects under stones were *Lepismids*, very appropriately in this land of ancient reptiles, for these are among the most primitive of insects, wingless because their race has never had wings, and we find them living today almost unchanged, going back as fossils in Baltic amber to the Lower Oligocene, a matter of about thirty-five million years. There were giant *Lepismids* on Seymour, one individual of which measured over three inches (78 mm.) from cerci to antennæ tips, but most of them were 50 to 60 mm. in total length. The body of the very large female was 26 mm. long. It was almost impossible to capture these entire, as they were injured at the slightest touch. This large species *Acrotelsa galapagœnsis*, was greyish black with a silvery sheen on the body scales, and with numerous tufts of short bristles.

I was attracted inland by a number of gulls and smaller birds which seemed to be feeding on a cloud of gnats. The latter proved to be grasshoppers, and the former lava gulls, purple martins and flycatchers. The giant grasshoppers had been in evidence since the first day I put foot on the Galápagos, but I had never seen so many as were gathered here on South Seymour. Hundreds flew up from the ground and grass as I walked along. They were feeding on the aromatic Bursera leaves, and many of these trees were completely denuded. On some branches there were many more insects than leaves and even the stems were lined. When I shook a limb, fifty or more would fly off into the air, and as many at once fly up to take their place. Hosts were continually flying high up into the air, and hovering like kestrels or fish hawks. Lava gulls were snatching the big females, while the lesser birds were taking the smallest males.

Except the dragonflies, and the feral so-called domestic animals such as pigs, donkeys, dogs and cattle, these grasshoppers were the wildest creatures living on the Galápagos. In early April at Harrison Cove they were courting, and every great female was the centre of three to five admiring males. I could not get near enough to observe much of the details of entreaty or selection, except that the males walked slowly back and forth in front of the females, on four legs, holding the third pair elevated to the fullest extent. The unusually brilliant colouring of this pair of legs may thus be of some secondary sexual significance. In this land so barren of insect voices, the clicking of the wings of these grasshoppers in flight was very noticeable.

These insects were as gay as butterflies, being intricately marked with yellow, scarlet and blue, the hind legs, as I have said, being especially brilliant, the upper leg banded red and black and the lower portions clear yellow. Only in the brightest coloured ones was the blue wholly dominant over the black, and equal in strength to the other primary colours. The females were often twice as large as the males, and at night, sometimes *against* a light breeze, the largest of the former sex would fly two miles out to the *Noma*, along the light beam from the searchlight, showing no signs of fatigue when they arrived. Although the males are full-winged I never knew one to come on board, even when we were anchored close to shore.

It was interesting to see the dominance of such strong fliers as these, for nearly eighty per cent of the orthopterous tenants of these islands are wingless or at least flightless. I did not find a single omnivorous or carnivorous reptile, bird or mammal which did not at one place or another have the remains of giant grasshoppers among its food. Even the mice fed on them and terns and gulls took them readily. Indeed, several times when an insect misjudged its direction and fell into the water, it was snapped up by some fish before it had a chance to struggle, and the tale of enemies seemed complete when one afternoon I saw two scarlet crabs pulling one apart between them.

Large dragonflies were hawking about, taking toll of the mosquitoes which I frightened out of the grass. I found it almost impossible to capture them in a net. They were far more difficult to approach than morpho butterflies in the jungle. So I resorted to an old method, and using a 22 calibre rifle, and shot cartridges, I shot at them at the outside range of effectiveness. My first attempt was amusing,—I only wounded my game, and after a somersault it regained its balance, and very much bent, flew very slowly to a high branch and there clung. The tip of the abdomen and one or two legs were injured. I had the same feeling that I have at the sight of a wounded animal, that I must put it out of its misery as soon as possible, so getting directly beneath I fired straight, and reduced it to a bedraggled pair of wings. My next half dozen shots were better, and I secured as many specimens almost uninjured, with perhaps a single wing broken, or some similarly unimportant injury. Going gunning for devil's darning needles seemed quite in place in this weird country, where seagulls fed on grasshoppers, and grasshoppers went to sea.

As if to emphasize this bizarre quality in the fauna, my next glance showed a full-sized, spotted-breasted, young mocking-bird, perched with fluttering wings, actually being fed by a slender-billed, black, female cactus finch. Twice as I watched, this unnatural parent made trips and brought food to the young bird who was larger than the finch. After the second trip another cactus finch appeared, sang several times over *chur-wee! chur-wee! chur-wee!* and both flew away, leaving the young mocker to his fate.

I now made my way to the high veldt land where I had already studied and captured Conolophus. The single species of Galápagos hawk was abundant here, and at one time I saw twelve at once. They were rather sociable and four or five sometimes perched close together on a boulder or low tree. By walking very slowly I could come within two feet, and even when I had flushed them several times, they exhibited no signs of increased timidity. The martins did not quite approve of them, and occasionally swooped twittering at them when in flight. There was at least a single basis for this sus-picion, for one hawk as I found by dissection had killed and eaten one of these birds. Although grasshoppers were flying about in all directions, the hawks seldom took one. Of four stomachs which I examined, only one had any remains of these insects. All had been feeding to repletion on giant red and black centipedes. I saw these striking creatures only rarely, and then it was the briefest glimpse as they disappeared into a pile of lava. Yet the hawks here and at least on two other islands, fed chiefly on a diet of centipedes, from six inches to a foot in length.

The martins to the number of about a dozen were hawking about, and occasionally swooping low down over the prostrate

Fig. 60

GALÁPAGOS HAWK

Buteo galapagensis (Gould)

So fearless were these hawks that it was easy to walk around them and choose the most artistic background for a photograph

FIG. 61

LAVA AND VEGETATION ON INDEFATIGABLE

The lava is scorching hot in the sun of mid-day, and sharp as a razor; the plants are armed with a host of thorns or, at touch, shed a mass of hooked seeds

yellow-flowered *Tribulus*. I could not discern what attracted them, until later when I examined several and found that they had been feeding upon moths, one bird having eaten twelve, and another, twenty-one of these fuzzy-winged insects, among which were a number of day-flying sphinx moths.

The hawks did little or no hunting in the heat of the day, and often sat on the same perch, either in or out of the sun, for hours, preening their plumage, or nibbling parrot-like at the bark or leaves. They showed considerable curiosity, and as we photographed or chased the big lizards, they followed us about, perching in nearby trees. As a rule the hawks appeared to keep in trios, and occasionally the immaturity of the third bird was apparent. Now and then they called, the usual staccato or rolling scream, but not nearly as loud or prolonged as the note of our northern birds. On these low, open islands, where distances are so short they can see too easily for loud tones to be necessary. Two distinct phases of these hawks were common, dependent neither on sex nor age; one was dark blackish brown all over, the other was pale buff, below, variegated with numerous large round spots.

The few sooty, ground finches present all flew up for a quick look at me, and then out of sight. They had almost no curiosity, and made their way about with a curious, slow, fluttering flight, the body held very steady and the wings beating with somewhat of an effort. Their notes were as simple as their garb, the usual song being *cu-wee! cu-wee! cu-wee!* Doves, small and short-tailed, flew past occasionally, less strongly and more slowly than doves in general, but with the same direct flight, and higher than any other land bird, twenty to fifty feet up.

A half hour after I reached the veldt I had a strange experience with a hawk. Desiring a specimen but not wishing to injure the skin, I backed away and at a considerable distance fired at it with fine shot in the third barrel of my gun. The bird turned a complete somersault, landed on the ground on its feet, lowered its head and ran full speed toward me and brought up exactly between my legs. I picked it up without resistance, placed it in a large basket and took it off to the yacht, where it readily took fish, preened itself and made no attempt to escape. Today, seven months later, it is living in good health at the New York Zoological Park.

I returned by way of the basaltic cliffs and found goats and small kids asleep here and there in the higher caves. On a flat rock I passed a big mother sea-lion nursing a thirty-pound youngster. She looked at me curiously, but made no movement until I rolled her off into the water, when she swam high out to see what next astonishing thing I would do. I took her pup and threw him as high into the air as I could. He fell into the water with a squawk and a splash and almost with a single movement of his tail shot himself out again upon the rock, looking at me and croaking, as if to say, "Please do it again." No tameness of horse or dog has ever impressed me with anything like the thrill which these wild creatures gave, in accepting the first human being they had ever seen as something which it was inconceivable could hurt them. I left the pup with his head bent over until it touched his back in an effort to watch me out of sight.

Returning to the *Noma* we trolled and caught two big, brown mottled groupers, and two of the very beautiful, gold-spotted Pacific mackerel *Scomberomorus sierra*, twenty inches long and fighters throughout their whole being.

Soon after lunch I went to the mainland of Indefatigable and made an impromptu attack on the interior. I had intended only describing a large half circle from one point of land to the next, but the going was so fair for the first half mile, that I took my bearings, lined up the first high craterlet with

the great central peak and set out doggedly to cover ground.
I have already described the gentle art of walking on Indefatigable on pages 67 and 79, so I need not repeat the details.
As soon as the small basins of red earth appeared, the soft,
waving grasses also materialized, and I began to gather the
spiny seeds of *Cenchrus*. Gray, in his botany, describes a
related species in our southern States, under the name of
hedgehog or bur-grass, and his final commentary is "a vile
weed." I called it this also, among other epithets, which
increased in power and sincerity as I encountered larger and
larger fields. Simultaneously, the level ground began to be
broken up into cross gullies and steep faults and canyons,
necessitating wide detours, and, in place of the rather solid lava
flows of the coast, there appeared acres of eternally balanced
and eternally sliding slabs.

After the first two miles I seemed to get in the lee of the
central ridge and every breath of air ceased. The clinker
radiated heat until it seemed as if it had not yet cooled from the
last eruption. Now and then I dipped down into a hollow
floored with stagnant water, choked with green slime, and
giving forth an aroma which seemed to thicken the still,
superheated air into a denser medium—an invisible heavy fog
of stench. At first I noticed occasional finches and doves
and even caught any unusual insect, but soon my mind drew
away from all casual objects, and my eyes, too tired to shift
about, watched only my next perilous step, and now and then
lifted to reorient my direction—headed ever for the mass of
purple raincloud which had eclipsed the mountains. This
was a better guide, quite as certain in direction and much
higher. For mile after mile I rose hardly at all, only rarely
from a red-hot knoll did I catch a glimpse of the bay behind me
and the ever lessening *Noma*. Its decrease in size was the only
warrant that I was making progress. The raincloud seemed as

far as ever, and I had long since lost sight of my craterlet guide.

Several times I almost broke my ankle on slipping masses of lava, some of which must have weighed two or three hundred pounds, and which, once overbalanced, rolled and clanged down to the lowest levels. I think that if we had planned to return straight to Panama on the morrow, I would have kept on, for I could then have recuperated at will, but with the thought of three days on the wonder island of Tower I did not dare cripple myself too much, and after what I conservatively estimated at five miles, I gave up and rested. I climbed a thorny tree on the highest hill at hand, and saw at least two more miles of almost level going between me and the rather abrupt first rise of land.

I rested, squatting on my heels, for the lava was too hot to sit on for a moment. My tongue seemed three times its usual size, and I watched the blood slowly drip from the big gouges in my legs resulting from frequent falls.

Hardly a living thing was in sight. Two little moths fluttered about a flowerless, grey-green amaranthus, several ants appeared and waved their antennæ inquiringly in the direction of my gory limbs, and a great centipede crawled out of one crevice and into another. No movement of wind, no rustle of leaf, no voice of insect or bird troubled the vibrating waves of heat, or the sound of my pulse throbbing in my ears. Then just before I shifted, stretched and started back, a sulphur butterfly—most appropriately coloured for this particular bit of hell—drifted near, brushed against my face, rested its tattered wings on my knee for an instant, and fluttered on, headed where I could not follow—inland.

Half-way back my fatigue was momentarily forgotten when I caught a glimpse of some animal moving slowly along on the opposite side of a screen of weeds and cactus. In another moment I saw two dogs emerge—police dogs in appearance, one larger, brown with some white markings, and the second half-grown. The first one snarled silently at me, turned at once, and both galloped off over the broken ground.

About a mile from the coast I left my back-trail and struck direct, and soon skirted a good-sized salt lagoon on which I saw five Galápagos pintailed teal, three adults, a full-sized but flightless young bird and a solitary downy duckling which followed me for some distance, keeping a few feet from the land, as I stumbled along over the rough boulders.

When at last I reached the shore I was going very slowly, and as often as not with the aid of my hands to relieve my feet which were in rather bad shape. I lay down in the low surf, promptly had a severe cramp in my leg and foot, and went aboard the yacht. By dinner time I was all right again, but no trip on any other island, James, Albemarle or Tower, nor any all-day hunt I have ever made in the high Himalayas, has equalled this for sheer uncertain frightfulness of one step after another.

After dinner we made a quick trip to the mainland again. With some bird lime, John Tee-Van almost at once caught a mockingbird which is still thriving in the Zoological Park. In three hauls of the seine we took many fish, once a pure culture of half-beaks, almost two hundred, varying from one and a half to seven inches in length. The last time we examined the net a beautiful little eel-like blenny was found entangled, which proved to be a new species, *Runula albolinea*.

A little before six o'clock the mosquitoes rose in a grey mist out of the grass, and whenever we stood still for a moment we were the centre of a hazy aura, an almost ectoplasm, from which a thousand little needles pricked us. The insects were small, and carried very little poison with their bites, and of course in this humanless land, there was no danger of malaria.

On other evenings as long as there was light enough for us to see, we never failed to observe a host of dragonflies hawking back and forth, while now and then little yellow-bellied fly-catchers would dive into the mass, their beaks snapping like castanets. This last night, eight mockingbirds made the most of this manna, so limited both temporally and spatially. When at last we pushed off and leaped into the big lifeboat, we thrashed about with branches, nets and coats, endeavouring to clear the cock-pit and not infest the *Noma.*

Climbing to the topmost lookout deck, I saw that the sun had just dipped behind the distant rugged peaks of James. This, our last Indefatigable sunset, was a skyful of purest gold, enamelling with equal glory the scarcely rippled water, while the air was like cool velvet on my face. With my whole heart I hated to leave these no-man's wonderlands.

I turned toward Indefatigable and watched the great central peaks, having shed their purple coat of rain, slip for a brief time into their evening wrap of gorgeousness. Then as the grey-black garb of night began to be drawn about their dread, mysterious shoulders, I shuddered. Looking back now I can appreciate so well the farewell words of my shipwrecked taxi-driver,—"We didn't mind the days so much, we was busy and there was turtles to get and things to see, but the nights got us. Then we wondered could we drink blood in place of water for another day and week—and month—and suppose we got sick like Fred Jeff, and suppose a ship didn't ever come. . . ."

For a few minutes the great searchlight played and flick-ered along the shore over two miles away, cutting a mighty swathe through the dusk. Even this brief minute of illumi-nation was enough to attract a number of tiny moths who, soon after they settled on our canvas, steamed away with us from their natal island. Overside, the tide rushed along with the cool current, and with it was borne a world of delicate beings, some glowing with phosphorescence, others dark and invisible, while others were clearly outlined because they had swallowed uncounted hosts of the still glowing ones. A fish—a shark—a dolphin—struck ten million sparks out of the watery life as they rushed along.

The afterglow shot up in perfect imitation of the glowing flame and light of an active volcano, perhaps prophetic of the end of these peaks as it was of their beginning—peaks so lonely in mid-ocean, with their strange beings of scales and feathers, all now asleep somewhere out in the darkness. Somewhere the sulphur butterfly, the one which brushed against me far inland half-way to those dread peaks, was clinging upside down to a thorn, a Galápagos butterfly—like the volcanoes—asleep.

I went down into the glare of laboratory lights, and did many and various things, mostly ineffectively, with microscope and vials and forceps, paper and ink and words. Hours afterward, a sudden thrill ran through the ship; for the first time in many wonderful days, the *Noma* had awakened and throbbed with life. On the same lofty deck, at midnight, I watched alone. The anchor came up, the yacht slowly circled and gathered speed on her straight course toward Tower Island far to the northward; the nineteen hundred and twenty-third twenty-sixth-of-April had passed.

Daphne Major slipped by, the very ghost of an islet, a mere blotter-out of stars beyond, and not a sound came from all the birds fast asleep,—on the highest rim of the volcano, as well as upon the deep, hidden, unforgettable floor of the dead crater.

So had Orizaba once faded from view, and so the jungle of Borneo and the lights of Rangoon; so had been erased the last silhouette of Kinchinjunga and of Fuji Tonight, with the passing of Indefatigable, there came the faint aromatic scent of Bursera leaves; but whether this, or the perfumed breeze which blows from the camphor groves of Kagoshima, or exciting odours drifting over Hoogly's waters from the Calcutta bazaars, or the scent of white jasmine in a Virginia garden, such memories are eternal, they are the saddest things in the world, and they pass all understanding.

CHAPTER XIV

THE SHIP-WRECKED TAXI-DRIVER

THE day after we reached New York, as I was overseeing the landing of some of the live animals and more delicate specimens from the *Noma*, a Checker Taxicab driver came up to me very much out of breath, and between gasps asked if I were one of the party who had been to the Galápagos. Then he asked if we had gone to Indefatigable.

I said, "What do you know about Indefatigable?"

He replied, "I was ship-wrecked there seventeen years ago."

I took his name and address and told him to come to my studio sometime and tell me his whole story. This he did and it proved to be well worthy of record.

Christiansen was a Dane, nineteen years old at the time of his great adventure, when he shipped as able seaman on the Norwegian bark *Alexander*. The crew consisted of Captain Petersen, the first mate, second mate Morrison, Nelson the cook, and sixteen men. They left New South Wales on November 6th, 1906, with a load of coal for Panama. Captain Petersen had made this voyage several times before and had never been longer than seventy days, but since it is the custom to carry twice as much food as can possibly be consumed, they had supplies on board for one hundred and forty days.

Christiansen's chum was Fred Jeff, a young American

from Tacoma, where his father was a cattle-man. The two boys had been shipmates on three previous voyages. A variety of nationalties were represented in the crew. There was Herman Schlesinger, a young German from Bremerhaven, a Frenchman whose name the taxi-driver had forgotten, the second mate from Glasgow, and an assortment of Scandinavians.

On May 8th the bark was still becalmed in the Pacific. For months there had hardly been breeze enough to stir her canvas and never enough really to fill her sails. For the past three or four weeks the crew had been on half rations, and had done no work. Christiansen said they lay about on deck and yarned, speculated on when the wind would spring up, and told each other the most minute details of their lives. He tells the story so well that his own words cannot be improved upon.

"The eighth day of May the captain called us all forward. 'Boys', he says, 'I want to put the case to you all. As I figure it we're about seven or eight hundred miles off the South American coast now. I don't see no sign of the weather changing, don't look like the wind is ever going to blow. Off here to the northwest of us there's some islands, the Galápagos. I don't know nothing about them, inhabited or not, but I think we can get there in the boats. Here's all the food and water we got left, piled up here in front of me. Now who wants to stay by the ship, and who says leave her and try for the islands?'

"Well, we looked at that pile of supplies and it looked pretty small, so we all decided we'd take a chance on the Galápagos. We didn't know what to do with the bark; nothing aboard to blow her up with, and we knew she'd be a nuisance to shipping, but we lit all her running lights and nailed notices to the cabin door, telling who we was and what had happened and where we was hoping to get. We run up the Norwegian flag and left her.

"The starboard watch went off in the Captain's boat, the port watch with the mate in the other. That made ten men in each boat. I was in the Captain's watch. We had less than a quart of water apiece when we went off. We had sixteen-foot oars and we rowed, four men off and four men on, two hours each, night and day. We rowed and rowed. Sometimes the mate's boat was ahead and sometimes ours. The third night we lost the mate's boat. When it come light we couldn't see him at all, and we never see him again.

"The third day the food give out, and the fourth day the water was gone. We rowed and rowed. We'd dip our hands in the sea and suck the water off our fingers. We knew it was bad for us, but we had to have something. What with the rowing and the salt water our hands got awful bad. I had cracks between my fingers that stayed running sores for months after.

"About the twelfth day we sighted land. The cook seen it first and told us, but we kept saying, 'No, that ain't land, its clouds.' And he says, 'No, its land,' and we'd say 'No, its clouds,' and the Captain sitting in the stern saying, 'Now boys, don't arger.' When it got a little lighter we seen it really was land, and I want to tell you, weak as we was, them oars *bent!*

"Along about noon we got up to it. We found out afterward it was Indefatigable Island. It looked like a tough place to land, no beach but just black rocks and a heavy surf on them. We was so crazy to get ashore we'd have run straight at a cliff, but the Captain kept saying, 'Keep her off, boys, you'll smash her sure if you try to land here.' We rowed along as near the shore as we could and every once in a while we'd get too close and a big wave would lift us nearly onto the rocks and we'd have to row hard to get back in still water. Finally we come to a little sandy beach,—no harbour or shelter, just a little piece of sand about thirty feet long. We didn't wait for the boat to touch, but jumped overboard and rushed up on shore. We remembered to pull the boat up a ways, but we wasn't thinking about anything but water.

"Well, you've been there, so you know the first thing we struck was lava, and nothing but lava. For a while it didn't look like we'd done ourselves much good by coming here, but we scattered and pretty soon we found a little rain-water held in the tops of rocks. There wasn't much in any one place and the spray had come over into it, so it was salty, but we drank off the top, and the salt and scum was mostly at the bottom.

"We spent a long time looking for water. Most of us felt pretty sick as soon as we landed, what with being weak and then drinking that queer-tasting water, so it was quite a while before we started back to the little beach where we left the boat. The cook was ahead of all the rest of us, and we saw him stop and look. Then he put both hands up to his head and stood there. We hurried to catch up with him and when we looked on the beach, say, there wasn't a piece of that boat left three feet long. It must have been dead low tide when we landed and the tide had come up and smashed the boat to bits on the lava. We didn't even see the oars again. Found the compass but of course it was smashed and spoiled. There was two or three pieces of our clothes sticking to a rock, and the empty water-cask with the top stove in.

"Well, there we was. We all sat down and there didn't seem to be much to say. Pretty soon we went off again to find some more little pools of water and then we was so tired we just stretched out, as comfortable as we could for the lava. Eat? No, we didn't seem to be very hungry then. All we thought about was drinking.

"The next day the Captain talked to us. He says, 'Boys, we are in a bad fix, but we got to do the best we can and try to be cheerful and keep up our hopes. This place don't look like a place anybody would live, but maybe up there in the hills there's people. It looks like trees and water might be up there, so we'd better try going inland as soon as we get something to eat and feel a little stronger. I can't help you any. I don't know any more about this place nor what to do than you do. On the ship we was captain and men, here we're all men, and one knows as much as the other.

"We looked around for something to eat. Pretty soon the Captain calls to me from the edge of the beach. He was crippled up pretty bad with rheumatism from the open boat and sleeping on the ground and all, and it was awful hard for him to get around over them rocks. We all went over and looked down and there was a turtle asleep on the sand. The Captain says, 'Who's got a knife?' Four of us had ours. 'That's mighty good eating,' he says, 'Here, Red, you and Jeff get down there and turn that turtle on his back.'

"Well, we hadn't ever seen one of those babies before and we didn't much like tackling him. We kinder walked around him for a while and wondered what he'd do when we took hold, but the Captain kept saying, 'Go on, turn him over, he can't hurt you,' so at last we done it. Then the Old Man yells to us to cut its throat and we did. When the blood spurted out, Fred Jeff got right down and drunk it, and he says to me, 'Boy, that's sweet.' It turned me. I didn't want any. But in a day or two when we got another, I tried it. I was so thirsty I'd a tried anything by that time. And it was sweet and after that we always drunk it.

"We cut off that turtle's under shell and ate the raw meat. There's a thick skin next to the top shell, and when you break it, there's a layer of sweet yellow fat underneath. So we scooped that out with our hands and ate it. We lived on raw turtle meat and blood, and a little brackish water once in a while for months and I guess it was that fat kept us alive. About all you can eat of a turtle raw is his legs and flippers. The shoulders are like hams,—the same shape.

"After we ate, we climbed up on a high place and looked around. The country along shore all looked the same, and there were no sails in sight. We decided we'd strike inland, and see if we couldn't find people. Well, you know how long our shoes lasted on that lava. It wasn't no time before we was about the same as bare-footed. We come back to the shore again. When we were first there, every few days one or two of us would think we'd try to get inland to the hilly country, where we could see the rain falling every day, but we always had to turn back after a mile or two, and after a while we got so we didn't even talk about it any more.

"We all hung round that first little beach for three days. It was a pretty quiet crowd. The cook was kind of an old man.

He had a wife and four children back in Stavanga, and it worked on him. He used to go off and sit by himself and worry about 'em. The second mate, Morrison, had a wife, but it didn't seem to bother him much. And the Captain's wife was dead. All the rest of us was single.

"Morrison was what you might call a regular old sea-dog. He had more push to him than any of the others, too. He wasn't a very young man either. Finally, one day he says, 'Well boys, we got to do something. Nobody'll ever find us here. We got to get moving and see what we can do. Maybe there's an anchorage where a ship might some day come.'

"So we started to work along the coast. We went as close in to where the thorny part begun as we could, and you bet we watched sharp for some sign that there was people on that island. Fred Jeff was feeling real sick all the time, and he stayed back with the Captain and the cook and the others. Morrison and a young Swede and me went on ahead, while the others stayed up on a high place to watch the sea for ships, and to see when turtles come ashore. Me and the Swede was the youngest of the crowd and we was the only ones that could swim. Morrison could swim just a little. After us three had gone as far as we could that day, we'd get up on a high place and hang up a shirt or something on a cactus, and then the others would come along and catch up with us, while we watched for ships and turtles. That way we stayed together. There was only a few nights when we wasn't all together.

"At first we could only get turtles when they came up on the beaches. When the sea was calm they didn't come much, but when it was kind of disturbed, they'd come ashore on a high tide and then the tide would go down and leave them asleep on the sand. Then we'd eat.

"We tried to get some seals but they always went into the water before we could reach 'em. Then we decided they must smell us, so the next one we see, we worked round down wind from him. He was laying asleep with his head resting on a ridge of rock. The Captain was there. He says, 'I don't know how much it takes to put one of them babies to sleep,' but Morrison had a big lump of rock and I had a big piece of turtle-shell. I crept up on all fours real close and then I rose up and hit him a clip on the head with the piece of shell and he rolled over and the Captain stuck a knife in him. We ripped off his skin, and say! it made the swellest shoes! One thickness didn't last long on the lava, so we'd put five or six thicknesses round our feet and punch holes and lace them on with thongs made out of long strips of skin. Our feet looked like ferry-boats, but it sure saved us a lot, for our shoes was all gone by that time. Yes, we tried eating the seal-meat, but it was awful tough and fishy, and we all decided we'd rather have turtle when we could get it.

"One day we was sitting up on the rocks looking down into a little cove and the turtles wouldn't come ashore. They just stayed out there and went up and down, up and down. I says to the Swede, 'Let's see if we can't get one of those boys in the water.' So the Swede walked out on one point of the cove and I went out the other as far as we could, and then we started to swim inshore. There was a turtle between us and the beach and the others walked into the water as deep as they could. Every time the turtle come up we'd splash and scare him in-shore till we got so far we could get a footing and then we grabbed him and turned him over. After that we didn't al-ways have to wait for them to come ashore. We got so we could turn them on their backs in deep water and swim in pushing them ahead of us. Sure it was hard to do, but you can do an awful lot when your food depends on it.

"Sometimes we wouldn't get any turtle for two or three days, but mostly we was pretty lucky. We tried eating some of the big birds like gulls, but they tasted bad raw and made us sick. We used to eat the big black lizards. We grabbed them by the tails and swung their heads against rocks. The tails was about all there was to eat of them. By the time you got the skin off the legs, most of the meat was stuck to it. Turtle was our best bet. We couldn't get any fish. Nobody had a watch or anything to make a burning-glass, so we never had a fire.

"It kept us busy, the Swede and me. We needed a seal almost every day, for our sealskin shoes wore out so fast on that rock, and there was ten men to provide shoes and turtle for. We had a talk with the others after a few days. We says, 'See here, we're willing to do all we can, but it ain't fair for us to do all the work. We being the only ones that can swim, we got to go after the turtles but after we get 'em ashore, you got to kill them and cut them up. That's your part of it.' And they all says sure they will. So that was the way it was divided.

"Nights was the bad time. All day we was pretty busy, and travelling along, climbing over the rocks and watching sharp both on the land and the sea sides of us. Once in a while Morrison and the Swede and I would think we sure could get inland, so we'd start off and pretty soon we'd be back, clothes torn off us, scratched and bloody, and just as wise as we was before. One day we thought we saw the smoke of a steamer, but it was far off on the horizon and it never come any nearer. Nights was the worst. We'd lie there and think about things, and wonder how much longer we could drink blood instead of water; then we'd get up and look at the sea, and think where we was, and suppose a ship didn't ever come. Nights was the time we felt it.

"Fred Jeff got worse and worse. He had dysentery or something, I guess. One day he laid down on a little beach, under a dead tree close to the water. We all stood around, the Captain and all. There wasn't anything we could do. He knew he was done for. 'Boys,' he says, 'I'm done for.' We didn't have anything to dig with. We carried him up off the beach and piled lava rocks over him. He was my shipmate on four voyages.

"Well, we kept on working round the shore. The Captain had six hundred pounds in English gold that he was carrying in a money belt. It was heavy, of course, and it didn't mean much to us then, so we buried it. I can see the place now as plain as if I was looking at it. There was a rocky place back from the shore where the salt water must have seeped up through the ground some way, for there was puddles of it down low between the rocks. There was a ledge about half way down to where the water was, and we pushed the gold back under the big boulders and left it there. I remember the place all right.

"One day we come to an inlet that was too deep to wade across. Afterward we found out if we'd waited for low tide we could have waded, but we didn't. We used to do things like that, I guess, that weren't real sensible, but our minds got kind of fixed on forcing ahead all the time, hoping to come to something better. So we never stopped to think about waiting for the tide. The Swede and I took the others across one at a time. I swum over with the Captain on my back, and the Swede helped the second mate over. This German boy, Herman Schlesinger, thought he could get over alone. We told him to wait, that we was coming back for him, but he thought he could make it. He drowned. We tried to get him but we couldn't. Three days afterward at low tide we got him. There wasn't anything left but his bones; the fish had nibbled them clean. The Swede dived and got his skull and ribs and I got his leg-bones. The tendons still held those together. We carried them up and piled them in the rocks. I suppose they're dust by this time.

"Our clothes was pretty far gone by this time, and sometimes we got sick on raw turtle and bad water. Of course we didn't get much water anyway; just once in a while we'd find a little scummy pool in a rock. But we were kind of taking things as they came. Nights was the worst.

"One day, I guess about two months after we landed, we three, going on ahead, come to a great high cliff that went straight down to the water. It was a bay, and way across on the other side was other cliffs, but up at the head was a smooth beach, and back of that it looked kind of green and nice. There was a little island at the opening of the bay, and lots of seals on it. I looked over the edge and down at the bottom of the cliff was a big man-eating shark. I says to the Swede, 'Boy, I'm not going down to that baby. It's bad enough up here.'

"We wanted to get down to that beach. Somehow it looked good to us. The cliff was too steep to climb along the face of it, so we had to strike back into the country and circle round to it. We waited for the others to catch up, and told them we were going down there, and they could stay there till we found out what it was like. It was awful going. When we got down, what was left of our clothes was in rags, and you couldn't have put your finger on us anywhere where we wasn't torn to pieces, what with thorns and cactus and lava. Some walking!

"When we got down on the beach, the first thing we see was four or five little fireplaces, some tumbled down. You better believe we rushed around looking for the people that built 'em. We found some soggy magazines, dated 1894, and two empty whiskey bottles and two pairs of lady's slippers and a pair of big felt slippers; on a bush was a pair of lady's stockings and a man's shirt. The minute we touched them they fell all to pieces. Rotted.

"And up at the head of the beach we found water, the first real drink of water we'd had in over two months. It wasn't exactly fresh, but it was water and there was plenty of it.

"Then we had to get the others down there. We left Morrison on the beach; he was pretty well done up with the trip down. We told him to stay there, and on our way up we'd come out on the top of the cliff once in a while and wave so he'd know we was all right, and for him to wave back so we'd know he was all right. We didn't know what there might be down there.

"Going back seemed easier than going down. I guess we'd made a sort of a trail. When we got to where the rest of them was, they was all sitting round in a circle. I looked and there was a little bit of smoke. They had a fire. I thought I'd gone crazy.

"'Where did that come from?' I says.

"Well, what do you think? That cook had had on two flannel shirts ever since we left the ship. While we was gone he thought he'd take off his top shirt. You see, if you'll excuse me mentioning it, we'd had plenty of company while we was on that island. Well, the cook started to pull off his top shirt and as he leaned over, something fell on the ground.

"The Captain saw it.

"'What's that?' he says.

FIG. 62

ABLE SEAMAN CHRISTIANSEN AND HIS NEW YORK
TAXI-CAB

FIG. 63

MALE SEA-LION BARKING

Christiansen killed one of these every day for its
hide, to protect his feet upon the lava

FIG. 64

THE NORTH SHORE OF INDEFATIGABLE

The castaways spent over two months in walking half around this island

Fig. 65

HARRISON COVE WITH EDEN IN THE DISTANCE

At the foot of one of these dead, stunted trees, Christiansen's pal, Fred Jeff, died

"The cook looked and he says, 'I dunno.'

"The Captain picked it up and it was a match.

"'Where did that come from?' he says.

"The cook looked and he says, 'I dunno.'

"Then the Captain looked at him, and sticking out under the flap of his second shirt pocket was another match.

"Well, sir, that cook had had a box of matches squashed flat in his underneath shirt pocket for two months and a half, and us living on raw turtle. Can you beat it?

"Some of them matches was damp, and talk about nursing! Baby! We laid them out in a row on a rock in the sun and we built a little wall of rocks around 'em, and then we all sat around and watched them dry. Once in a while someone would reach out and turn one over when he thought one side was done.

"Then we all went down to the beach. We had six turtle legs to carry but we threw away two of them on the way. It was hard going. That was the time we saw a land turtle, the only one we ever did see. She was a pretty big one, about four feet long.

"We found Morrison and we told him about the matches and we made a fire. We scattered and looked all round but we didn't find any more signs of people. The Captain says, 'Well boys, if we're ever found, we'll be found here. I don't believe there's anybody lives inland, or we'd have seen a trail or some signs in all the ways we've been. This is the only harbour we've seen, there's water here, and we might just as well stay here and wait.'

"So we stayed. We built up the fireplaces and cooked our meat, and it sure tasted good that way. It was a good place for turtle, too. A good many used to come in that bay, and now we wasn't moving on all the time, we didn't have to worry much about getting seals for shoes. But it was worse staying in one place than when we was moving. You see, when we was travelling, we had something to think about and somewhere to try to get to, but when we sat down and just waited, it was hard. It worked on us.

"Funny, after we got to that beach, we hardly talked at

all. We'd already told each other everything about ourselves, and now we just sort of grunted and held out our hands when we wanted something, or made some kind of a signal. Seemed like it was too much trouble to talk. I tried to talk to the Captain a couple of times, asking him about what he thought our chances was of getting away, and things like that, but he was queer. I don't mean he was crazy but losing his ship was on his mind, of course, and he was kind of an old man, and had been through a lot. You couldn't talk to him. For hours and hours he'd walk up and down and up and down the beach and once in a while he'd stop and sort of shudder all over, and then start walking again.

"We stayed on that beach about three months. One day we was all eating. I can see us now. I was lying back beside the fire-place with my arms under my head, stretched out. All I had on was a pair of pants; one leg was torn off at the knee and the other was even shorter than that. My whiskers and my hair was long, and I was grease all over. Of course we was all like that.

"All of a sudden this young Dane started to yell, 'Ship! Ship! Ship!' He was young and didn't seem to realize how we was all fixed, and he'd done that a couple of times before, thinking he was funny. So this time when he done it, Morrison jumped up and gave him a belt over the head that knocked him down. And then the old cook started to yell, 'Ship! Ship! Ship!'

"We all looked up, and there coming round the east point of the bay was a sail. For a minute we just sat there. Then we jumped up and began to yell. We expected she'd come about and into the bay, but she kept on going past. Say, we nearly went crazy. We ran up and down and screamed and waved and cried, but she kept right on and went out of sight behind that island in the middle of the bay. Seemed like she was out of sight behind it for an hour, and we put things on the fire to make smoke, and we tried to think of some way we could make a big noise so she'd hear us way out there. Of course there wasn't a thing we could do. And then she come out from behind the island, and stood straight across for the other point. We thought she was gone. I remember the cook was rolling in the sand and crying, and I guess none of us was much better. But the captain of that sloop knew what he was doing. He'd been in that bay before and he knew how to make it. Just as she had almost passed the western point, she come about and made a long tack into the bay. Then she come round on the other tack, and before she could make the third one, the Swede and me was in the water swimming out to her. We swum quite a long ways before they got a boat over the side.

"What do you think was the first thing we says to that captain? Say, that's right! How'd you know? Sure we asks him for tobacco. He gives us two cigars but we didn't stop to smoke. We just crammed 'em in our mouths and pretty near eat 'em.

"We almost tore that captain to pieces, we was so glad to see him. When he got ashore, the first thing Captain Petersen says to him is, 'Did you ever hear anything of our other boat with the mate and the rest of the crew?'

"And this old captain says, 'Why, Lord bless you, Petersen, they're home long ago!'

"It seems they was picked up by a fisherman about sixty miles off Guayaquil the same day that we got ashore on Indefatigable. One current took us northwest and another carried them northeast. The limit, wasn't it?

"The mate told the authorities about us and an Ecuadorian gunboat went out to look for us. They found the ship all right. She was wrecked on the south end of Albemarle, down by the bow, her stern straight up, with the Norwegian flag still flying. So they reported back the wreck found, but no survivors. Then a British gunboat went out from somewhere and made the same report,—wrecked bark but no us. There was a relative of Captain Petersen's working down in Chile in the saltpetre mines, and he heard about it, so he hired this old German captain to look for us. He had cruised round the Galápagos a good deal and knew all the places where his small ship could go and a gunboat couldn't. He'd cruised round three or four islands, and this day he was asleep in his cabin when one of the black deck-hands rushed in, grabbed him by the collar and dragged him up on deck. This black man was so excited he couldn't speak but he hung on to the skipper with one hand and kept pointing and pointing, and there was our smoke.

"Well, that German skipper planned to sit down and hear all we had to say, and spend the night there and start for Guayaquil the next morning, and I want to tell you, it's God's truth, we got down on our knees to that man to beg him to start *then*. He didn't want to, but he did.

"The gold? Why, we never even thought of it till we was almost to port. I suppose it's still there, don't you? Not very likely anybody would find it; one chance in a million, maybe. I sure would like to go back some day and look for it. I've heard since that there may be pirates' gold on some of those islands, but I know there's sailors' gold there.

"When we got back to Guayaquil the British consul took us on; he was attending to Norway's business in Ecuador. You know what we looked like. You ought to have seen us when they trailed us into the best hotel in town, with all the dressed-up ladies and gentlemen standing around. Some of them squealed and ran away from us, and some squealed and ran at us. They had a couple of policemen to keep people away and wouldn't let anyone speak to us till we'd had a bath. Then they gave us each a suit of overalls and sent us up to Panama on a cattle-boat. That was a swell trip. Cattle on one side of the deck, and cattle on the other side, and us sleeping in the middle on straw.

"When we got to Panama, the Norwegian consul was going to give us half pay for the time we was on that island. He didn't, though. We got our full pay. They sent us over to Colon, and was going to ship us up to Galveston on another cattle-boat. By that time I was fed right up to the neck with the Norwegians.

"'Here,' I says, 'give me my pay. Here's where I take the air.'

"And I went to work dredging the Panama Canal. Then I got fever and beat it for the States."

CHAPTER XV

SOME of the most wonderful things in life have come to me as by-products, obliquely, casually. Tower Island is the last, and very far from the least of these. We were headed toward the Galápagos for the second time, and on the seventeenth of April realized that if we kept on at the present rate of speed we should reach Indefatigable at dusk on the following day and have to lie off shore all night. So at three o'clock we shifted our direction slightly to the northwest and headed for Tower, of which we knew nothing, save that it was the most northeasterly island of the archipelago. The weather was perfect with the usual long, low swell of the Pacific.

An hour later we passed an enormous school of dolphins travelling steadily almost in our direction, southwest, but more slowly. They must have been going about eight knots, all in a line, ten or fifteen abreast, and half a mile long, probably three thousand individuals. The glass showed a nearly solid mass of churning backs, so there were as many again beneath the surface. Now and then one would shoot up, turn over several times, and send up a shower of spray in which a rainbow was born, to die in the same second. They were migrating, perhaps because of some change in current and temperature, and consequently in food. A single frigate-bird hovered over them.

The next morning, birds were seen more frequently, and several unconnected banks of cloud on the horizon became more and more suspicious, until out of the one ahead appeared a low, dark line of land, topped with trees. Soon the whole long extent of Tower Island was visible, and at three o'clock we were within a mile of the shore. The surf was so high along the entire length that landing on this side was unthinkable and we steamed slowly around the northern end. By this time the air was full of birds and over the wireless a perfect maze of man-o'-war birds and boobies milled around, each of the former giving the tip of the mast a sound peck as the bird passed it. There were at least five hundred man-o'-wars all trying to alight on the flimsy cross-piece which held the wireless. On each side of the yacht hundreds of red-footed, brown boobies soared just out of arm's reach, and shearwaters skimmed close to the surface. Giant sharks passed and repassed, and now and then a devil-fish leaped out, and fell with the impact of a barn door dropped on the surface.

A low promontory seemed a bit more sheltered, and we put off in small boats. As we neared the island the cliffs rose higher and higher, and both ledges and cliff top were seen to be covered with nesting birds. Over the centre of the island a great cloud of birds reached upward as high as they could be seen, all soaring slowly and gracefully in the same direction. Although this was on an infinitely larger scale, it reminded me of the cylinder of soaring pelicans in Florida, or occasional vast flocks of vultures over a tropical city.

The promontory looked very unlike the view seen through glasses from the yacht. The waves did not break over it, but they surged ten feet up its jagged front, hung a moment, then sucked away sideways and under the rocks, and finally broke, in such a manner that a misjudgment of a few inches would mark the difference between a safe landing and a crushed boat. I could have made the leap ashore, but the return would have been very risky both for me and the boat, so we kept on around the island. The *Noma*, two miles out, saw our movements and followed parallel with us. On and on we went, the line of cliffs continuing high and inaccessible. The air overhead was filled with gulls, frigate-birds, boobies, tropic-birds, shearwaters, petrels and terns. Never have I seen so many birds in so limited an area.

Once more we pulled in, this time toward a sandy beach, but while two great waves passed beneath, we backed barely in time to escape the crest of a third, at least six feet high, which would have smashed us into chips.

The sun was low, the first hint of dusk approaching, and there, within stone's throw, was the wonderful island, the summit dotted with brown and white boobies and with the scarlet throat pouches of nesting man-o'-war birds. I was just about to give up and go back to the ship, when, having been three-fourths of the way around, we saw a slight break in the cliffs. This indentation grew to an inlet, with green water, then a cove, a sound, and when we were actually in the mouth, it opened out into the most magnificent bay I have seen in many years. On the map Tower is a rough, solid rectangle, about two by five miles in area. Here in the very heart of the southern portion was a bay, over a mile across, with a very narrow opening, almost hidden from passing ships. There seemed but one name for such a find as this, so then and there I called it Darwin Bay. (Fig. 67). We rowed across it as rapidly as possible, and landed with only a half hour of daylight left. But this was sufficient to show how interesting was the island and its life, and I made up my mind to cut short my visit to Indefatigable and Seymour, and spend three days here on the return trip. My description of the island begins with our return narrative.

Nine days later, on April twenty-seventh, the first light of dawn showed Tower plain ahead. To the northwest Abingdon had a curious warm light over her as of fire, a bit of the after-glow-like sunrise. As a matter of fact, the sun never actually came out but remained pleasantly cooled by a cloud layer all the morning, which concentrated in a short, heavy shower about noon, followed by an afternoon of glorious sunshine.

A few minutes after six, standing forward close to the star-board rail, several of us saw a very large ray or devil-fish struck a glancing blow by the yacht, and go rushing off along the surface. Realizing what a splendid chance this was for definite observation, I concentrated on the colour and pattern, and when it vanished, I sketched it at once, the details being verified by my companions (Fig. 81).

From tip to tip of wings it was at least ten feet, of somewhat the usual manta or devil-fish shape, except that the wings were not noticeably concave behind, and the lateral angles were not acute. The cephalic horn-like structures were con-spicuous and more straight than incurved. In general the back was dark brown, faintly mottled, while the most con-spicuous character was a pair of broad, pure white bands extending halfway down the back from each side of the head. The wing tips also shaded abruptly into pure white.

From dawn until seven o'clock, dozens and dozens of flocks of boobies of two species came flying from the island directly toward and past us, southwest, in the general direction of James. Remembering that the sea-birds of Daphne all flew due north every day, I realized that the intersection of the Tower and Daphne birds probably indicated an area of un-usually good fishing ground, almost exactly on the equator and 90° 30′ west longitude. The same instinct which impels the Florida pelicans to go forty or fifty miles to their fishing grounds takes these birds far out to sea, while, as we were to

FIG. 66

MODEL OF TOWER ISLAND, SHOWING DARWIN BAY

FIG. 67

DARWIN BAY LOOKING SOUTH FROM THE LANDING BEACH

Fig. 68

Frigate-bird

Fregata aquila (Linné)

Male on Nest with the brilliant Scarlet Throat Pouch fully distended

Fig. 69

Green-footed Booby

Sula dactylatra granti Rothschild

The rarest of the Galápagos Boobies

see later, individuals could find an abundance of food close at hand near their nests.

The boobies flew steadily and rapidly, usually close to the water, and in small flocks of eight to forty. Their flight, like that of the pelicans, was the wonderful rhythmic *flap-flap-flap-flap soarrrrrrrrrr, flap-flap-flap-flap soarrrrrrrrrr*. By far the greater number were red-footed, and these showed two colour phases, brown and white. The latter seemed to be about one to forty of the former. With them were a few of the larger, white, green-footed species. At a distance I could not tell the difference between these and the white red-foots, but closer, the size and the colours of the face, beak and feet were good characters. The green-foots were usually single birds, either flying with a flock or alone. Once only I saw three—two adults and a bird of the year—flying together.

We approached Tower from the southwest, with the narrow opening of Darwin Bay plainly visible. Had we not known it was there, even at a mile distant we might not have suspected its magnitude. The scarlet throat pouches of hundreds of frigate-birds were distinct half a mile away, looking like an abundance of giant red flowers scattered along the top of the cliff.

Half a mile off shore our longest wire found no bottom, then suddenly we struck twenty-five, twenty-one, sixteen and twelve fathoms. The yacht backed slowly away, and we sent First Officer Healy on to make soundings. Midway between the two arms of the bay he reported four and a half fathoms, but to the east this deepened considerably. This was the submerged wall of the great crater which formed the bay.

We all piled into boats and went ashore, and an hour later, when we looked back, the *Noma* was well within the entrance, only a quarter of a mile away, steaming slowly to an anchorage, close under the northern cliffs in seventeen fathoms of water.

The bay was roughly round, with the narrow eight-fathom navigable part of the opening to the southwest. On all sides it was surrounded by precipitous cliffs fifty to one hundred feet high, and on the northern or main island side these took the form of successive terraces of lava, each broken sharply off. Only low rollers entered the bay and broke so gently against the cliffs that we could land almost anywhere on isolated rocks or tidal platforms. Hidden behind a low reef in the extreme inner, northeast portion, was a charming beach of pure white sand, a cove within a bay, and this gave access to a quiet, tidal lagoon and a great expanse of low, level land. Here too, by surmounting a series of boulders, which in many places were almost as regular as the steps of Cheops, one could easily reach the summit of the cliffs. The colour scheme, as usual, was astonishing—the clear emerald water, white sand, black lava, and scattered green mottlings of cactus and other plants, while everywhere glowed bright yellow flowers of a half dozen species —*Opuntia, Cordia, Gossypium, Mentzelia* and others.

This single page would easily contain all the biological notes ever made upon Tower Island. Few parties have even approached it, and still fewer landed. A chest with more than three hundred thousand dollars' worth of gold was buried and less than a score of years ago was salvaged from one of the cliffs of Tower. Twenty-four species of birds, twenty-one plants, one lizard, two ants, and a grasshopper and a cricket have been recorded.

The wonder of the island was the birds. The sky was a closely-woven mesh of flying forms, a frigate-bird warp with a woof of boobies and gulls. The bushes and the rocks and the sand were alive with them, sitting upon their nests and eggs.

The dominant forms were the man-o'-war or frigate-birds. During the day, or the morning at least, it seemed to be the custom for the males to sit, while their mates were off fishing.

When we first approached the island, hundreds of female frigate-birds soared over our wireless, not a male among them. On the other hand it may be that curiosity is a characteristic of the female alone.

The inter-relationships of the various species were intricate and of great interest. I could have spent weeks in the nesting colony just back of the sand beach without exhausting the dramatic happenings constantly going on. The ground was covered everywhere with a dense growth of *Mentzelia aspera*, which, although it belongs to a nettle family, did not sting. It was two feet high, and produced a maze of crooked, interlocked branches topped by a veneer of dusty-green leaves. The hairs which covered stems, twigs and foliage were recurved and gummy, so that they stuck tightly. If one sat down on a section of the springy, mattress-like growth, the entire foliage came away upon one's clothing, and in walking through this vegetable quicksand my skin and garments soon became plastered with the leaves. The flowers were yellow, but small and inconspicuous. This was the foundation for the homes of many hundreds of frigate-birds and red-footed boobies, nesting close together, side by side. But propinquity did not of necessity mean good fellowship, as I found after some minutes of observation.

A few concrete notes, covering a half hour, on my first day on Tower will illustrate the conditions. I penetrated deep into the colony, stepping high over the *Mentzelia*, and sinking through to the ground at each step. I found a composite armchair of lava and sand, and settled down. There were eight nests within twenty feet, three frigates and five red-footed boobies. Underneath the maze of branches, fork-tailed gulls and occasional doves crept about, while now and then a mockingbird ran toward me and peered curiously, singing to itself as it wondered.

Why the boobies do not choose the ground, as the doves and gulls and their cousins the blue-foots, I do not know. Half their troubles in life would disappear if they did not insist in building a very poor nest, a foot or two above the sand. As I shall show, there is more excuse for the frigate-birds.

Within reaching distance from my seat a male frigate-bird sat patiently on his rough stick nest. By putting my head close to the ground and looking up I could see that there was no egg beneath him. His plumage was dull brown with a mantle of glossy green hackles. Eyes, beak and feet were dull, but out of this sombreness, like fire out of lava, billowed the burning scarlet of the enormous breast pouch. When distended with air this was like a huge bladder, completely hiding the bird. Its distension was not dependent upon conscious muscular action for I saw birds quite sound asleep, with their beaks resting upon the top of this balloon as if on a pneumatic pillow. They even soared with it blown up, although in the wind they were put to constant effort in balancing, compared with their pouchless fellows. The colour is at its best when the sun is beyond the bird and shining through the skin tissue—the scarlet being intensified, and colouring by reflection both the leaves of the plant and the eye and feathers of the bird itself.

I reached over and stroked it with but slight protest, the bird turning away from me, but not snapping or pecking. Taking the tip of one great wing in my hand I raised it up as high as I could reach. The bird spread the other and, lightly as thistle-down, lifted and drifted away. Its ascent could not have been more effortless had the pouch been filled with hydrogen.

I sank back in my ground chair, and instantly there came a metallic twang of pinions—a loud *wonk ! wonk !*—and another frigate-bird swooped, caught up a twig, and as a polo player at full gallop swings at a ball, so the bird reached, plucked, and was off. Another and another followed, and before the owner returned a half dozen sticks had been purloined by its neighbours. Down on the rumpled nest sank the first bird and began rearranging the ruins.

Then another emotion obsessed him; he bent his head back until it sank between his shoulders, the red balloon projecting straight upward, and the long angular wings spread flat over the surrounding bushes. The entire body rolled from side to side, as if in agony, while the apparently dying bird gave vent to a remarkably sweet series of notes, as liquid as the distant cry of a loon, as resonant as that of an owl. In our human, inadequate, verbal vocality, I can only record it as *kew-kew-kew-kew-kew-kew!* In a higher tone the female answered him from the sky, *oo-oo-oo-oo-oo-oo!*

She now descended in a narrow spiral, so small in diameter that the end of the left wing seemed almost a stationary point. For a few minutes, the birds sat close together, going through various forms of dying ecstasies. She had no red throat pouch, but, instead, a breast of solid white feathers. Again the male let the breeze lift him, and I watched him rise without a single flap or movement of wings or feathers. Up and up he rose until the red became a faint spot of colour, then with half-closed wings he dropped like a plummet and was lost in a crowd of half a dozen of his fellows who were mobbing a brown booby. The unfortunate bird had broken off a ridiculously small bit of twig not over three inches in length, but clung to it tenaciously while attacked from rear and sides. At last a frigate-bird met him full head on, and to avoid a collision both birds banked sharply upward and for a moment hung suspended, feet together. The booby gave up and dropped the twig, another assailant saw it, and before it reached the ground had dived and caught it. To my surprise I found it was my particular bird, who returned and joined his mate. Now followed as much celebration as married couples of another kind would bestow on a lifted mortgage. After each had fingered or rather billed the addition, and had *kewed !* his or *oood !* her appreciation, the negligible plunder was pushed down into the substance of the frail platform, and promptly fell through to the sand beneath. Fifteen minutes later, when watching a pair of boobies near by, I glanced back just in time to see a mockingbird skulking along the ground. It seized the self-same twig, and with considerable effort rose with it to its round nest ín a neighbouring cactus. This ended the tale of one twig.

I noticed that about one frigate-bird in every five was crouched down low on its nest, with its balloon punctured, or at least shrunken to an unlovely dark crimson mass of folded skin. I investigated and found that these were the males who had begun the drab cares of domestic life in earnest. Beneath each was a single white egg. No more for him were the scarlet joys of puffed up pouches, no more rollicking looning to greet the returning mate. Now when she came back, he shifted over, gave a perfunctory rattle, and settled down again.

I noticed another thing which may or may not have a very interesting interpretation. In three instances, two with eggs and one with a young chick, when the parent left the nest, there was no attempt made on the part of the frigates to steal twigs. Yet on all sides, wholesale robberies of empty nests were going on. Observers of colonies in other parts of the world report that not only are the nests thus stolen, but the young birds snapped up and devoured by unnatural neighbours of the same species. Whether my three examples of immunity represented an honour-among-thieves characteristic of this Tower colony I cannot say. I liked to think it did, just as females with young are often respected by members of the herd or flock. At any rate, I could tell by a glance over the colony how many eggs there were compared with empty nests, using the formula,—

inflated pouch = empty nest,

deflated pouch = egg in the nest.

Toward noon many of the male frigates were joined by their mates. In place of the scarlet pouch this sex showed a front of white plumage, and often there was a hint of immaturity in a wash of rufus over these feathers. Around her eyes was a broad, conspicuous band of pink skin, a shade of pink which fairly screamed aloud at the scarlet of her husband's waistcoat, but love in the Galápagos, as elsewhere, was colour-blind. An unexpected phase of life of these singular creatures was the presence of last year's birds. These were easily distinguished by their white head and neck. As in the plumage of the females, there seemed to be a possibility of acquisition of absolutely clear colour, and the pure white head probably represented the perfected juvenile plumage. But there were all sorts of intermediate conditions, from a faint wash to a deep tinge of rufus over the light areas.

In at least three-fourths of all the nesting frigates, a solitary youngster in this plumage sat perched close to the new home of his parents. He fed apart from them, and usually soared with fellows of his own age, but when he alighted, as I proved again and again, the same individual always perched as closely as possible to one certain pair of old birds. They would not permit him too near the nest, and they never showed any hint of affection or even recognition, except that no third adult or other young frigate was permitted even this intimacy of association. Lonesomely he perched, hunched up, for hours, sometimes all day, content to be near his parents, who now were absorbed in one another, in a renewed courtship, or in a new egg or chick. These juveniles appealed to me as a pitiful sight,—a hobble-de-hoy age, too young to mate and found a home, too old to claim parental affection, just waiting and watching.

The relationship of the frigate-birds and the red-footed boobies was invariably that of robber and victim, parasite and host. It would be interesting to know which first established the colony; probably the boobies, followed by the well-named man-o'-war birds. The nests of the two were thoroughly inter-mingled, and every trip of the boobies was made through a gauntlet of enemies. Yet the beak of the booby was a much more effective weapon than that of the frigate-bird. It could draw blood or make a serious stab at an approaching hand, and I once saw one strike a full-grown frigate-bird of the year and kill it instantly. But with numbers roughing it in mid-air, the booby had little chance, and usually lost its twig, and at least two or three of the fish it had captured, lucky if it got through with a single fish. The rejoicing of two boobies when this was once accomplished could be heard over all the colony din. Even with its pouch full of fish, and a green-leaved twig in its beak, it could scream its rage loudly at being attacked. It turned and twisted and performed marvels of aircraft, but the frigates were like its shadows and anticipated every effort.

The boobies were of a beautiful gull brown with much made-up bill and face. When one of them looked down its long beak at me it reminded me more than anything else of a circus clown in full regalia. The bill was greenish-yellow shading into blue at the tip, the base of the bill and narrow forehead pink, set off by jet black pigment behind. The skin around the eye was bright blue grey, the eye itself cadmium-yellow, framed by eyelids of clear forget-me-not blue. When the bright red feet were added to this, the harlequin effect was most striking. And with it all, the bird wore an air of anxious sobriety which heightened the bizarreness of the colour scheme.

The voices of the boobies were harsh, a series of raucous squawks, like a whirling rattle. The bird on the nest greeted his mate with an outcry registering joy, which for a moment outsounded the whole colony.

These red-footed boobies showed two adult colour phases —one brown, one white. Half a mile away, in the interior of the island, I came across a level area bounded by steep lava terraces. Here was a pure culture of the white phase of redfoots, fifteen pairs altogether, sitting on their eggs, both sexes being present in the middle of the day. Along the edge of this area, on the top of the terraces at least on two sides, were four nests with one brown and one white bird sitting, and two other nests with no birds present. Beyond this zone were nothing but brown birds, except far away along the shore where an occasional pied pair was seen. This inland, local concentration of a dichromatic phase, which as far as we know has no basis in age, sex or season, is particularly significant. It looks like the beginning of a real segregation, almost specific, arising from a mutation which to our coarse appraisal seems to have no selection value.

Although boobies and frigate-birds have been found nesting together on islands in many parts of the tropics, and much has been written of their habits, yet no one has ever made anything like an intensive study of them. We had no time to do this, but every observation showed how interesting it would be. The frigates were essentially birds of the air—almost as much so as hummingbirds. Their wings were astonishingly long, very narrow and strong in proportion to their size and weight. Where a gull is provided with thirty flight feathers, a frigatebird has forty.

It is interesting to compare average male frigate-birds and red-footed boobies. Both may weigh two and a half pounds and the actual length of the head, neck and body is much the same in the two birds. The other dimensions are of considerable significance. If we consider each of the dimensions of the frigate-bird as 100%, the booby shows the following relative proportions:

Weight, 100%	Extent, 57%
Total length, 76%	Tail, 42%
Culmen, 77%	Tarsus, 128%
Wing, 60%	Middle toe, 120%

Thus, though the birds are of equal weight, the booby is far smaller in length, beak, wing, tail, and extent, but well ahead in development of feet and toes. The shorter, sharp bill of the booby is adapted for grasping fish at the completion of a terrific dive, while the hooked beak of the frigate-bird is rather for daintily picking up fish from near the surface, or catching them in mid-air when dropped by the persecuted boobies. With vigorous kicking of the feet the boobies can easily take flight from the water, while only a cliff edge, or an elevation of a few feet in a bush enables the frigates to rise in flight.

As compensation for the 140% advantage of wing of the frigates, we must count the weak, almost useless legs and feet. These birds must nest above the ground, or else in a period of protracted calm, they might starve to death before they could get on the wing.

I found the most decided difference between the young, newly-hatched chicks of the various sea-birds. The young of the red and of the blue-footed boobies might be indistinguishable as far as external characters went. Held on the hand, however, and encouraged to move, there was at once apparent very significant distinctions in activity. The chicks of the red-footed booby and of the frigate-bird climbed with astonishing strength and success from finger to finger and hand to hand, hooking the bill and neck up over a finger, and reaching out

FIG. 70

RED-FOOTED BOOBY

Sula piscatrix websteri (Rothschild)

A Tree-nesting Booby on Tower Island

FIG. 71

TOWER ISLAND COLONY OF BOOBIES AND FRIGATE-BIRDS

FIG. 72

GALÁPAGOS DOVE

Nesopelia galapagænsis (Gould)

Female on the lookout above her nest, Tower Island

FIG. 73

NEST OF DOVE, WITH THE HALF-GROWN SQUABS ON THE GROUND, UNDER A SLAB OF LAVA

with wings as well as feet. But chicks of equal age of blue-footed boobies and tropic-birds merely squirmed and tried to push down or along on a level toward any available shadow. This reflected perfectly the habits of nidification of the species, the first two nesting always on bushes or trees, the two latter on the ground or in shallow crevices.

The surprising, precociously developed, negatively geotropic activity was probably a very important factor in keeping the young red-foots and frigates on their flimsy nests, or enabling them to climb up again, if accidently fallen to a lower twig.

I rose to my feet at last, to the great interest but not the fear of all the birds nearby, and walking westward along the shore I climbed slowly up the bouldered cliffs to a wonderful look-out, the entrance to an ideal pirate's cave, with all the great expanse of Darwin Bay spread out before me. I lay flat on a gently sloping, shadowed slab of lava and watched another hour of Tower's life. In the distance was the *Noma*, with her wireless cross-pieces covered with birds, a booby on each mast top and a mist of frigates, dense as gnats, revolving slowly overhead.

A tide pool near the base of my cliff was half emptied and along its shrinking rim stalked immature yellow-crowned night herons and lava gulls, eagerly snatching at stranding fish.

A pair of little blue-eyed doves came trotting along and pattered over my shoes before they sensed me as a living creature. With a crackle of wings they were off, but as if drawn, magnet-wise, they swung around to a neighbouring rock and, with bobbing heads, put their whole souls into a scrutiny of me. I sat up and the world-old tragedy of fear began. I did not understand. This was Galápagos, where fear had not entered and yet here the female was using all her wiles to tempt me away. She threw herself prone upon her breast and flapped helplessly, moving along by spasmodic jerks of both feet at once. Her head swayed from side to side, her beak opened as if in dying gasps, and she uttered low, undovelike sounds, *wow ! wow !* almost under her breath. Ceasing for a moment, she joined her mate and both crept about like mice, over and under the jagged lava.

I watched for the focus of their efforts, and finally, six feet away I lifted a small piece of clinker. There was the home of the Galápagos dove, a few wisps of grass stems, a twig or two, forming a vague mat on which squatted a pair of squabs, fat and scantily downed in white. Although dazed by the flood of light which devastated the darkness of their little lava cave, they showed no fear: this was only the latest incident in the unknown out-of-door life which was ahead of them. The mother dove fairly brushed against my hand in her appeal for her precious twins, so I lowered the lava slab and went back to my couch. She took this as good faith, and no longer played her litle deception, but gave me one last glance and slipped down upon her babies.

During the first day I saw a half dozen of these doves here and there, but not until we began to search carefully did I realize how numerous they were. We found eight nests within a hundred yards of our landing beach, four with eggs, and four with young birds.

To the left, beneath my look-out, was a small colony of seven pairs of fork-tailed gulls, standing and preening their feathers, or now and then floating gently up from their harmoniously coloured eggs. It was a study in contrasts,—the gnarled and twisted ledge of sharp-pointed lava and tortured, firewoven filaments of stone, run together in a maze of cruel clinker, and then, from the heart of this, a gull, the softest, gentlest being in the world, to rise quietly, and leave behind it a grey chick clad in plumage of thistle-down.

TOWER—AN ISLAND SANCTUARY

Between my present roost and the bay was a bad-lands of miniature canyons and reefs and slips and faults of lava, jagged pinnacles and diminutive mountain ranges alternating with pools and wells of clearest emerald sea-water. This water rose and fell through mysterious subterranean channels, and though far from the shore, held fish of the most wonderful colours and remarkable variety. These were down near the lower levels. High up on the cliff's foot were ancient pools, dark and scum-covered, in tanks and troughs filled only by the waves of fiercest storms. And here were further proofs of the great tempests—barrels, boards, carved wooden posts, wicker baskets, pieces of chests, cabin doors, and companion ladders—wreckage of old and long-forgotten ships.

The lower pools gleamed in the sunlight like precious stones set in dull platinum. The eye could make of them at will a mirror of perfect reflection, or a medium all but clear as the air itself. Thinking mirror, arresting the focus of one's glance on the surface, there was no need to raise the eyes to the sky to tell every faintest cloud, every missing feather in the wings of scarlet-throated frigate, or booby or graceful tropic-bird. Then when one thought farther down, the clouds melted away, the birds became unfocussed motes, and there swam into ken angel-fish, blue and purple, with slashes of white over their shoulders, zebra fish with awning bands of brillance, others with yellow tails and scarlet heads, and now and then one which beggared description, a swimming prism, a finnv spectrum (Plate V).

I watched several of the party catching these fish with bits of dough for bait. One was pulled out all green and scarlet and violet, with spots of yellow gold and fins of blue and orange, and laid out on a slab of lava to be painted. Within the period of a minute it had turned a dark sooty black, but when placed in a pail of water, every tint and shade and hue returned as if a black veil had been withdrawn. As they fished, a mocking-bird flew and ran about, stealing bits of dough from their box and their hooks and paying for it between swallows with short bursts of melody.

From behind a crag of grey-blue tufa, came one of the girls, swimming lazily, surrounded by all the marvellous fish, with herons and gulls watching her from the nearby lava, and frigates dipping low in flight to see what new fellow islander this was. It was a surprise to realize that she was a mere human, and not what the pool demanded—a mermaid.

Wherever I looked, strange things took form. A yard out of reach there was a great flat mat of white-flowered *Convolvulus*—an island morning-glory which in leaf and blossom might have clung to any New England trellis. But over it, casting a mist of tendrils everywhere, was something very different, at first glance a maze of dried grass, yet all tender and alive. Its flowers also were white, but very tiny, cup-shaped and extremely sweet of odor. (As I scrawled this last word a mockingbird alighted on my elbow and shadowed my paper with his little form.)

The curious dead and alive plant was *Cuscuta*, a native dodder or hell-bind, a parasite belonging to the same family as its victim, the morning-glory, whose life it was draining through a myriad suckers. The unnatural plant—leafless, rootless, wholly without green chlorophyll—seemed worthy of this unreal world of Tower and I pictured in my mind the sprouting of the young seedling—as strange a proceeding as exists in the whole realm of plants. Out from the round seed grows a club-shaped end which thrusts itself deep down into the ground, not from any normal reason of root or tuber for nourishment or moisture, but simply as a steadying point; it might as well be a motile telegraph pole or fence post. Then the opposite end lengthens, twists about, feels here and there, disregarding solid bodies but with leech-like certainty grasping any growing bit of vegetation. Up this it twines, the weird plant growing from the middle, not from either end, and using only the cellular reserve food within its mass, provided by its parent. A speeded-up motion picture of the young plant at this time would seem to show it actually climbing, its posterior end leaving the ground and the leafless, branchless tendril become to all appearances an active scansorial worm. But the actuality is not quite as revolutionary as this, for the apparent movement is due to the progressive death, shrivelling and disappearance of the posterior end, while the middle still provides the means for anterior growth. If it has by ill luck grasped a dead stem, it relaxes in time and falls to the ground. All growth ceases, but for many weeks the tendril keeps a dull glimmer of life, and if within this period some tender plant sprouts near by, the last spark of vitality of the *Cuscuta* is turned into movement. If it once reaches such a sprouting seedling, a tremendous flying wedge of cells presses against the tissues of the stem, breaks through and with the first draught of alien sap derives renewed life, strength and safety.

My reverie was broken by a tiny black wasp which appeared near my face, lugging along a spider quite as large as itself. With all my skill I strove to net him, but the gauze caught on a half dozen needles of lava and the little hymenopteron flew off. No wasp had as yet been recorded from Tower and only one of this group from any of the islands. I always accept such an occurrence as proof positive that some day I shall return and certainly capture my little wasp.

I walked a mile along shore, occasionally on the bay itself, or skirting inland estuaries. Every hundred feet the entire scenery changed, first a mass of tumbled lava boulders, then a curved cove with a windrow twenty feet high of a pure culture of wave-pounded, short stubs of coral, a very rare sight in these volanic islands. Next a jungle of fifty-foot mangroves, with a score of boobies' nests high above my head; beyond lay a sheltered inlet of white sand with a dozen sea-lions basking in the shallows. All were blissfully upside down, with flippers stuck helplessly straight out of the water.

I skirted some shallow water bounded by small, smooth rocks and was attacked for the first time on the islands. As I leaned over to examine a great sea-snail, something hurtled out of the water, a full foot into the air, twisting and turning near my face. I leaped back startled and saw that my assailant was a moray eel, not more than two feet in length, dark olive brown in colour, minutely dotted with white. I threw a stone at it, and it leaped and leaped again with open mouth. I walked along and it followed, darting from pool to pool and springing whenever near enough. Never have I seen such vicious ferocity, such astounding courage in attacking so relatively enormous an opponent. For a moment the old Spanish name for Tower seemed appropriate—*Quita Sueño*—Nightmare Island.

After the moray had disappeared, I watched the tide-pools and found them battle-fields of hosts of hermit crabs. At first glance all the mollusks of the coast seemed to have gone mad, and were leaping and running about at an unheard-of molluscan rate of speed. Within were usurping crabs, and dozens were fighting with each other, antennæ and claws being chopped off, crabs hauled bodily out of their shelters, and a constant, forced exchange of houses taking place. Peace and kindliness had no place in the Tower tide-pools.

On the lava reefs Amblyrhynchus lizards were common, but on all the coast I never saw a large one. All were less than fifteen inches in length, most of them under a foot,— apparently an incipient, insular, dwarfed race.

My last memory of Tower is of a long walk into the heart of the island, where, a mile away from the shore, all thought of the sea passed, except when I saw the webbed feet of the boobies which nested everywhere. In a little heated amphitheatre I rested in the shade of a twenty-foot *Cordia,* which burned with the brillance of its hundreds of big yellow trumpet blossoms. The heat, the flowers, the blue sky with scudding, white clouds, the distant song of a mockingbird, all these were as they should be, but over all lay a film of magic, of strange lack. This was not a normal country, it was something more and less than other tropical lands. I looked at the leaves and they were all perfect. No caterpillar had bitten them, no carpenter bee had cut circles, no leaf-miner had scrawled its hieroglyphics across their surfaces. I listened; not a hum of bee or fly came from the host of flowers, not a flitting wing of butterfly or dart of wasp, or metallic boom of flying beetle. There was absolute silence in the world of insects, except rarely a whirr from a passing grasshopper. No galls or insect webs were visible, no larvæ; the few caterpillars of moths were nocturnal, hiding under lava from the birds and the heat.

By diligent search I found a few pale centipedes under bark, a scattering of tiny grey moths and black and brown crickets under stones. Here also were small grasshoppers with reduced, useless wings, a few wingless beetles and lava-coloured spiders. Isopods, strangely out of drawing in this arid place, were the most abundant invertebrates. As far as my knowledge went, there were only two mosquitoes on Tower Island, both of which bit me and one of which I captured.

Wherever I went I was the centre of an admiring, or at least curious group of birds, and whenever I squeaked or chirped I added more to the ensemble. But the best plan of all to attract as many small birds as possible I now put into practice. A pair of red-footed boobies overhead was having some terrible family disturbance. Squawk after squawk rang out, and attracted no attention whatever; finches and mockers called to one another and went about their business as though the silence of the insect world extended to the realm of birds. I reached up and seized the foot of one of the boobies, and both forgot their matrimonial woes and concentrated on me. My ear could detect no difference in the squawks—they were as rasping and raucous as before, but in a few seconds every small bird within hearing was at hand, some within arm's reach peering down at me with concentrated interest. I had been able to change the whole tenor of the boobies' outcries, therefore for reasons of their own I was worthy of long continued scrutiny by all the glade's inhabitants. The mockingbirds were the most abundant, the little brown *Certhideas* next, then the sooty ground finches, the great-billed ones, doves, and sixth, the little cone-billed finches.

There was great individuality in the songs of the Tower mockingbirds, but the basis of all was simple; *chuc-chuc! ca-ca-kee! ca-ca-koo! ca-ca-kaw!* varied, interspersed, accented and reversed.

As everywhere in the archipelago, the mockers were essentially ground and rock feeders and walkers. Their favorite mode of feeding was to flick the leaves and dirt with closed beak, first to one, then to the other side.

The *Certhidea* was a drab little warbler, worthy in colour of this lava isle. It had a sweet, simple warble, eight notes run and jumbled together, no one distinct, and impossible to transcribe. Rarely a variation was given—a long-drawn-out wheezy note, followed by the usual vocal tangled knot. So tame were they that when I wanted a specimen, it took me quite five minutes to get thirty feet away. The great-billed *Geospizas*, or black finches, sang from the tree-tops and were the only birds which did not readily respond to my squeaking. Their utterance, like that of all the finches, was monotonous, a series

Fig. 74

FORK-TAILED GALÁPAGOS GULL

Creagrus furcatus (Néboux)

One of the most beautiful of the native birds, common on Tower Island

Fig. 75

A MONTH OLD FORK-TAILED GULL

Caught as a young chick, and living out a long life in the Zoological Park

Fig. 76

Young Galápagos Brown Pelican

With a large number of flat feather flies on its down

Fig. 77

Mockingbird About to Puncture and Devour the Egg of a Sea-bird

of double syllables, two notes apart, the first often being *C*. Transcribed, it sounded like *kah-lee! kah-ee! kah-lüh!* or another bird would pronounce it *ker-dee! ker-dah!*

The second time I shook up a booby for the purpose of making a census of small birds, a short-eared owl flew silently to a branch overhead and regarded me with great, yellow eyes. The small birds gave him plenty of room, but mingled a slight appearance of timidity with a casualness which indicated that his food seldom included birds. This individual, however, had in his stomach three large grasshoppers, two spiders, two six-inch centipedes, one roach and a caterpillar, and in addition the remains of a sturdy black finch.

.   .   .   .

We pulled away from the beach late in the afternoon of the twenty-ninth of April. The bay was as beautiful as when we first saw it, the flocks of birds as dense overhead, and the last noteworthy thing was a four-foot golden grouper which for some time swam like a molten fish between our two small boats.

As Tower sank below the horizon, a magnificent sunset, like a volcanic outburst, glowed behind Bindloe. Later the full moon rose from the heart of a jet cloud, showing first as a tiny spark, and bursting at once into full orange-silver glory. The glowing path on the water stretched back to the horizon— to where there floated quietly in the midst of the great ocean Eden, Guy Fawkes, Daphne, Seymour and Indefatigable. The moon-path was the colour of the head of Conolophus, of angel-fish fins, of cactus blossoms, of pirates' gold, of our taxi-driver's buried sovereigns; it reached the *Noma* and extended a little ahead—a prophecy perhaps of other golden days to come in these islands of enchantment.

CHAPTER XVI

MAN AND THE GALÁPAGOS

By Ruth Rose

WHEN Robinson Crusoe was rescued by privateers from his desert island, he turned privateer himself, and was made captain of a captured ship. After helping his freebooting friends in the sack of Guayaquil, they all sailed with their plunder to islands so superlatively desert that in comparison his lonely Juan Fernandez must have seemed a Paradise.

This group of extinct volcanoes known as the Galápagos Archipelago remained uninhabited for three hundred years after their discovery, and a malign influence seems to have been over every attempt to colonize them in modern times. Perhaps it was the sense of a spell, of something more than the mystery of isolation and desolation, that led the earliest Spanish navigators in these seas to call them *Las Islas Encantadas*, —the Enchanted Isles. They are one of the few places on earth where aboriginal man never existed; there surely is no other place where man has so much adventured, and which has yet remained for the most part a desert,—unmapped and unexplored.

Over a stretch of almost four hundred years the meagre history of the Galápagos can be traced; the mere enumeration of the visitors to their shores would hardly fill a half-page of type. But it is again the story of quantity versus quality. Think of the happy state of that person or household or city that could boast of having never received a dull visitor! And this country,—if we may join these scattered islands by a name, as they were probably once joined by solid land—has been the scene of dramas, comedies, tragedies and mysteries that in variety and interest can hardly be surpassed.

A royal funeral procession paused here; one of a fleet of prizes captured by buccaneers vanished from the face of the calm waters and was never seen again; a lonely grave marks the site of a duel fought here by two American naval officers; and on these shores ten shipwrecked sailors kept life in their suffering bodies for two and a half months by means of raw turtle meat, when all the time one of their number carried a forgotten box of matches!

Reading the history of the Galápagos is like beginning at the hub of the wheel and following each spoke back to an encircling rim of widening interest that is far removed from the starting point. That is the penalty and the advantage of never having a dull visitor. Captain Benjamin Morrell, for instance, visited the archipelago in 1825, but no one could be satisfied to read his account of what happened there only. His particular spoke leads back to his birthplace in Rye, New York, to his running away to sea during the War of 1812 and his earnest endeavours to be a combatant, which resulted only in a series of sojourns in British prisons; and to his discovery of islands in the South Seas where he was the helpless witness of a cannibal feast in which fourteen of his crew were devoured by treacherous natives.

We began to trace the history of the Galápagos backward from modern times; each record contained a reference to some earlier record, until we topsy-turvily reached the end at the beginning,—at a record which is hardly more than a rumor.

Before Pizarro's shadow fell across Peru, even before Columbus discovered his New World, there was an Inca king, Tupac Yupanqui, who made himself ruler over all the land and then turned ambitious eyes toward the great unknown sea. To go out of sight of land was an unventured peril to his subjects, but their great king dared, in a fleet of wide-sailed balsas, which tradition says were made of inflated sealskins fastened together, and a number of rafts. As Sarmiento says, "He resolved to challenge a happy fortune and see if it would favour him by sea."

Years passed and the conquistadores came. The grandson of the great Tupac Yupanqui was the ruling Inca when for the first time the fierce and greedy eyes of the Spanish invaders gazed on Chimborazo's slopes above the jungle and saw only gold and the means of extorting it. The Inca king, Atahualpa, was a means; he was thrown into prison, insulted, tortured, and at each new outrage his persecuted people sent a fresh wave of gold to the feet of his captors in the frenzied hope of ransoming their beloved ruler. But Atahualpa died at last, a shameful death, by order of Pizarro, his insatiable jailer.

All this time and for long afterward, the Spaniards heard over and over again the native legend of Tupac Yupanqui's voyage, and how he had found strange lands far out in the western sea. He had been gone a year, they said, and had then made a triumphal entry into Cuzco, with a great throne of copper, negro prisoners, and the skins of "strange animals like horses," as trophies of his travels. They told of the two islands that he had named Nina-chumbi and Hahua-chumbi, which is translated as the Island of Fire and the Island of Beyond. From the first name it seems likely that Tupac Yupanqui's were the first human eyes to see a volcanic eruption on the Galápagos. Perhaps it was such a spectacle that turned him back before he had seen more than two islands of the archipelago.

The Inca left no written record of this enterprise, but much accepted history is founded on similar word-of-mouth legends. Pedro Sarmiento de Gamboa is the earliest authority for the story. He wrote a history of the Incas which he obtained from forty-two Inca witnesses, and this manuscript was sent to Spain in 1572.

Tupac Yupanqui's booty could not have come from the Galápagos, for copper thrones and negro prisoners are not commodities to be found on desert islands. Perhaps this daring amateur navigator penetrated to archipelagoes further west, from which he brought strange plunder. But the invaders were not to be distracted from this Land of Gold by tales of distant chimerical isles, and so through lack of credulity or interest or enterprise, the Galápagos remained terra incognita yet a while, and were only discovered by accident, a fashion of discovery which was not infrequent in those days when even the most skilled navigators had perforce to "set their course by God and by guess."

A note which is in all probability a reference to the Galápagos is found in an old work by P. Matire, who says:

"There came to me the day before the ides of October, this year, 1516, Rodriguez Colminares and Francisco de la Puente, who affirmed, the one that he had heard of, the other that he had seen, divers islands in the Mare del Sud to the west of the Pearl Islands, in which trees are engendered and nourished, which bring forth aromatic fruits as in India and therefore they conjecture that the land where the fruitfulness of spice beginneth, cannot be far distant. And many do only desire that leave be granted them to search farther, and they will of their own charges, frame and furnish ships and adventure the voyage to seek these islands and regions."

These aromatic trees are very likely bursera, with cherry-like fruit, which we found so common; the crushed leaves give out a clean, pungent-sweet odor, resembling sweet-fern.

In 1535 Fray Tomás de Berlanga set out on the Pacific. He was the third Bishop of Panama, and it is to him that southern America is supposed to owe the introduction of the plantain, a fruit so flourishing in our tropics that it is seldom thought of as originating in India and Southern Asia. The Bishop had been ordered by his Emperor to go to Peru and report on conditions there under Pizarro. It is a far cry from the court of Charles V, Emperor of Spain, Germany, Austria, the Netherlands, to the jungles and peaks of Darien, but communication was maintained, and the square-rigged Spanish ships brought the imperial commands to these savage domains.

Fray Tomás set out from Panama on his coastwise voyage with a considerable retinue, and with horses aboard, for their convenience in seeing Peru. After a week of sailing with favourable wind a dead calm fell and prevailed for eight days. They drifted helplessly in the grip of the strong currents which were the affliction of many succeeding mariners in these waters. Their supplies were running low and so when land was sighted, they were greatly rejoiced, for they had only a small amount of water remaining, scarcely enough for two days.

So the first Spaniards arrived at the Galápagos, giving us the first authentic record of their discovery. In a letter to the Emperor, Fray Tomás afterward described their sufferings in search of water. At the first island where they landed, which was probably Barrington, not a drop of water was found, and there was no pasturage for the horses. Deciding at length that this desert place afforded nothing but turtles and iguanas, they re-embarked and made sail for another island which was much larger and of a more promising appearance. However, they were again becalmed and for three days the ship lay motionless between the two islands, while men and horses were tortured by thirst. At length they landed on what we must suppose to have been Charles, and again scattered in search of a stream or spring. Finding nothing of the kind, they attempted for two days to dig a well, but after much exhausting labour, the small quantity of liquid that they obtained was undrinkable, "being as bitter as the sea." In their extremity they had recourse to that plant which has saved many a thirsty man and animal, and by chewing the fruit and juicy stalks of the cactus, they kept themselves alive. Not all withstood these privations; one man and two horses died of thirst, and so the first grave on the Galápagos was dug.

Sunday came, and the Bishop celebrated high mass on those desolate shores where, as he said, "It looked as though God had caused it to rain stones." Sea-lions slipped through the creaming surf and looked curiously at the little group of strange beings, and long-winged sea-birds poised and swooped above the cross and the Spanish flag, planted side by side among the cactus and thorny scrub.

After the service, the despairing search for water recommenced, and as though their efforts had been blessed by their devotions, they now found, here and there, cupped in hollowed rocks, a quantity sufficient to fill all the vessels they had on board. They then resumed their interrupted voyage, but their vicissitudes were not over. Another man died, they lost their course more than once, and were so long on the way that they were suffering from thirst again before they sighted the mainland. Two days of calm kept them tantalizingly off the shore, and the Bishop relates that having no water left, they were forced to drink wine. This seems to have been the least of the hardships encountered. What the horses drank in this case, we are not told.

At length they arrived at Puerto Viejo, not far from the

place where Tupac Yupanqui was supposed to have embarked on his legendary voyage.

In the Bishop's letter to his Emperor, he dwells upon the tameness of the birds of the Galápagos, "which did not fly from us but allowed themselves to be taken," and speaks of the giant tortoises, capable of carrying men upon their backs. The Spanish eye of those days was ever alert for gold and precious stones, and Fray Tomás thought that even here he had found riches, for he reported that in the sand of the beach on the second island he saw small stones, crystalline and glittering, which he at first thought might be diamonds, and other yellow ones which resembled amber.

After hearing of the treasure of the Incas, Spanish voyagers in this part of the world were probably more than half expecting to find beaches strewn with precious stones.

The Bishop made a careful report and obtained very good observations, for those days, of the latitude of the islands, but he modestly refrained from giving his own name to his discovery.   Indeed he did not name them at all, so perhaps his neglect was from indifference rather than modesty.

A dozen years passed before the Galápagos had another visitor.   The Spaniards, having conquered Peru, were now engaged in civil war in the vanquished country.   In 1546 Diego Centeno, opposing Gonsalvo Pizarro, found himself with a handful of men in flight before the forces of Carvajal, Pizarro's fierce old lieutenant, who had been close at his heels for many days.   Centeno, seeing that he was being forced to the coast, instructed his loyal captain, Diego de Rivadeneira, to detach himself from the main body of troops and hurry on to get a ship in which they might all escape to New Spain.   The port of Quilca was the rendezvous to which the vessel was to be brought for the others to embark.

Rivadeneira, with a dozen soldiers, reached the coast after

an exhausting forced march. At Arica he found two ships. One of them was useless, being stranded on the shore. Rivadeneira sent off a raft with two or three men to tell those aboard the other ship that Carvajal was dead, and that Don Diego Centeno, now victorious, commandeered their vessel. News travelled slowly in those days and in such a country. The credulous crew immediately went ashore to see what they were required to do, and the resourceful Captain profited by their absence to seize the ship, which he thus obtained without meeting with the least resistance. Having taken aboard as many provisions as they could get, they sailed for the port which Centeno had designated.

But while Rivadeneira was employing stratagem, Centeno had been so incessantly harried by Carvajal that his small forces finally broke, and he led them in precipitate flight down to the coast in search of the rescuing ship. They arrived at Quilca before Rivadeneira, and finding no means of escape, Centeno gave up the struggle and sorrowfully dispersed his followers.

Very soon after him Carvajal arrived. He heard the fate of Centeno's army, and also that Rivadeneira had seized a ship at Arica. Determined to capture at least some of the rebels, he had a number of balsas lying in wait for Rivadeneira. When the captured ship appeared, the men on the rafts made signals purporting to be from Centeno. The crafty Captain, however, did not perform his part in the plan; instead of sailing his ship into the midst of the fleet of rafts, he cannily sent a small boat to investigate them, and thus discovered the trick.

He then made haste to get away, and finding that there was no hope of getting his chief and no army left to rescue, he sailed out on the Pacific, without map or compass, hoping to reach that part of New Spain which we know as Nicaragua.

After twenty-five days they sighted land, an unfamiliar

country rising to high cloud-capped peaks. They had at first no notion that it was an island, and since they were quite ignorant of their position, they were thrown into a panic, thinking that it was Puná or Túmbez, on the mainland. Rivadeneira thought that the pilot had betrayed them by bringing them back into the clutches of Pizarro, but after the poor pilot pleaded that he was doing his best with neither instruments or charts, Rivadeneira spared his life. For three days they sailed round this land, thus discovering it to be a great island. They could not land because of heavy seas, but they cruised about in these waters until they had seen twelve more islands, all smaller than the first. At one of these a party succeeded in landing. Hardly were they ashore when the silence and the desolation appalled them; they suddenly feared lest their companions on board the vessel should sail away and leave them on these barren shores. That such an idea should occur to them seems a striking commentary on the comradely spirit that must have prevailed. They hastily returned to the beach, without having made more than a cursory search for needed water, and taking nothing from the island but a few birds, they made another attempt to reach New Spain.

Everyone who ever came to the Galápagos arrived thirsty; most voyagers left the islands in the same condition. Rivadeneira and his men, twenty-two in all, suffered for days until a rainstorm quenched their thirst. Their food was consumed, and then these soldiers so recently turned sailors, stripped off their spurs and from them fashioned clumsy harpoons, with which they speared sharks and other fish. Once as they lay becalmed they saw a great sea-turtle floating on the surface; a boy from the crew was thrown overboard to grapple with the animal, which they knew to be delicious eating. Suddenly the wind sprang up, the boat slipped through the water, and, as casually set forth, "the poor boy was devoured by sea-creatures."

This incident possibly shows that the men of this same party who feared to be marooned by their comrades were not the victims of an altogether groundless panic.

They arrived at last at San José de Istapa, which is now a part of Guatemala. Rivadeneira reported his discovery and described giant tortoises, iguanas, sea-lions, and flamingos, owls and other birds, but like the Bishop, he gave no name to the place he had so briefly visited.

Various Spanish voyagers visited the islands during the last half of the 16th century, and called them *Las Islas Encantadas*, partly because they were absolutely solitary and uninhabited, and partly because of the difficulty of finding them by the methods of navigation of those days. The means of calculating distances sailed were uncertain; the strong unknown currents made calculations even more like guesswork, so that many pilots swore that they had sailed directly over the places where the islands had formerly been seen, without glimpsing so much as one rock above the surface. Here was reason enough for calling them Enchanted.

This difference in reckoning also gave rise to the belief that there were two groups of islands in this parallel, and there are charts as late as 1750 on which two archipelagoes are set down.

On a map in Ortelius' *Theatrum Orbis Terrarum* of 1570, the archipelago appears as *Insulæ de los Galopegos*, so by that time it had evidently become the namesake of the gigantic reptiles that were so numerous there at that time and for many succeeding years. Each island has been named and renamed half a dozen times, but the archipelago itself has remained the Islands of the Tortoises. To be sure, Ecuador, in a recent commemorative frenzy, renamed it the Archipelago of Colon, to celebrate the quattro-centenary of Columbus' discovery, but no one seems to have taken this change seriously. The Republic of the Equator presumably has the right to name its possessions to please itself, but a habit of three hundred and fifty years' standing is not so easily changed. As the White Knight would say, "Galápagos isn't its *name,* it's only what it's *called.*"

Hawkins, in 1593, disposed of the Galápagos with one line. By that time the islands must have been sufficiently well-known to have dispelled the idea that here Nature would furnish forth shiploads of spices, for his brief description merely says: "Some fourscore leagues to the westward of this cape (Cape Passaos) lyeth a heape of Ilands the Spaniards call Illas de los Galápagos; they are desert and beare no fruite."

That eminent and indefatigable historian, Señor Jiménez de la Espada, recently unearthed a mention of the Galápagos which still further enriches the list of diversified happenings there. It is related by León Pinedo that a legate of Santo Domingo, Fray Martín Barragán, was shipwrecked on one of these islands about 1602, and that he afterward said his conversion to the Church took place during his three-year sojourn there. Details are aggravatingly missing, but the imagination supplies a picture of the tattered, bearded survivor of disaster, crying in a wilderness his conviction of sin.

In the latter part of the 17th century the buccaneers came to the Pacific,— those men whose names never fail to thrill the hearer, whether he regards them as dashing freebooters or as bloodthirsty outlaws. Many of the islands of the Caribbean had become depressingly law-abiding, and a filibuster's lot was no longer such a happy one. Henry Morgan had blazed a trail to plunder across the Isthmus of Darien, and it seemed feasible to capture the Spanish ships as they came up the coast with rich cargoes from Peru, instead of seizing the loot after the Spaniards had transported it across the Isthmus.

For more than a century the natives of Peru and Darien

had cowered before Spanish persecution; it was now the Spaniards' turn to cower before the onslaughts and depredations, actual or threatened, of the buccaneers. The Spaniards knew that the imaginations of these freebooters had been diverted to the South Seas, and they were properly filled with forebodings. Dampier relates:

"Before my going to the South Seas, I being then on board a privateer off Portobel, we took a packet from Carthagena. We opened a great many of the merchants' letters, several of which informed their correspondents of a certain prophecy that went about Spain that year, the tenor of which was, that the English privateers in the West Indies would that year open a door into the South Seas."

In 1675 the Viceroy of Peru heard a rumour that pirate ships had been seen off Chile, and he sent Antonio de Vea to investigate. The rumour proved groundless, but the Spanish continued in a bad state of nerves until at last in 1680 the thing they feared came upon them.

The first buccaneer expedition into the Pacific had for leaders John Coxon, Richard Sawkins, John Watling and Bartholomew Sharp. Lionel Wafer, "chyrugeon," author and pirate, has left a vivid account of his adventures. William Dampier, at that time a seaman of no celebrity, was among these buccaneers, and Basil Ringrose published a journal concerning this series of voyages. The only bearing that this has on the Galápagos is a negative one. Certain of the buccaneers under Sharp tried to stop at the islands, but were prevented by unfavourable winds.

The second irruption of filibusters into the Pacific was in 1684. Several of the men who had painfully toiled across the Isthmus to reach the South Sea, were of this second party that no less painfully won to their destination round the Horn.

John Cook, not to be confused with James Cook, the famous navigator, was their captain. Dampier and Wafer, Ambrose Cowley and Edward Davis were among his colleagues. They sailed from the Chesapeake, stopped at several of the Cape Verde Islands and finally reached the coast of Guinea, near Sierre Leone. Here they captured a large Danish vessel of thirty-six guns. Having burned their old vessel, "that she might tell no tales," they embarked in their prize, which they rechristened the *Batchelors' Delight,* and directed their course across the Atlantic for the Strait of Magalhanes.

Any writer of fiction would be unmercifully jeered who depicted a crew of desperate pirates, on slaughter and rapine bent, pausing in their gory musings to keep Valentine's Day. The appended quotation from Cowley's *Voyage Round the World,* should encourage authors to let imagination do its worst, untrammelled by probability:

". . . Then haling away S.W. we came abreast of Cape Horn the 14th day of Feb. (1684) where we chusing of Valentines and discoursing of the Intrigues of Women, there arose a prodigious Storm, which did continue till the last day of the Month, driving us into the lat. of 60 deg. and 30 min. South, which is further than ever any Ship hath sailed before South; so that we concluded the discoursing of Women at Sea was very unlucky and occasioned the Storm."

"Cherchez la femme" has a wide application.
Another interesting bit from Cowley's journal:

"Towards the beginning of the Month of March, the Wind coming up at South, we were soon carried into warm Weather again; for the Weather in the lat. of 60 deg. was so extream cold that we could bear drinking 3 quarts of Brandy in 24 hours each Man, and be not at all the worse for it, provided it were burnt."

MAN AND THE GALAPAGOS

After rounding the Horn despite Woman and Wine, the *Batchelors' Delight* fell in with another piratical ship called the *Nicholas*, with Eaton as commander. They agreed to voyage together for a time, and a few days later stopped at the island of Juan Fernandez. After rescuing a Mosquito Indian who had been abandoned here by Captain Watling three years before, the two ships sailed to the islands Lobos de la Mar, where they "scrubbed the ships" and were glad to stay awhile, having a good many sick. Captain Cook was one of those ill. After a week, three Spanish ships came in sight, which the buccaneers chased and captured, apparently meeting with no resistance. They were not laden with the conventional treasure of plate and bullion. Dampier says that in one ship were eight tons of quince marmalade, surely strange pirate booty. He also says that a large part of the cargoes was "flower," and that in the largest ship was "a stately mule sent to the president (of Panama) and a very large image of the virgin Mary in wood . . . sent from Lima by the viceroy."

The adventurers had barely missed much richer plunder, for when this ship first sailed from Lima, she had carried 800,000 pieces of eight, but stopping at Guanchaco, the presence of strange ships in the South Seas was reported and so the treasure was there put ashore.

The buccaneers learned from their prisoners that preparations had been made to give them a warm welcome on the coast, so they decided to take their three prizes to the Galápagos, which they sighted after twelve days.

Ambrose Cowley made the first chart of the Galápagos during this visit, and he did a very good piece of work, considering the limited opportunity he had to survey the archipelago. His chart is infinitely better than the one made one hundred and nine years later by the Captain of the Royal Spanish Armada, who came to make a reconnaissance of the

Galápagos and stayed three days.   The two maps are reproduced on another page and afford an interesting comparison.

Cowley's account of the island is appended:

". . . Thereupon we stood away to the Westward, to
try if we could find those Islands which the Spaniards call
Gallappagos or Enchanted Islands, when after 3 weeks sail
we saw Land, consisting of many islands, and I being the first
that came to an anchor there, did give them all distinct
Names.

"The first that we saw lay near the lat. of 1 deg. 30 min.
South; we having the Wind at South, and being on the North
side thereof, that we could not sail to get to it, to discover what
was upon it.   This Island maketh high Land, the which I
called King Charles's Island: And we had sight of three more
which lay to the Northward of this, that next it I called Cross-
man's Island: The next to that Brattle's; and the third, Sir
Anthony Dean's Island. We moreover saw many more to the
Westward. . .   Then we came to an Anchor in a very good
Harbour, lying toward the Northernmost end of a fine Island
under the Equinoctial Line: Here being great plenty of Provisions, as Fish, Sea and Land Tortoises, some of which weighed
at least 200 Pound weight, which are excellent good Food.
Here are also abundance of Fowls, viz., Flemingoes and
Turtle Doves; the latter whereof were so tame, that they
would often alight upon our Hats and Arms, so as that we
could take them alive, they not fearing Man, until such
time as some of our Company did fire at them, whereby
they were rendered more shy.   This Island I called the Duke
of York's Island: there lying to the Eastward of that (a fine
round Island) which I called the Duke of Norfolk's Island.
And to the Westward of the Duke of York's Island, lieth another curious Island, which I called the Duke of Albemarle's;
in which is a commodious Bay or Harbour, where you may ride
Landlock'd: And before the sad Bay lieth another Island, the
which I called Sir John Narborough's: And between York
and Albemarle's Island lieth a small one, which my fancy led

me to call Cowley's enchanted Island; for we having had a sight of it upon several points of the Compass, it appear'd always in as many different Forms, sometimes like a ruined Fortification; upon another Point, like a great City, etc. This Bay or Harbour in the Duke of York's Island I called Albany Bay; and another Place York Road. Here is excellent good, sweet Water, Wood, etc, and a rich Mineral Ore. From thence we sailed to the Northward, where we saw three more fine Islands. (Abingdon, Wenman, Culpepper) All of them that we were at, were very plentifully stored with the aforesaid Provisions, as Tortoises, Fowls, Fish and Alguanoes, large and good, but we could find no good Water on any of all these places, save on that of the Duke of York's Island. But at the north end of Albemarle Island there were thick, green Leaves of a thick substance which we chewed to quench our thirst; and there were abundance of Fowls on this Island which could not live without Water, tho we could not find it.

"After that we had laid up, and put on shoar at Albany Bay and other Places, 1500 Bags of Flower, with Sweetmeats, etc. we sailed to the Noathward again to try a second time amongst the Islands, if we could find any fresh Water, if ever we should have occasion to touch hereafter amongst thcm; but it happened so, that we fell in with such a very strong Current that when we would have sailed back again to the Duke of York's Island to have watered our Ship, we could not stem it. This made us steal away N.N.E. and the first Land that we made upon the Main was Cape Trespontas."

King Charles II died soon after Cowley named the islands, so the Duke of York's Island became King James' Island, and so appears on Cowley's chart.

Their first anchorage was probably at Indefatigable, called Duke of Norfolk's by Cowley. Here they stayed only one night, and then sailed to James, where they made a tent ashore for Cook, their sick captain, and busied themselves in getting a supply of fresh meat, which they found to be an easy task. Dampier was a man of remarkable powers of observation and description, and while Cowley was chiefly interested in noting down the positions of the islands and in baldly stating what was to be found on them, no detail of botany, zoology, or meteorology was too small to receive Dampier's painstaking attention. With the true spirit of adventure and of scientific curiosity, he was far more the explorer and naturalist than he was the buccaneer.

His description of the Galápagos is in part as follows:

"The Spaniards who first discovered them, and in whose drafts alone they are laid down, report them to be a great number, stretching northwest from the line, as far as five degrees north, but we saw not above fourteen or fifteen. They are some of them seven or eight leagues long, and three or four broad. They are of a good height, most of them flat and even on the top; four or five of the easternmost are rocky, barren and hilly, producing neither tree, herb, nor grass; but a few dildo trees, except by the sea side. The dildo tree is a green prickly shrub, that grows about ten or twelve feet high, without either leaf or fruit. It is as big as a man's leg, from the root to the top, and it is full of sharp prickles, growing in thick rows from top to bottom; this shrub is fit for no use, not so much as to burn. Close by the sea there grows in some places bushes of Burtonwood, which is very good firing. This sort of wood grows in many places in the West Indies. . . . I never saw any in these seas but here. There is water on these barren islands, in ponds and holes among the rocks. Some other of these islands are mostly plain and low, and the land more fertile; producing trees of divers sorts, unknown to us. Some of the westermost of these islands, are nine or ten leagues long, and six or seven broad, the mould deep and black. These produce trees of great and tall bodies, especially mammee-trees, which grow here in great groves. In these large islands, there are some pretty big rivers, and on many of the lesser islands, there are brooks of good water. The Spaniards when they first discovered these islands, found multitudes of guanoes, and land turtle or tortoise, and named them the Gallipagos islands. I do believe there is no place in the world that is so plentifully stored with these animals. The guanoes here, are as fat and large, as any that I ever saw; they are so tame, that a man may knock down twenty in an hours time with a club. The land-turtle are here so numerous that five or six hundred men might subsist on them alone for several months, without any other sort of provision: They are extraordinary large and fat, and so sweet, that no pullet eats more pleasantly. One of the largest of these creatures will weigh one hundred and fifty to two hundred weight, and some of them are two feet, or two feet six inches over the callapee or belly. I never saw any but at this place, that will weigh above thirty pounds weight. . . . The air of these islands is temperate enough considering the clime. Here is constantly a fresh sea breeze all day, and cool refreshing winds in the night; therefore the heat is not so violent here, as in most places near the equator. The time of the year for the rains is in November, December and January; then there is oftentimes excessive dark tempestuous weather, mix'd with much thunder and lightning. Sometimes before and after these months there are moderate refreshing showers; but in May, June, July and August the weather is always very fair.

"We staid at one of these islands, which lies under the equator, but one night, because our prizes could not get in to an anchor. We refresh'd ourselves very well, both on land and sea turtles, and the next day we sailed from thence. The next island of Gallipagos that we came to is but two leagues from this; it is rocky and barren like this, it is about five or six leagues long, and four broad. We anchored in the afternoon, at the north side of the island, a quarter of a mile from the shore, in sixteen fathom water. It is steep all round this island and no anchoring only at this place. Here it is but ordinary riding, for the ground is so steep, that if an anchor starts it never holds again, and the wind is commonly off from the land, except in the night, when the landwind comes more from the west; for there it blows right along the shore, though but faintly. Here is no water but in ponds and holes of the rocks. That which we first anchored at hath water on the north end, falling down in a stream from high steep rocks, upon the sandy bay, where it may be taken up. As soon as we came to an anchor, we made a tent ashore for captain Cook, who was sick. Here we found the turtle lying ashore on the sand; this is not customary in the West Indies. We turned them on their backs that they might not get away. The next day more came up; when we found it to be their custom to lie in the sun, we never took care to turn them afterwards, but sent ashore the cook every morning, who killed as many as served for the day; this custom we observed all the time we lay here, feeding sometimes on land turtle, sometimes on sea turtle, there being plenty of either sort. . . . The sea about these islands is plentifully stored with fish, such as are at Juan Fernandes. They are both large and fat and as plentiful here as at Juan Fernandes; here are particularly abundance of sharks. The north part of this second isle we anchored at lies 28 minutes north of the equator, for I took the height of the sun with an astrolabe. These isles of the Gallipagos have plenty of salt. We stayed here but twelve days, in which time we put ashore 5000 packs of flour for a reserve, if we should have occasion for any before we left these seas."

From the Galápagos, the *Batchelors' Delight*, the *Nicholas* and their prizes proceeded northward to the Island of Cocos, but missed it altogether through unfavourable winds, and so continued to New Spain. Off Cape Blanco Captain Cook died, "though," says Dampier, "he seemed that morning to be as likely to live, as he had been some weeks before; but it is usual with sick men coming from the sea, where they have nothing but the sea air, to die off as soon as ever they come within view of the land."

Edward Davis, quartermaster of the *Batchelors' Delight*, was unanimously elected to succeed in command.

During these adventures of the *Batchelors' Delight*, many more buccaneering expeditions had found their way across the Isthmus and, seizing Spanish ships whenever opportunity

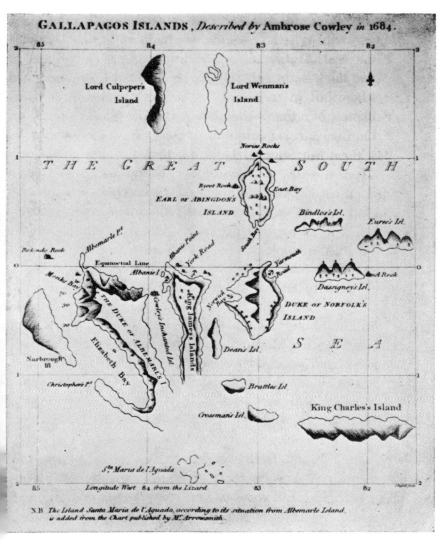

FIG. 78

FIRST CHART OF THE GALÁPAGOS ISLANDS

Made by Ambrose Cowley, the Buccaneer, in 1684

Mapa trazado en 1793 por los marinos españoles de la fragata "Santa Gertrudis".

FIG. 79

CHART MADE BY ALONZO DE TORRÉS, CAPTAIN IN THE ROYAL SPANISH ARMADA

Although made one hundred and nine years after that of Captain Cowley, it is wholly inferior

offered, had enlarged their piratical fleet to considerable numbers. Eaton in the *Nicholas* parted company with Davis and the *Batchelors' Delight* to cruise to the Ladrones and the Bashee Islands, and Ambrose Cowley threw in his fortunes with Eaton.

When Davis established himself at the island Taboga in the Bay of Panama, he was in company with Captain Swan in the *Cygnet*, and with a small barque, also manned by buccaneers. They were soon joined by two parties, one of about three hundred men, both English and French, commanded by Grogniet and L'Escuyer, and the second headed by Rose, Le Picard and Desmarais. A member of the second party was Raveneau de Lussan, whose journal contains revolting details of incredible cruelty both to men and animals.

The buccaneer forces were now composed of almost a thousand men in ten ships, and they presently encountered the Spanish fleet of fourteen sails which had put ashore its treasure and come out to give battle. This was an almost bloodless affair; neither side appeared over-enthusiastic. The buccaneers knew the treasure was no longer aboard the Spanish ships, and the Spaniards were evidently not in an impetuous frame of mind. Pirates and dons spent two rather absurd days in circling round each other and then the Spanish fleet was overcome with discretion and incontinently fled. The only loss the buccaneers suffered was one man killed, six wounded and half a rudder shot away.

This futile affair led to reproaches and quarrels among the filibusters and the upshot of it was that the French and English separated. Davis was re-enforced by the arrival of William Knight in command of about forty Englishmen.

After taking the towns of Ria Lexa and Leon, the English confederacy was dissolved. Swan and Townley proposed to go to the East Indies, and Dampier, desirous of seeing that part of the world, embarked with Swan. Lionel Wafer remained with Davis and the record of the next two visits to the Galápagos is from his account.

Davis and Knight were now associated in various enterprises and after some months, during which they suffered from an epidemic of spotted fever and lost several men, they returned to the Galápagos, probably late in 1684 or early in 1685, to seek some of their stores of flour. Of this they took away about five hundred sacks, some of which had been eaten by birds. They watered also, but unfortunately it is impossible to tell at which island they found the means. Wafer only says, "From Cocos we came to one of the Galápagos Islands. At this island there was but one watering place and there we careened our ship." He also speaks of a tree, "low and not shrubby, very sweet in smell and full of very sweet gum."

Dampier's manuscript journal refers to this second visit by Davis:

"Part of what I say of these islands I had from Captain Davis, who was there afterwards, and careened his ship at neither of the islands that we were at in 1684, but went to other islands more to the westward, which he found to be good, habitable islands having a deep fat soil capable of producing anything that grows in those climates; they are well watered, and have plenty of good timber. Captain Harris came hither likewise, and found some islands that had plenty of mammee trees, and pretty large rivers. They have good anchoring in many places, so that take the Galápagos by and large, they are extraordinary good places for ships in distress to seek relief at."

This optimistic version of the water question can only be explained by the visit having been made in a *very* rainy season.

The struggles of some of these men with natural history classification makes interesting reading. Dampier devotes considerable space to disproving the current opinion that crocodile and alligator were male and female of the same species, and a few pages further on attempts to show that the tapir and the hippopotamus were one and the same. He had never seen either creature. The following extracts are taken from "Captain Woods Voyage Through the Streights of Magellan," published in 1669, showing an effort to place two denizens of the Galápagos as Fish, Flesh or Fowl:

"We have already mentioned those Birds called Penguins to be about the bigness of Geese; but upon second thoughts to call them Fowls I think improper, because they have neither Feathers nor Wings, but only two Fins or Flaps, wherewith they are helped to swim. . . .

"We have also mentioned the great number of Seals found here, the same being a great creature that feedeth in the Sea and swims like a Fish, but its similitude is like a Beast; they take their rest, sleep and bring forth their Young on the Shoar, and I think they may for all that be called Fish. Some of them are as big as the largest Horses and will keep good in salt several months."

In succeeding years, the Galápagos became for marauders in the Pacific a refuge similar to Tortuga in the West Indies. They were so far from the routes followed by Spanish shipping that the pirates felt no fear of sudden assault; tortoises, birds and fish were wonderfully plentiful and easily caught; and having found watering places, which they did on the islands now known as Charles and James, it was a perfect place for them to recuperate and re-fit after long cruises, interspersed with attacks on the settlements of the mainland.

Here there was none to dispute their right to all that the country afforded, and here they once swarmed on beaches which are today as devoid of human traces as though that lively procession of pirates had never troubled these placid waters.

GALAPAGOS

In 1687 Davis visited the Galápagos for the third time, after the sack of Guayaquil. The plunder had been divided "by a very ingenious and unobjectionable mode of distribution. The silver was first divided; the other articles were then put up at auction and bid for in pieces of eight, and when all were so disposed of, a second division was made of the silver produced by the sale."[1]

Lionel Wafer was still with Davis, and "appears to have been one of those to whom fortune had been most unpropitious."[2] He says:

"I shall not pursue all my coasting along the shore of Peru with Captain Davis. We continued rambling about to little purpose, sometimes at sea, sometimes ashore, till, having spent much time and visited many places, we were got again to the Galápagos from whence we were determined to make the best of our way out of these seas."

From the meagre information available, it seems probable that Davis spent most of his time in the archipelago in a harbour at Charles Island. On this, his third and last visit, he again careened and re-victualled, taking aboard a large supply of flour and tortoise meat. It is related that they considered the oil from the tortoise as not inferior to fresh butter and that they saved of it sixty jars of eight gallons each for their homeward voyage. They arrived in the West Indies in 1688, just in time to avail themselves of the free pardon offered in King James' proclamation to all buccaneers who would abandon their filibustering ways.

In 1708, two ships, the *Duke* and the *Dutchess*, were fitted out for privateering by the merchants of Bristol. By this time, England and France, those periodical allies, were again arrayed against each other in the War of the Spanish Succession,

<hr/>

[1] *History of the Buccaneers of America*, James Burney. [2] *Idem.*

so this expedition was directed against both Spanish and French.

Captain Woodes Rogers was commander in chief, and sailed aboard the *Duke*. He was a man of great authority and resource, and he had need to be, for his motley crews planned mutiny more than once.

Stephen Courtney was captain of the *Dutchess;* he was a gentleman of Bristol who had contributed largely toward fitting out the ships.

Thomas Dover, of Dover Powder fame, was third in command, though he was a physician by profession. He too had a financial interest in the expedition. It is said of him that "he was a man of a rough temper, and not easily pleased; but as he had not the chief command, this was of the less consequence."[1]

The pilot was that man who perhaps contributed more than any other of his time to the knowledge of strange countries, their people and their natural resources. This voyage is the last record we have of William Dampier's adventurous life. Like most men whose lives have been spent in the pursuit of pure knowledge, he was poor and unsuccessful in a material way. He had sailed unknown seas for many years, and had been commander of his own ship on divers voyages, but in 1708 he was obliged to accept a subordinate position to cruise again those waters where his very name struck terror to the Spanish.

About the middle of December they rounded the Horn into the Pacific, and a few weeks later they added to their numbers the famous castaway, Alexander Selkirk, prototype of that fictional character now known in every civilized country. The two ships reached Juan Fernandez, where they were eager to land, as scurvy was making havoc among the crew,

[1] *All The Voyages Round The World,* collected by Capt. Samuel Prior, London, 1827.

and fresh provisions and water were the only hope of cure. While they were preparing to land in the evening, they saw a fire inshore, and knowing the island to be uninhabited, they thought this light proceeded from French vessels at anchor, so they waited until the morning. Seeing no ships by daylight, Dover in the pinnace went ashore. Towards evening he returned, bringing with him (according to Woodes Rogers)

" . . . a Man cloath'd in Goat-skins, who look'd wilder than the first Owners of them. He had been on the Island four Years and four Months being left there by Capt. Stradling in the *Cinque-Ports;* his name was Alexander Selkirk a Scotch-Man, who had been Master of the *Cinque-Ports*, a Ship that came here last with Capt. Dampier, who told me that this was the best Man in her; so I immediately agreed with him to be a Mate on board our Ship. 'Twas he that made the Fire last night when he saw our Ships, which he judg'd to be English. . . . The reason of his being left there was a difference betwixt him and his Captain; which, together with the Ships being leaky, made him willing rather to stay here, than go along with him at first; and when he was at last willing, the Captain would not receive him. He had been in the Island before to wood and water. . . .

" He had with him his Clothes and Bedding, with a Firelock some Powder, Bullets, and Tobacco, a Hatchet, a Knife, a Kettle, a Bible, some Practical Pieces, and his Mathematical Instruments and Books. He diverted and provided for himself as well as he could; but for the first eight Months had much ado to bear up against Melancholy and the Terror of being left alone in such a desolate place. . . . At first he never eat anything till Hunger constrain'd him, partly for grief, and partly for want of Bread and Salt; nor did he go to bed till he could watch no longer; the Piemento Wood, which burnt very clear, serv'd him both for Firing and Candle, and refresh'd him with its fragrant Smell.

" . . . his way of living and continual exercise of walking and running, clear'd him of all gross Humours, so that he ran with wonderful Swiftness thro the Woods and up the Rocks and Hills, as we perceived when we employed him to catch Goats for us. We had a Bull-dog, which we sent with several of our nimblest runners, to help him in catching Goats; but he distanc'd and tir'd both the Dog and the Men, catch'd the Goats, and brought 'em to us on his Back.

"He told us that his agility in pursuing a Goat had once like to have cost him his Life; he pursu'd it with so much Eagerness that he catch'd hold of it on the brink of a Precipice, of which he was not aware, the bushes having hid it from him; so that he fell with the Goat down the said Precipice a great height, and was so stunn'd and bruis'd with the Fall, that he narrowly escaped with his Life, and when he came to his Senses, found the Goat dead under him. He lay there about 24 hours, and was scarce able to crawl to his Hutt, which was about a mile distant or to stir abroad again in ten days.

" . . . After he had conquer'd his Melancholy, he diverted himself sometimes by cutting his Name on the Trees, and the Time of his being left and Continuance there. He was at first much pester'd with Cats and Rats, that had bred in great numbers from some of each species which had got ashore from Ships that put in there to wood and water. The Rats gnaw'd his Feet and Clothes while asleep which oblig'd him to cherish the Cats with his Goats flesh; by which many of them became so tame, that they would lie about him in hundreds, and soon deliver'd him from the Rats. He likewise tam'd some Kids, and to divert himself would now and then sing and dance with them and his Cats; so that by the Care of Providence and Vigour of his Youth, being now but about 30 years old, he came at last to conquer all the Inconveniences of his Solitude, and to be very easy. When his Clothes wore out, he made himself a Coat and Cap of Goat-Skins, which he stitch'd together with little Thongs of the same, that he cut with his knife. He had no other Needle than a Nail; and when his

Knife was wore to the back, he made others as well as he could of some Iron Hoops that were left ashore, which he beat thin and ground upon Stones. . . .

"At his first coming on board us, he had so much forgot his Language for want of Use, that we could scarce understand him, for he seem'd to speak his words by halves. We offer'd him a Dram, but he would not touch it, having drank nothing but Water since his being there, and 'twas some time before he could relish our Victuals.

. . . .

"This Morning we clear'd up Ship, and bent our Sails, and got them ashore to mend and make Tents for our sick men. The Governour, for so we call'd Mr. Selkirk, caught us two Goats, which make excellent Broth, mix'd with Turnip Tops and other Greens, for our sick Men, being 21 in all, but not above two that we count dangerous; the *Dutchess* has more Men sick, and in a worse condition than ours."

Owing to Dampier's previous acquaintance with this man and knowledge of his seamanship, he was put in command of one of the prizes captured soon after and he assisted the privateers in their enterprise of taking Guayaquil. Dampier commanded the artillery during this adventure, which was moderately successful, the privateers demanding thirty thousand pieces of eight for the ransom of the town, and obtaining besides quantities of provisions and wines, and as much jewelry as the inhabitants had not managed to conceal. The control exerted by Rogers over his turbulent crew is shown by the statement that only one man was drunk during their occupation of the town.

On the 8th of May the *Duke* and the *Dutchess*, with four prize vessels bore away for the Galápagos, carrying with them some Spanish hostages, since the last installment of the town's ransom had been short three thousand pieces of eight. Three days later a malignant fever broke out among the crew and spread so alarmingly that by the time they sighted the islands, which was on the 16th, one hundred and forty men were on the sick list. This fever they supposed was contracted at Guayaquil, where it had raged shortly before their arrival.

Woodes Rogers' account of the Galápagos is in part as follows:

"May 19th. Yesterday in the afternoon the Boat return'd with a melancholy Account, that no Water was to be found. The Prizes we expected would have lain to Windward for us by the Rock about 2 Leagues off Shore; but Mr. Hatley in a Bark, and the *Havre de Grace*, turn'd to Windward after our Consort the *Dutchess;* so that only the Galleon and the Bark that Mr. Selkirk was in staid for us. . . . At 5 in the Morning we sent our Boat ashore again to make a further search in this Island for Water. About 10 in the Morning James Daniel our Joiner died. We had a good Observation, Lat. 00° 32' S.

"May 20. Yesterday in the Evening our Boat return'd but found no Water, tho they went for 3 or 4 miles up into the Country. They tell me the Island is nothing but loose Rocks, like Cynders, very rotten and heavy, and the Earth so parch'd, that it will not bear a Man, but break into Holes under his Feet, which makes me suppose there has been a Vulcano here; tho there is much shrubby Wood, and some Greens on it, yet there's not the least Sign of Water, nor is it possible, that any can be contain'd on such a Surface. At 12 last Night we lost sight of our Galleon, so that we have only one Bark with us now.

"May 21 Yesterday in the Afternoon came down the *Dutchess* and the French Prize. The *Dutchess's* Bark had caught several Turtle and Fish, and gave us a Part, which was very serviceable to the sick Men, our fresh Provisions that we got on the main Land being all spent. They were surpriz'd as much as we at the Galleon, and Hatley's Bark being out of sight, thinking before they had been with us. We kept Lights at our Top-mast's Head, and fir'd Guns all Night, that they might either see or hear how to join us, but to no Purpose.

"Capt. Courtney being not yet quite recover'd, I went on board the *Dutchess*, and agreed with him and his officers, to stay here with the *Havre de Grace* and Bark, whilst I went in Quest of the missing Prizes. At 6 in the morning we parted, and stood on a Wind to the Eastward, judging they lost us that way. Here are very strange Currents amongst these Islands, and commonly run to Leeward except on the Full Moon I observed it ran very strong to Windward; I believe 'tis the same at change.

"May 22. Yesterday at 3 in the Afternoon we met with the Galleon under the East Island, but heard nothing of Mr. Hatley's Bark. At 9 last Night Jacob Scronder a Dutch-man and a very good Sailor, died. We kept on the Wind in the Morning to look under the Weather Island for Mr. Hatley, and fir'd a Gun for the Galleon to bear away for the Rendezvous Rock, which she did.

"May 23. Yesterday at 3 in the Afternoon we saw the Weather Island near enough, and no sail about it. We bore away in sight of the Rock, and saw none but our Galleon; we were in another Fright what became of our consort, and the 2 Prizes we left behind; but by 5 we saw 'em come from under the Shore to the Leeward of the Rock. We spoke with 'em in the Evening; we all bewail'd Mr. Hatley and were afraid he was lost; We fir'd Guns all Night, and kept Lights out, in hopes he might see or hear us, and resolv'd to leave these unfortunate Islands, after we had viewed two or three more to Leeward. We pity'd our 5 Men in the Bark that is missing, who if in being have a melancholy Life without Water, having no more but for 2 Days, when they parted from us. Some are afraid they run on Rocks, and were lost in the Night, others that the 2 Prisoners and 3 Negroes had murder'd 'em when asleep; but if otherwise, we had no Water, and our Men being still sick, we could stay little longer for them. Last Night died Law. Carney of a malignant Fever. There is hardly a Man in the Ship, who had been ashore at Guiaquil but has felt something of this Distemper, whereas not one of those that were not there have been sick yet. Finding that Punch did preserve my own Health I prescrib'd it freely among such of the Ship's Company as were well, to preserve theirs. Our Surgeons make heavy Complaints for want of sufficient Medecines, with which till now I thought we abounded. . . .

"May 21. Yesterday at 5 in the Afternoon we ran to the Northward and made another Island . . . and this Morning we sent our boat ashore, to see for the lost Bark, Water, Fish or Turtle. This Day Tho. Hughes a very good Sailor died, as did Mr. George Underhill, a good Proficient in most parts of the Mathematicks and other Learning, tho not much above 21 years old. He was of a very courteous Temper, and brave, was in the Fight where my Brother was kill'd, and serv'd as Lieutenant in my Company at Guiaquil. About the same time another young Man, call'd John English, died aboard the *Havre de Grace*, and we have many still sick. . . .

"May 25. Yesterday at 6 in the Evening our Boat return'd from the Island without finding any Water, or seeing the Bark. . . . Last Night Peter Marshal a good Sailor died. This morning our Boat with Mr. Selkirk's Bark went to another Island to view it. . . .

"May 26. Last night our Boat and Bark return'd, having rounded the Island, found no Water but Plenty of Turtle and Fish. This Morning we join'd the *Dutchess*, who had found no Water. About 12 a Clock we compar'd our Stocks of Water, found it absolutely necessary to make the best of our way to the Main for some, then to come off again; and so much the rather, because we expected that 2 French Ships, one of 60, and another of 40 Guns, with some Spanish Men of War, would suddenly be in quest of us.

"May 30. . . . Had we supplied ourselves well at Point Arena, we should, no doubt, have had time enough to find the Island S. Maria de l'Aquada, reported to be one of the Gallapagos, where there is Plenty of good water, Timber, Land and Sea Turtle, and a Safe Road for Ships. . . . Its probable

there is such an Island, because one Capt. Davis, an Englishman, who was a buckaneering in these Seas, above 20 Years ago,
lay some months and recruited here to Content; He says
that it had Trees fit for Masts; but these sort of Men, and
others I have convers'd with, or whose Books I have read,
have given very blind or false Relations of their Navigation,
and Actions in these Parts, for supposing the Places too remote to have their Stories disprov'd, they imposed on the
Credulous, amongst whom I was one, till now I too plainly
see that we cannot find any of their Relations to be relied on;
Therefore I shall say no more of these Islands, since by what I
saw of 'em, they don't at all answer the Description that those
Men have given us."

The *Duke* and the *Dutchess* returned to the mainland, and
on September first set sail for the Galápagos again, where they
spent almost two weeks searching for Hatley, the missing
mate, his men and the inexplicably vanished ship. No trace
of him was ever found, though the rudder and "boltsprit" of
a small bark lying on a beach led them to suppose at first
that the mystery was explained. However, on examination these relics were found to be too old to be a part of the
lost ship.

Rogers speaks of the tales that Dampier used to tell them
—reminiscences of his buccaneering days round about the
Galápagos. They took on board quantities of turtle and tortoise, and the Captain concludes his observations of the archipelago with these remarks:

"I saw no sort of Beasts but there are Guanas in abundance
and Land Turtle almost on every Island: 'Tis strange how the
latter got here, because they can't come of themselves, and
none of that sort are to be found on the Main. Seals haunt
some of these Islands, but not so numerous, nor their Fur so
good as at Juan Fernando's. A very large one made at me
several times, and had I not happen'd to have a Pike-staff

pointed with Iron in my Hand, I might have been killed by him; (one of our Men having narrowly escap'd the Day before.) I was on the level Sand when he came open-mouth'd at me out of the Water, as quick and fierce as the most angry Dog let loose. I struck the point into his Breast, and wounded him all the three times he made at me, which forc'd him at last to retire with an ugly Noise, snarling and showing his long Teeth at me out of the Water: This amphibious Beast was as big as a large Bear."

Of the men concerned in this voyage, Dampier and Selkirk are the most interesting personalities. The *Duke* and the *Dutchess* did not reach England until 1711, and after that absolutely nothing further is known of Dampier, navigator, philosopher, naturalist and buccaneer. He was no longer a young man, having devoted forty years of his life to the most strenuous of callings, and he died in such complete obscurity that even the most meagre record of his passing has not been found. His published *Voyages* were popular but apparently brought him little money. He lived the most conspicuous and adventurous of lives, and died unnoticed, perhaps in some peaceful village.

Selkirk arrived in England practically penniless. Presumably he shared in the spoils of the expedition but a privateer and his plunder were soon parted. In his necessity, it occurred to him that the story of his four years' solitary exile might have a money value, so he wrote a long account of his adventures and showed it to a friend. The friend recommended him to seek the advice of a rising young writer of the day, and thus did Robinson Crusoe knock at Daniel Defoe's door. Defoe knew Opportunity when he heard her. He kept the sailor's manuscript for some time and finally returned it, saying he could make nothing of it, and discouraged Selkirk from further effort. Not very long after, *Robinson Crusoe* appeared; it was the best seller of that day and many succeeding ones, but Selkirk died in poverty.

No desert island would be complete without its story of buried treasure, and the Galápagos furnish many such tales. Most of them deal with the loot of the buccaneers, but there is even a legend that the Incas brought a quantity of gold here when Pizarro invaded Peru; their ignorance of navigation, however, would make this seem wildly improbable. Concerning the buccaneers, at least, we have definite facts upon which to base tales of concealed booty. Both Davis and Rogers came here soon after acquiring rich plunder and some re-division of the spoils took place on the islands. Davis was here three times, and Rogers left the first time with the intention of soon returning, so to cache a certain amount of treasure in this isolated spot until their hazardous marauding was over, would not have been an unnatural thing to do.

In *The Cruise of the Dream Ship*, Ralph Stock says that twice of recent years treasure has been unearthed in the archipelago. One fortunate discoverer built an hotel in Guayaquil and still flourishes; the other drank himself to death as fast as possible.

As for Nature's buried treasure, there is some cannel coal of poor quality on Chatham; there is also a theory that the oil belt of Ecuador extends submarinely to the Galápagos, so an immense reservoir *may* exist there. Perhaps some day a modern kind of oil-hunter will replace those wandering groups who for years have made fitful onslaughts on the giant tortoises for the sake of their rich yellow fat, which is tried out into a valuable food product. At this safely remote day, buccaneers seem much more desirable than oil-hunters of any sort. No romantic soul could without a pang see oil-derricks on the beaches where pirate ships once went through that unspeakably picturesque process known as careening, and surely the wine-jars of the buccaneers are more interesting and worthy relics than the hundreds of enormous empty shells that in some places strew the ground in mute witness to the oil-hunters' ruthless slaughter of an extraordinary form of life.

The buccaneers' day had passed, the privateers' was passing. The whalers were coming, whose calling was almost as picturesque as the pirates', and sometimes nearly as gory. The British, ever alert for industrial and mercantile possibilities, sent out Colnett in 1793 to explore the whaling grounds and report on islands and harbours where ships employed in this industry might lie for repairs. The Galápagos were already somewhat vaguely known as a possible resort for refitting, and were in the midst of some of the best whaling waters. Captain Colnett can speak for himself and the Galápagos, which he does at some length in his book entitled *A Voyage to the South Atlantic and Round Cape Horn into the Pacific Ocean, undertaken and performed by Captain James Colnett, of the Royal Navy, in the ship Rattler.*

"On the same day (June 19, 1793) I took my departure from Cape St. Helena for Gallipagos Isles. . . . On the twenty-fourth, at four A.M., we made one of the Gallipagoe Isles, bearing West by North six or seven Leagues. . . . The land, toward the East, was covered with small trees or bushes without leaves, and very few spots of verdure were visible to us; a few seals were seen on the shore. The land rises at short intervening distances in small hills or hillocks of very singular forms, which, when observed through a glass, and at no great distance from the shore, have the appearance of habitations, while the prickly pear trees and the torch thistles, look like their owners standing round them. In other parts, the hills rise so sudden on the low land, that, having a small offing, they appear to be so many separate islands. About four miles off the North East end, there is a small islet, which is connected by a reef with the main isle; it is covered with seals, and the breakers reach some distance from the shore. The highest land, at this part of the isle, is of a very moderate height, descending gradually to the shore, which consists, alternately of rocks and sand; some of the rocky parts being much insulated, they form winding inlets of two or three miles in depth and from one to two cables in breadth.

"At the distance of two or three miles to the westward of the islet, I hove to and sent the chief mate on shore to sound and land. At eight P.M. he returned with green turtle and tortoises, turtle doves and guanas; but they saw no esculent vegetable, nor found any water that was sufficiently palatable to drink.

" This island contains no great number, or variety, of land birds, and those I saw, were not remarkable for their novelty or beauty; they were the flycatcher and creeper, like those of New Zealand; a bird, resembling the small mockingbird, of the same island; a black hawk, somewhat larger than our sparrow hawks, and a bird of the size and shape of our black bird. Ringdoves, of a dusky plumage, were seen in the greatest number; they seldom approached the sea till sun-set, when they took their flight to the Westward, and at sun-rise returned to the Eastward; so that if there is any water on the isle, I should suppose it would be found in that part. Besides, it is the highest land, and a small quantity of water, lodged in the hollow of a rock, would supply these birds for a considerable time. My second visit, to these isles, confirmed, my supposition, as small oozings were then found, at the foot of two or three hills, which may be occasioned by pools of rain water collected on the tops of them, as is frequently seen on the North West coast of America. An officer and party, whom I sent to travel inland, saw many spots, which had very lately contained fresh water, and about which, the land tortoise appeared to be pining in great numbers. Several of them, were seen within land, as well as on the sea coast, which if they had been in flesh, would have weighed three hundred weight, but which were now scarcely one third of their full size.

"I was very much perplexed to form a satisfactory conjecture, how the small birds, which appeared to remain in one spot, supported themselves without water; but the party on their return informed me, that having exhausted all their water, and reposing beneath a prickly pear-tree, almost choked with thirst, they observed an old bird in the act of supplying three young ones with drink, by squeezing the berry of a tree into their mouths. It was about the size of a pea, and contained a watery juice, of an acid but not unpleasant taste. The bark of the tree produces a considerable quantity of moisture and on being eaten allays the thirst. In dry seasons, the land tortoise is seen to gnaw and suck it. The leaf of this tree is like the bay-tree, the fruit grows like cherries, whilst the juice of the bark dies the flesh a deep purple, and emits a grateful odor; a quality in common with the greater part of the trees and plants in this island; though it is soon lost, when the branches are separated from the trunks or stems. The leaves of these trees also absorb the copious dews, which fall during the night, but in larger quantities at the full and change of the moon; the birds then pierce them with their bills, for the moisture they retain, and which I believe, they also procure from the various plants and evergreens. But when the dews fail in the summer season, thousands of these creatures perish; for on our return hither, we found great numbers dead in their nests, and some of them almost fledged. It may, however, be remarked that this curious instinctive mode, of finding a substitute for water, is not peculiar to the birds of this island; as nature has provided them with a similar resource in the fountain-tree that flourishes on the Isle Ferro, one of the Canaries; and several other trees and canes, which, Churchill tells us in his voyages, are to be found, on the mountains of the Phillipine Islands.

"There is no tree in this island, which measures more than twelve inches in circumference, except the prickly pear, some of which were three feet in the girth, and fifty feet in height. The torch thistle, which was the next in height, contains a liquid in its heart, which the birds drank when it was cut down. They sometimes even extracted it from the young trees, by piercing the trunks with their bills.

"We searched with great diligence for the mineral mountain, mentioned by Dampier, but were not so fortunate as to discover it; unless it be that from which the heavy sand and topazes were collected, and of which I ordered a barrel to be filled and brought it away.

". . . The various kinds of sea-birds which I had seen on the coast of Peru, we found here, but not in equal abundance. There were also flamingoes, sea-pies, plovers and sand-larks. The latter were of the same kind as those of New Zealand. No quadruped was seen on this island, and the greatest part of its inhabitants appeared to be of the reptile kind, as land tortoises, lizards, and spiders. We saw also dead snakes, which probably perished in the dry season. There were, besides, several species of insects, as ants, moths, and common flies in great numbers; as well as grass-hoppers and crickets.

"On the shore were sea guanas and turtle; the latter were of that kind which bears a variegated shell. The guanas are small and of a sooty black, which, if possible, heightens their native ugliness.[1] Indeed, so disgusting is their appearance, that no one on board could be prevailed on to take them as food.

"We saw but few seals on the beach, either of the hairy or furry species. This circumstance, however, might be occasioned by its not being the season for whelping. . . .

"Dampier mentions that there is plenty of salt to be obtained here at this season, but I could not find any. . . .

"The rocks are covered with crabs and there are also a few small wilks and winkles. A large quantity of dead shells, of various kinds, were washed upon the beach, all of which were familiar to me; among the rest were the shells of large crayfish, but we never caught any of them alive. On several parts of the shore there was driftwood of a larger size than any of the trees that grow on the island; also bamboos and wild sugar canes, with a few small cocoa nuts at full growth though not larger than a pigeon's egg. We observed also some burnt wood, but that might have drifted from the continent, been thrown overboard from a ship, or fired by lightening on the spot.

[1] The sea-guana is a nondescript; it is less than the land guana and much uglier; they go to sea in herds a-fishing, and sun themselves on the rocks like seals, and may be called alligators in miniature.

". . . As I could not trace these islands by any accounts or maps in my possession, I named one Chatham Isle, and the other Hood's Island, after the Lords Chatham and Hood. . . .

"The Rodondo is an high barren rock, about a quarter of a mile in circumference, and is visible as far as eight or nine league, has soundings round it at the distance of a quarter of a mile thirty fathom. Here our boats caught rock-cod in great abundance. I frequently observed the whales leave these isles and go to the Westward and in a few days, return with augmented numbers. I have also seen the whales coming, as it were, from the main, and passing along from the dawn of day to night, in one extended line, as if they were in haste to reach the Galipagoes. It is very much to be regretted that these isles have to this period been so little known but only to the Spaniards. . . .

"On reaching the South point of James's Isle, I got sight of three other isles which I had not seen before, nor can I trace them in the Buccaneers accounts, no more than the isle which we saw to Westward, when at anchor in Stephen's Bay, Chatham Isle. These three isles now seen, I named after the admirals Barrington, Duncan, and Jarvis. The two Northernmost, which are nearest to James's Isle, are the highest, and presented the most agreeable appearance, being covered with trees. The Southernmost, which I named Barrington Isle, is the largest and was the greatest distance from me, it is of a moderate height, and rises in hummocks; the South end is low, running on a parallel with the water's edge. We did not land on either of them. In this expedition we saw great numbers of penguins, and three or four hundred seals. There were also small birds, with a red breast, such as I have seen at the New Hebrides; and others resembling the Java sparrow, in shape and size, but of a black plumage; the male was the darkest, and had a very delightful note. At every place where we landed on the Western side, we might have walked for miles, through long grass and beneath groves of trees. It only wanted a stream to compose a very charming landscape. This isle appears to have been a favourite resort of the Buccaneers, as we not only found seats, which had been made by them of earth and stone, but a considerable number of broken jars scattered about, and some entirely whole, in which the Peruvian wine and liquors of that country are preserved. We also found some old daggers, nails and other implements. This place is, in every respect, calculated for refreshment or relief for crews after a long and tedious voyage, as it abounds with wood, and good anchorage, for any number of ships, and sheltered from all winds by Albemarle Isle. The watering-place of the Buccaneers was entirely dried up, and there was only found a small rivulet between two hills running into the sea; the Northernmost of the hill forms the South point of Fresh-water bay. Though there is a great plenty of wood, that which is near the shore, is not large enough for any purpose, but to use as fire-wood. In the mountains the trees may be of a larger size, as they grow to the summit of them. I do not think that the watering-place which we saw, is the only one on the island; and I have no doubt, if wells were dug any where beneath the hills, that it would be found in great plenty: they must be made, however, at some distance from the sandy beach, as within a few yards behind them, is a large lagoon of salt water, from three to eight feet in depth, which rises and falls with the tide; and in a few hours a channel might be cut into it. The woods abound with tortoises, doves, and guanas, and the lagoons with teal. The earth produces wild mint, sorrel, and a plant resembling the cloth-tree of Otaheite and the Sandwich Isles, whose leaves are an excellent substitute for the China tea, and was indeed pre-ferred to it by my people as well as by myself. There are many other kinds of trees, particularly the moli-tree, mentioned by Mr. Falkner, and the algarrooa, but that which abounds, in a superior degree, is the cotton tree. There is a great plenty of every kind of fish that inhabit the tropical Latitudes; mullet, devil-fish, and green turtle were in great abundance."

The Galápagos became an even more popular resort for whalers than it had been for buccaneers. One of the favourite anchorages was in the immensely long, crescent bay formed by the curve of Albemarle round Narborough; in this bay is a particularly sheltered nook that the whalers called Port Rendezvous. This is now known as Tagus Cove, a small, narrow haven, with deep water almost to the very foot of the high black cliffs that hem it in.

Strangely enough, the first establishment of any kind on the Galápagos was a post-office, though there were no inhabitants. It was a simple edifice. It consisted of a cask fastened to a tree at an anchorage on Charles Island which still bears the name of Post-Office Bay. Here whaling ships that had just rounded the Horn on voyages that might last two, three or even five years, left letters for their families, and here the homeward-bound ships stopped to gather up such mail as they might find deposited. This was a wholly unofficial, entirely friendly service.

The supply of fresh meat afforded by the islands proved a godsend to the whalers, for they discovered that the giant tortoise would live for months without either food or water and suffer astonishingly little loss of flesh. Scurvy was the ever-present menace to those sea-farers on their interminable voyages, and the tortoise was the equivalent of a cold-storage room in those days, since there was no other animal that they could keep alive for the sake of having fresh meat during the months when they never even sighted land. A stop at the islands became a regular feature of most whaling trips, and every ship would take away as many tortoises as could be stowed in the hold, where they continued to exist after the un-emotional manner of their kind, without apparent distress. There are accounts of ships that took five and six hundred of these creatures at a time, a larder that could be drawn upon at will. No wonder that in 1923 the *Noma* expedition found but one solitary tortoise which, for all that we can assert to the contrary, may have been the last of a great race.

The story of the first resident of the Galápagos is worth

telling. It is quoted from Captain Porter's *Journal of a Cruise:*

". . . on the east side of the island (Charles) there is another landing, which he calls Pat's landing; and this place will probably immortalize an Irish man, named Patrick Watkins, who some years since left an English ship, and took up his abode on this island, and built himself a miserable hut, about a mile from the landing called after him, in a valley containing about two acres of ground capable of cultivation, and perhaps the only spot on the island which affords sufficient moisture for the purpose. Here he succeeded in raising potatoes and pumpkins in considerable quantities, which he generally exchanged for rum, or sold for cash. The appearance of this man, from the accounts I have received of him, was the most dreadful that can be imagined; ragged clothes, scarce sufficient to cover his nakedness, and covered with vermin; his red hair and beard matted, his skin much burnt, from constant exposure to the sun, and so wild and savage in his manner and appearance, that he struck everyone with horror. For several years this wretched being lived by himself on this desolate spot, without any apparent desire than that of procuring rum in sufficient quantities to keep himself intoxicated, and, at such times, after an absence from his hut of several days, he would be found in a state of perfect insensibility, rolling among the rocks of the mountains. He appeared to be reduced to the lowest grade of which human nature is capable, and seemed to have no desire beyond the tortoises and other animals of the island, except that of getting drunk. But this man, wretched and miserable as he may have appeared, was neither destitute of ambition, nor incapable of undertaking an enterprise that would have appalled the heart of any other man; nor was he devoid of the talent of rousing others to second his hardihood.

"He by some means became possessed of an old musket, and a few charges of powder and ball; and the possession of this weapon probably first stimulated his ambition. He felt himself strong as the sovereign of the island, and was desirous of proving his strength on the first human being that fell in his way, which happened to be a negro, who was left in charge of a boat belonging to an American ship that had touched there for refreshments. Patrick came down to the beach where the boat lay, armed with his musket, now become his constant companion, directed the negro, in an authoritative manner, to follow him, and on his refusal, snapped his musket at him twice, which luckily missed fire. The negro, however, became intimidated, and followed him. Patrick now shouldered his musket, marched off before, and on his way up the mountain exultingly informed the negro he was henceforth to work for him, and become his slave, and that his good or bad treatment would depend on his future conduct. On arriving at a narrow defile and perceiving Patrick off his guard, the negro seized the moment, grasped him in his arms, threw him down, tied his arms behind, shouldered him, and carried him to his boat, and when the crew had arrived he was taken on board the ship. An English smuggler was lying in the harbour at the same time, the captain of which sentenced Patrick to be severely whipped on board both vessels, which was put in execution, and he was afterward taken on shore handcuffed by the Englishmen, who compelled him to make known where he had concealed the few dollars he had been enabled to accumulate from the sale of his potatoes and pumpkins, which they took from him. But while they were busy in destroying his hut and garden, the wretched being made his escape, and concealed himself among the rocks in the interior of the island, until the ship had sailed, when he ventured from his hiding-place, and by means of an old file, which he drove into a tree, freed himself from the handcuffs. He now meditated a severe revenge, but concealed his intentions. Vessels continued to touch there, and Patrick, as usual, to furnish them with vegetables; but from time to time he was enabled, by administering potent draughts of his darling liquor to some of the men of the crews, and getting them so drunk that they were rendered insensible, to conceal them until the ship had sailed; when, finding themselves entirely dependent on him, they willingly enlisted under his banners, became his slaves, and he the most absolute of tyrants. By this means he had augmented the number to five, including himself, and every means was used by him to endeavour to procure arms for them, but without effect. It is supposed that his object was to have surprised some vessel, massacred her crew, and taken her off. While Patrick was meditating his plans, two ships, an American and an English vessel, touched there and applied to Patrick for vegetables. He promised them the greatest abundance, provided they would send their boats to his landing, and their people to bring them from his garden, informing them that his rascals had become so indolent of late, that he could not get them to work. This arrangement was agreed to; two boats were sent from each vessel, and hauled on the beach. Their crews all went to Patrick's habitation, but neither he nor any of his people were to be found; and, after waiting until their patience was exhausted, they returned to the beach, where they found only the wreck of three of their boats, which were broken to pieces and the fourth one missing. They succeeded, however, after much difficulty, in getting around to the bay opposite to their ships, where other boats were sent to their relief; and the commanders of the ships, apprehensive of some other trick, saw no security except in a flight from the island, leaving Patrick and his gang in quiet possession of the boat. But before they sailed, they put a letter in a keg, giving intelligence of the affair, and moored it in the bay, where it was found by Captain Randall, but not until he had sent his boat to Patrick's landing, for the purpose of procuring refreshments; and, as may be easily supposed, he felt no little inquietude until her return, when she brought him a letter from Patrick to the following purport, which was found in his hut.

"SIR,

"I have made repeated applications to captains of vessels to sell me a boat, or to take me from this place, but in every instance met with a refusal. An opportunity presented itself to possess myself of one, and I took advantage of it. I have been a long time endeavouring, by hard labour and suffering, to accumulate wherewithal to make myself comfortable; but at different times have been robbed and maltreated, and in a late instance by captain Paddock, whose conduct in punishing me and robbing me of about five hundred dollars, in cash and other articles, neither agrees with the principles he professes, nor is it such as his sleek coat would lead one to expect.[1]

"On the 29th of March, 1809, I sail from the enchanted island in the *Black Prince*, bound to the Marquesas.

"Do not kill the old hen; she is now sitting and will soon have chicks.

(Signed)   FATHERLESS OBERLUS."

"Patrick arrived alone at Guayaquil in his open boat, the rest who sailed with him having perished for want of water, or, as is generally supposed, were put to death by him on his finding the water to grow scarce. From thence he proceeded to Payta, where he wound himself into the affection of a tawny damsel, and prevailed on her to consent to accompany him back to his enchanted island, the beauties of which he no doubt painted in glowing colours; but, from his savage appearance, he was there considered by the police as a suspicious person, and being found under the keel of a small vessel then ready to be launched, and suspected of some improper intention, he was confined in Payta gaol, where he now remains, and probably owing to this circumstance Charles Island, as well as the rest of the Gallipagos, may remain unpopulated for many ages to come. This reflection may naturally lead us to a consideration of the question concerning the population of the other islands scattered about the Pacific ocean, respecting which so many conjectures have been hazarded. I shall only hazard one, which is briefly this; that former ages may have produced men equally as bold and daring as Pat, and women as willing as his fair one to accompany them on their adventurous voyages. And when we consider the issue which might be produced from a union between a red-haired wild Irishman and a copper-coloured mixt-blooded squaw, we need not be any longer surprised at the different varieties in human nature.

"If Patrick should be liberated from durance, and arrive with his love at this enchanting spot, perhaps (when neither he nor the Gallipagos are any longer remembered) some future

[1]Captain Paddock was of the Society of Friends.

navigator may surprise the world by a discovery of them, and his accounts of the strange people with which they may probably be inhabited. From the source from which they shall have sprung, it does not seem unlikely that they will have one trait in their character which is common to the natives of all the islands in the Pacific, a disposition to appropriate to themselves the property of others. From this circumstance, future speculators may confound their origin with that of all the rest."

In 1800 Amasa Delano, stout old Boston sea-captain, visited the Galápagos and made some interesting observations there. He was evidently one of those fortunates who arrived at the islands during or immediately after the rainy season, for he says that at James Bay "we likewise found fresh water here as Captain Colnett mentions, which was very good and filled 18 to 20 of our butts."

He made several natural history notes, and I quote his remarks anent that "innocent animal," the iguana:

"The Land Guana is very similar in shape to the lizard or alligator, having four legs, and is about two feet and a half long. Their shape is like a short, thick snake with four legs; but it is a very innocent animal. Its colour is like that of burned rocks or cinder, and their skin looks almost as coarse and rough. They are tolerably good eating and would be made use of for food were there not so many terrapins and sea-turtles to be got at this place.

"The sea guana resembles the land guana in its shape, being about the same size but its back and head go up to a sharp ridge on the top, and a comb runs from near the nose over the top of its head to near the end of its tail on the top of this ridge, which gives it the most disagreeable appearance of any animal to be found here. The colour of the skin is nearly black and has as rough and coarse an appearance as the land kind. It obtains its being entirely out of the sea. The other kind feeds upon the same vegetable substance as the terrapin.

"The largest kind of lizards found here resembles the land guana in everything except size, they being only a little more than half the length. Their colour and coarse appearance are the same with the exception of a bright vermilion red throat, which makes it appear as if bloody. There are to be found here also two smaller kind of lizards. The smallest is not much longer than a man's finger. The size of the other kind is between the two. There is no particular difference in the shape of the three kinds; but the colour of the two latter is gray."

A chapter could almost be written about the names applied by sailors to the strange animals they met in their voyages. They are sometimes particularly interesting as reflecting the history of the time. So in Delano's *Narrative* we find a reference to the fact that in his day the sailors drew upon the Napoleonic wars in naming two of the denizens of the bird rookeries. The large solemn pelicans parading the barren shores were called the Russian Army, while the boobies, present in even larger numbers, received the name of Bonaparte's Army. The lovely snow-white bird, known to us as the tropic-bird, was then called the Boatswain's Mate, because of his clear, sweet whistle.

Delano wrote two excellent descriptions of the difference in manner of diving between the Russian Army and Bonaparte's Army. Concerning the pelican he says:

"They are the most clumsy bird that I ever saw. When in the act of diving they make the most awkward appearance that can be imagined; which cannot be better described than by comparing it to the manner in which a sailor washes his clothes, by making them fast to the end of a rope and throwing them from the forecastle into the sea; when they strike the water, they spread out, with the trowsers in one direction, the shirt in another and the jacket in a third. The pelican makes a plunge into the water for the purpose of obtaining its food in a similar manner, its wings being extended, its mouth open and its bill expanded; and with two enormously large feet spread out behind exhibits itself in a very ludicrous and sprawling manner."

Indefatigable seems to be the least known of all the islands even to the present day. I have searched in vain to discover when, by whom and after what it was named. Cowley called it the Duke of Norfolk's Island, but so late as 1813 Captain Porter, failing to recognize it from any position laid down in the chart, sent his chaplain to examine it, by whom it was named Porter's Island. Delano refers to it as follows:

"There is one very large island to the S E of James's Island, 8 or 10 leagues distant, which I know but very little of, as I never visited it, neither have I seen anyone that did. Its appearance . . . was more favourable than any one of the cluster."

What is probably the first description of an eruption in the Galápagos is found in Delano's *Narrative:*

"The most extraordinary phenomenon happened while we were riding in James's Bay in the year 1800 that I ever witnessed in my life; and as I do not remember to have ever heard or read of anything like it either before or since, I will here insert an extract from the ship's Journal, describing it:
" ' . . . As our boat was coming from the watering place on the evening of the 21st of August between sunset and dark, with a load of water, we saw a large black cloud gathering over the highest mountain on Albemarle Island, which was the same place where one of the men on board our ship had asserted that he had seen a volcano burning in 1797; soon after the cloud gathered, it formed a spire or piked end similar to that of a cloud when about to meet a water-spout. It descended to the top of the mountain, with a body of fire following it apparently of the size of the largest part of the steeple of a meeting-house. Its illumination was so great that it at-

tracted the attention of all the people in the boat although they were at the time rowing with their backs toward it. After the fire had descended to the top of the mountain it continued some seconds, when it broke like a water-spout, and left a streak where it had passed, which appeared as brilliant as a column of fire and continued for nearly half an hour before it wholly disappeared.'"

One incident in the *Narative* is too amusing to omit, even though it is entirely irrelevant to the Galápagos; later in this same voyage Delano touched at the Sandwich Islands, and when he left took with him two native boys, one of whom he called Bill, and the other a natural son of the King of the Islands, who called himself Alexander Stewart, after a white man to whom he took a fancy. Bill went with Delano to Boston, where "he performed on the stage several times, in the tragedy of Captain Cook, and was much admired by the audience and the publick in general!"

Captain Porter's *Journal of a Cruise* is a most delightful book, abounding in vivid description and humour. It also contains more real information about the Galápagos than is to be met with elsewhere. During the War of 1812 he rounded the Horn in the frigate *Essex*, with the intention of clearing the Pacific of British whalers. The whalers from Nantucket and New Bedford had fared badly at the hands of the British and their Peruvian allies until Porter's advent, when he succeeded in making prizes of nearly the whole British whaling fleet.

After a frightful passage round the Horn, "we began to form our projects for annoying the enemy." Obtaining information that at least twenty British ships were cruising in the vicinity of the Galápagos, Porter sailed from Payta in company with the *Barclay*, an American whaler, and sighted Chatham Island on April 17th, 1813. Finding no ships either there or at Hood, he stood over for the harbour of Charles Island. Here they perceived no vessels, but this landing-place was the well-known Post-office Bay, and from the cask used as a repository for letters, Porter took papers which confirmed the information he had already received concerning the presence of the enemy in these waters. He was thus enabled to make up a list of the ships he might expect to encounter. He says:

" . . . The landing here is very good; and, at the time lieutenant Downes was on shore, a torrent of very fine water many feet deep, discharged itself near the beach; but as it was raining constantly while he was on shore, and the mountains were completely capped with the clouds, added to which, as the banks of the deep ravine, worn away by the stream, clearly showed that the torrent had subsided ten feet within a very short period, it was evident to us, that this stream owed its existence to temporary rains alone. This opinion was not only confirmed by those on board the *Essex* who had been there before, but by some person who had bountifully left on the island, near the post-office, several articles for such persons as might be there in distress, among which was a cask of water. It is known that in the centre of the island is a small spring of water, which a stranger might not be acquainted with or, if he had a knowledge of it, might not have the strength to reach; but if the stream in question existed constantly, where would be the necessity of leaving this cask of water along side of it?

" . . . for although it seldom rains on shore and never at sea here, yet the tops of the mountains are almost constantly covered with thick clouds, great part of the moisture from which, instead of being soaked up by the light and spongy soil of the mountains, would find its way in running streams to the sea, were the islands sufficiently furnished with trees to condense more constantly the atmosphere, and interlace their roots to prevent its escape into the bowels of the mountains. . . . I made all sail to make the south head of Albemarle.

When we arrived within eight or nine miles of a point, which I have named Point Essex . . . I took my boat and proceeded for the aforesaid point, where I arrived in about two hours after leaving the ship, and found in a small bay, behind some rocks which terminate the point, a very good landing, where we went on shore. . . . The rocks were everywhere covered with seals, penguins, guanas, and pelicans, and the sea filled with green turtle, which might have been taken with the greatest ease, had we been enabled to take them into our boat; for we sometimes rowed right against them, without their making any exertion to get out of our way. Multitudes of enormous sharks were swimming about us and from time to time caused us no little uneasiness, from the ferocious manner in which they came at the boat and snapped at our oars; for she was of the lightest construction, with remarkably thin plank, and a gripe from one of those would have torn them from her timbers. But we guarded as much as lay in our power against the danger, by thrusting boarding-pikes into them as they came up."

Lieutenant Downes was sent in a small boat to reconnoitre in Banks' Bay for enemy ships, but perceived none. However,

"we did not wish to believe that the bay was destitute of vessels; and while there was room to build a hope of meeting the enemy, we kept our spirits up with the expectation of finding them, either in the bay, or at anchor in a cove called the Basin, on the Albemarle side of the passage between Elizabeth and Banks' Bay, where the whalers frequently go to refit and wood, and get tortoises. Here, at times, a small quantity of fresh water may be obtained, but never more than sixty gallons per day, and seldom so large a quantity, and this only after heavy rains . . . so fully was I of the belief that I should fall in with an enemy that would offer some resistance that I considered it most prudent to clear away the guns every night, and keep the hammocks stowed in the nettings, so as to be prepared for any force that might be assembled."

GALAPAGOS

Captain Porter visited the Basin, which was Tagus Cove, and said that "the art of man could not have formed a more beautiful basin . . . surrounded by high cliffs, except at the very bottom, where is the only landing for boats at a small ravine." He speaks of

"trees of a considerable size, which would afford wood for shipping, and among them a species from which oozed a resinous substance, in very large quantities, dripping from the trunk and limbs. This tree produces a fruit nearly as large as a cherry; it was then green, and had a very aromatic smell and taste. . . . On the side of a rock . . . we found the names of several English and American ships cut, whose crews had been there; and but a short distance from thence was erected a hut, built of loose stones, but destitute of a roof. In the neighbourhood of it were scattered, in considerable quantities, the bones and shells of land and sea tortoises. This I afterwards understood was the work of a wretched English sailor, who had been landed there by his captain, destitute of everything, for having used some insulting language to him. Here he existed near a year on land tortoises and guanas, and his sole dependence for water was on the precarious supply he could get from the drippings of the rocks; at length, finding that no one was likely to come to take him from thence, and fearful of perishing for want of water, he formed a determination to attempt at all hazards getting into Banks Bay, where the ships cruise for whales. With this in view he provided himself with two sealskins, with which, blown up, he formed a float; and, after hazarding destruction from the sharks, which frequently attacked his vessel, and which he kept off with the stick that served him as paddle, he succeeded at length in getting along side an American ship early in the morning, where his unexpected arrival not only surprised but alarmed the crew. His appearance was scarcely human; clothed in the skins of seals, his countenance haggard, thin, and emaciated, his beard and hair long and matted, they supposed him a being from another world. The commander of the vessel where he arrived felt a great sympathy for his sufferings, and determined for the moment to bring to punishment the villain who had, by thus cruelly exposing the life of a fellow-being, violated every principle of humanity; but from some cause or other he was prevented from carrying into effect his laudable intentions, and to this day the poor sailor has not had justice done him."

A fortnight of fruitless cruising, alternating with fishing and turtle hunting, found them almost discouraged. Several times the cry of "sail ho!" had inspirited them, but in each case the lookout had been deceived by the likeness to a sail of a rock or sand-bank.

"There were few on board the ship who did not now despair of making any captures in the Gallipagos Islands; and I believe that many began to think that the information that we had received respecting the practice of British vessels frequenting those islands, as well as the flattering expectations which this information had given rise to, had been altogether deception. But I could not so lightly lay down the opinions, which had caused me to visit those islands, and had been formed on information that could not be doubted. I determined not to leave the Gallipagos so long as there remained a hope of finding a British vessel among them. . . .

"At daylight on the morning of the 29th, I was roused from my cot, where I had passed a sleepless and anxious night, by the cry of "sail ho!" "sail ho!" which was re-echoed through the ship and in a moment all hands were on deck. The strange sail proved to be a large ship, bearing west, to which we gave chase; and in an hour afterwards we discovered two others, bearing southwest, equally large in their appearance. I had no doubts of their being British whale-ships; and as I was certain that toward mid-day, as usual, it would fall calm, I felt confident we should succeed in taking the whole of them. I continued my pursuit of the first discovered vessel, and at nine o'clock spoke her under British colours. She proved to be the British whale-ship *Montezuma*, captain Baxter, with one thousand four hundred barrels of spermaceti oil. I invited the captain on board. I took his crew on board the *Essex*, put an officer and crew in the *Montezuma*, and continued in pursuit of the other vessels, which made all exertions to get from us. At eleven A.M. according to my expectation, it fell calm; we were then at the distance of eight miles from them. . . . Thick and hazy weather is prevalent here, and as there was every indication of it, I was fearful that, in the event of a breeze, one or the other of them might make its escape from us, as I had understood that they were reputed fast sailers. I therefore thought it advisable to attempt them in our boats, and with this in view had them prepared for the purpose, and in a few minutes they departed in two divisions. . . . I had given the most positive orders that the boats should be brought into action all together, and that no officer should take advantage of the fleetness of his boat to proceed ahead of the rest, believing that some of them, from their extreme anxiety to join with the enemy, might be so imprudent as to do so. At two o'clock the boats were about a mile from the vessels (which were a quarter of a mile apart) when they hoisted English colours and fired several guns. The boats now formed in one division, and pulled for the largest ship, which, as they approached, kept her guns trained on them. The signal was made for boarding; and, when lieutenant Downes arrived within a few yards of her gangway, and directed them to surrender, the colours were hauled down. They now proceeded for the other vessel, after leaving an officer and some men on board, and as soon as she was hailed, she followed the example of the first in striking her colours. Shortly afterwards a breeze sprang up, the prizes bore down for us, and we welcomed the safe return of our shipmates with three hearty cheers. The captured vessels proved to be, as I had expected the *Georgiana*, captain Pitts, of two hundred and eighty tons, and the *Policy*, of two hundred and seventy-five tons; and these three vessels, which we had taken with so little trouble, were estimated to be worth in England upwards of half a million dollars.

". . . The possession of these vessels, besides the great

satisfaction it produced, was attended by another advantage of no less importance, as it relieved all our wants except one, to wit, the want of water. From them we obtained an abundant supply of cordage, canvas, paints, tar, and every other article necessary for the ship, of all of which she stood in great need, as our slender stock brought from America had now become worn out and useless. Besides the articles necessary for the ship, we became supplied with a stock of provisions, of a quality and quantity that removed all apprehensions of our suffering for the want of them for many months, as those vessels, when they sailed from England, were provided with provisions and stores for upwards of three years, and had not yet consumed half their stock. All were of the best quality; and were it only for the supplying our immediate wants, the prizes were of the greatest importance to us. We found on board of them, also, wherewith to furnish our crew with several delicious meals. They had been in at James Island, and had supplied themselves abundantly with those extraordinary animals the tortoises of the Gallipagos, which properly deserve the name of elephant tortoise. Many of them were of a size to weigh upwards of three hundred weight; and nothing, perhaps, can be more disagreeable or clumsy than they are in their external appearance."

From that time Porter's fleet steadily increased. Some of his prizes he sent to the mainland, some he equipped and armed from other vessels. The greatest problem was that of manning the captured ships. Before Porter's cruise was finished, the only men left on the *Essex* above the rank of seaman were himself and the surgeon's mate. All the other officers, including the chaplain and the doctor, were in command of prize vessels. A twelve-year-old midshipman was his last resort as an officer to put in command,—a midshipman whom he had adopted three years previously in New Orleans, and who was to be the first full Admiral of the United States Navy. This boy was David Farragut, and half a century later he and

Captain Porter's own son co-operated at the taking of New Orleans.

The difficulty of obtaining crews to man the prizes was overcome in a rather surprising way. Porter says, "A number of seamen captured in the prizes had already proffered their services to us; and on inquiry I found many of them to be Americans . . ." and numbers of the British were willing to give their paroles and stay to work their own ships.

From captured vessels Porter obtained materials to repair the old *Essex*, which had suffered severely in the passage round the Horn; he says, ". . . for although we had had it in our power to supply ourselves at Valparaiso, I did not procure them (materials) there, confidently believing that the enemy would, in due time, furnish us with what we wanted." His confidence was amply rewarded. He was an adept at disguise and he constantly changed the appearance of the *Essex* by building false parts and re-painting her, so that British ships should not recognize her as the frigate that had worked such havoc in their fleet.

The invariable Galápagos water problem was pressing. He returned to Charles Island with all the ships except the *Georgiana*, a prize in command of Lieutenant Downes, who was sent to look for enemy ships round other parts of the archipelago, while Porter occupied himself in a search for water.

"I had heard of a spring in the interior, which could be approached from a beach on the west side, about six miles distant from the ship. To this place I proceeded next morning, taking with me two ten-gallon kegs to make the experiment with, in order to estimate the quantity we could procure from thence daily. We found the spring at the distance of three miles from the beach, and the water, after clearing it out, proved excellent. But it was found to be extremely laborious work getting it down to the beach, as our stoutest men were exhausted after taking down one keg each; and it was found that each man could not carry any more than three kegs in twenty-four hours, owing to the distance, the badness of the roads, and the excessive heat on shore. I concluded, however, on attempting to get some water to answer our present purpose, notwithstanding the difficulties which opposed us, and with this view returned to the ship to make the necessary arrangements, and on my way loaded my boat with some excellent fish.

"On landing at the beach leading to the spring, we found fresh embers, and a tortoise, which had not been killed apparently more than two days. On our way to the spring we found other testimonies of persons having been recently there; such as a pair of mockasons, made of English canvas, and a tortoise shell containing about two quarts of English barley.

"This part of the island abounds with tortoises, which frequent the springs for the sake of the water, and upwards of thirty of them were turned on their backs by us, as they came down to drink during the short time we remained there, which was not more than an hour and a half. But we were enabled to bring down only one, and he was selected more for his antiquated appearance than for his size or supposed excellence. His weight was exactly one hundred and ninety-seven pounds, but he was far from being considered of a large size.

"As I returned from the spring, I could not help reflecting on the extraordinary scheme that I was about attempting to procure water, and was almost appalled by the obstacles which presented themselves. In addition to the difficulties of getting it down to the beach, it would be necessary there to put it into large casks and from thence raft it to the ship, a distance of six miles, through a high sea, and sometimes against rapid currents. To this must be added the danger and inconvenience of having one half of my crew at least separated from the ship, thus leaving not only her but our prizes exposed, in a defenceless state, to the attacks of an enemy. As water was to be procured in that part of the island, I thought it not unlikely that it might be found near the bay in which we lay and well knowing the roving disposition of seamen, I determined to let a party go on shore to amuse themselves, confidently believing, if water was to be found within two or three miles of us, it would be discovered by them. On their return at night I was not disappointed, for they informed me that they had found upwards of forty or fifty barrels of water lodged in the different hollows of the rocks about a mile and a half from the shore; that the difficulties of getting to it were very great, but they did not doubt that each man would be enabled to bring down, in ten-gallon kegs, forty gallons per day. I immediately caused casks to be landed, and by sending parties on shore daily, procured while we lay there two thousand gallons, much of it, to be sure, of a filthy appearance, having a bad taste and smell, and filled abundantly with slime and insects. But to us it was a treasure too precious to lose, and the greatest industry was used to save every drop of it, for fear that the sun, which was evaporating it rapidly, should cheat us of our prize.

"In order that no means should be left untried to procure a large supply of water, I caused two wells to be dug in the most likely places for finding it; but, after digging a considerable depth, salt water flowed in and disappointed our hopes. I also sent on shore a wooding party, which soon procured us as large a supply of fuel 'as we stood in need of.

"Early in the morning of the third day of our arrival, a sail was discovered to the westward, standing in for the island. I immediately caused preparation to be made for sending the boats after her, as the wind was very light; but on her nearer approach, when she made her private signal, discovered it to be the *Georgiana*. Her arrival, although unexpected, gave me much pleasure, and on lieutenant Downes coming on board . . . I despatched him to Albemarle, in pursuit of the stranger who had touched at the island before us, directing him to stop at Charles' Island as soon afterwards as possible, and, should he not find me there, to search at the foot of the stake to which the letter-box is attached, where I should bury a bottle containing instructions for him. . . . We occupied ourselves in painting our ship's bends and upper-works, keeping parties every day on shore bringing down to the beach tortoises for the ship's company, of which they succeeded in getting on board between four and five hundred. Although the parties on this employment (which were selected every day, to give all an opportunity of going on shore) indulged themselves in the most ample manner in tortoises meat (which by them was called Gallipagos mutton) yet their relish for this food did not seem in the least abated, nor their exertions to get them on board in the least relaxed, for everyone appeared desirous of securing as large a stock of this provision as possible for the cruise. They were brought the distance of from three to four miles, through thorns and over sharp rocks; yet it was no uncommon thing for them to make three and four trips a day, each with tortoises weighing from fifty to a hundred weight. We were enabled to procure here, also, in large quantities, an herb in taste much resembling spinage, and so called by our people; likewise various other pot-herbs, and prickly pears in great abundance, which were not only of an excellent flavour, but a sovereign anti-scorbutic. It afforded me great pleasure to observe that they were so much relished by our people.

"The cotton plant was found growing spontaneously, and a tree of a very aromatic flavour and taste, which was no other than the one formerly mentioned, found on the island of Albemarle, and producing in large quantities a resinous substance. This Mr. Adams declared was the alcornoque, so famous for the cure of consumptions and is probably the same as that mentioned by Colnet, and called by him the algarrooa.

"The only quadrupeds found on the island were tortoises, lizards, and a few sea guanas; the land guana was not to be found. Doves peculiar to these islands, of a small size, and beautiful plumage, were very numerous, and afforded great amusement to the younger part of the crew in killing them with sticks and stones, which was nowise difficult, as they were very tame. The English mockingbird was also found in great numbers, and a small black bird, with a remarkably short and strong bill, and a shrill note. These were the only birds except aquatic found here; the latter were not numerous, and consisted of teal, which frequented a lagoon on the east part of the bay, pelicans, boobies, and other birds common to all the islands of these seas. Sea turtles and seals were scarce and shy.

"That every person might be employed to the most advantage, I directed that those having charge of prizes should paint them, and otherwise put them in good order as to appearance, in the expectation that they would bring a higher price among the Spaniards, to whom I intended offering them for sale at the first opportunity. They were noble ships, and a little paint added greatly to the beauty of their appearance. . . . The appearance of the *Essex* had been so frequently changed, that I had but little apprehension of her being known again by those who had seen her before, or from any description that could be given of her. While we lay here, I permitted all the prisoners to go on shore whenever they wished it, as many of them were afflicted with the scurvy. One in particular was so bad with it as to be scarcely able to move. But on getting him on shore, where he could procure a kind of sorrel and the prickly pear and burying his legs in the earth every day, he was so far recovered before our departure, as scarcely to complain of his disease, and could walk as briskly as any among us, assisting frequently in bringing down water and tortoises from the rocks and mountains.

"We here found the tomb of a seaman who had been buried five years before, from a ship called the *Georgiana*, commanded by captain Pitts. Over it was erected a white board, bearing an inscription, neatly executed, showing his age, etc., and terminating with the following epitaph, which I insert more on account of the extreme simplicity of the verse, and its powerful and flattering appeal to the feelings, than for its elegance, or the correctness of the composition:

> "'Gentle reader, as you pass by,
> As you are now, so wonce was I;
> As now my body is in the dust,
> I hope in heaven my soul to rest.'"

"The spot where his remains were deposited was shaded by two lofty thorn bushes, which afforded an agreeable shade and fragrance, and became the favourite resort of our men at their meals. The pile of stones (which had been piously placed over the grave by his shipmates) served them both for table and seat, where they indulged themselves amply in their favourite food, and quaffed many a can of grog to his poor soul's rest!"

The chaplain, Mr. Adams, who explored Indefatigable, and named it Porter's Island, reported that the tortoises there were "generally of an enormous size, one of which measured five feet and a half long, four feet and a half wide, and three feet thick, and others were found by some of the seamen of a larger size."

Mr. Adams also reported having passed a ship which he believed to be English, and which appeared to be bound for Albemarle. Porter thought that she might be visible from one of the adjacent hills, so.

"landed on the western point of the bay, and, in company with lieutenant Gamble of the marines, and Mr. Shaw, purser, proceeded to ascend a high and rugged mountain there situated, which did not appear to us a difficult task to attempt. But we were soon convinced of our error, for it was not without great labour and fatigue, and at the risk of our lives, that we succeeded in reaching the top of it, after crawling through thorn bushes, wounding ourselves by the prickly-pear trees, and scrambling over loose lava which tore our shoes and was every moment giving way under us. We at length, however, arrived, exhausted with thirst, heat, and fatigue, at the summit, where we had an extensive view of the islands, but could perceive no vessels in the offing. Our descent was no less hazardous; and on our way back we found a large tortoise, which we opened with some difficulty, with the hope of finding some water to allay our thirst. But we were disappointed in only finding a few gills, of a disagreeable-tasted liquid. This our stomachs revolted at; we therefore had recourse to sucking the leaf of the prickly pear, which we found to serve our purpose."

Porter was among the islands from April 18th to June 8th, a great part of the time struggling against currents which seemed invariably to set in the direction in which he did not wish to go. He witnessed a not very spectacular eruption on Albemarle on June 6th, and on June 8th passed to the north of Abingdon, bound for the mainland. He had buried two men in the waters of the archipelago, one a seaman, the other Dr. Miller, who had been ailing for a long time. The water question he sums up as follows:

"I have no doubt but the spring formerly mentioned at Charles' Island is a never-failing one, where water may at all times be had; the distance from the sea, to be sure, is great and but few would attempt to water a ship of war from it; it may, however, be of use to those who are really suffering for water. Colnet and others mention streams of water at James' and Chatham Islands, but I am induced to believe, from what I have learnt from my prisoners, that they owe their existence to temporary rains, and are similar to the place I visited near the basin in Albemarle, where it is said water has been obtained formerly. Supplies from them, however, are too precarious to place any dependence on, and it is advisable for every vessel visiting the Gallipagos, to lay in a good stock of that necessary article, as they may not be so fortunate as myself in capturing vessels with a large quantity on board, which, although contained in the oily casks of a whale-ship, and from them, as may be supposed, derived no very agreeable taste or smell, but on the contrary, produced nausea when drank; yet we considered it the most valuable part of our prize."

Porter stood over to Túmbez on the mainland, and after taking three more prizes, sent Downes to Valparaiso with four British ships to sell, and with instructions to rejoin him at the Galápagos. He planned to return there because of information that three armed English ships had gone to their favourite fishing-ground a fortnight before. He captured them all in

Banks' Bay in a few hours, but even in the excitement of the pursuit he was observant of other matters:

"Notwithstanding the great interest I felt for the critical situation of my prizes, as well as that which every officer must feel when in pursuit of an enemy, I could not help remarking the operations of Nature on the south side of Narborough and on the southern part of Albemarle. Narborough appeared to have undergone great changes since our last visit, by the violent irruptions of its volcanoes; and at this time there were no less than four craters smoking on that island, and one on the south part of Albemarle. I should have before mentioned, that a few hours after leaving Charles' Island, a volcano burst out with great fury from its centre, which would naturally lead to the belief of a submarine communication between them."

The largest ship captured in this engagement

"proved to be the *Seringapatam*, which had taken the letters, wood, kegs, etc., from Charles' Island. The capture of this ship gave me more pleasure than that of any other which fell into my hands; for besides being the finest British ship in those seas, her commander had the character of being a man of great enterprise and had already captured the American whaleship *Edward*, of Nantucket, and might have done great injury to the American commerce in those seas. Although he had come into the Pacific on a whaling voyage, he had given but little attention to that object while there was a hope of meeting American whalers. On requiring of this man that he should deliver to me his commission, he, with the utmost terror in his countenance informed me that he had none with him, but was confident that his owners had, before this period, taken out one for him, and he had no doubt would send it to Lima, where he expected to receive it. It was evident that he was a pirate, and I did not feel that it would be proper to treat him as I had done other prisoners of war. I therefore ordered him and all his crew in irons; but after enquiring of the American prisoners, whom I found on board the prize, as to the manner in which they had been treated by the crew of the *Seringapatam*, and being satisfied that they, as well as the mates, were not to blame for the conduct of their commander, I liberated them from confinement, keeping Stavers only in irons."

A strange sea fight took place off Albemarle, when they sighted a strange sail, which hoisted the American colours, but at the same time made all sail away from them.

"Every exertion was made to come up with her, as she was evidently a whale-ship; and from every appearance I had no doubt of her being English. The winds becoming light, inclinable to calm, we made use of our drags, and found considerable advantage from them; but from the constant labour requisite to work them, our people became very much harassed, and finally worn out with fatigue. We had, however, by the greatest exertions, approached within four miles of the chase, and were enabled, by the assistance of our glasses, to see all his movements. He now got his boats ahead to tow his ship, with a view, as I supposed, of running her on shore on the island of Abingdon, which was not far distant. To prevent his effecting this object, I despatched the gig and whale-boat, the first under command of lieutenant M'Knight, the other under Mr. Bostwick, clerk, with a few good marksmen to drive them from their boats, but with the most positive orders to make no attempt on the ship. They soon succeeded in driving the boats alongside the ship, but found great difficulty in keeping out of the range of his shot, as he had mounted two guns on his forecastle, with which he kept up a constant fire on our boats, having hauled down his American colours and hoisted English. At four o'clock on the afternoon of the 30th, both ships were perfectly becalmed, at the distance of three and a half miles from each other, our two boats lying ahead of the enemy, and preventing his boats from towing; my crew so worn out with fatigue as to be incapable of working the drags to any advantage; the enemy with English, and we with American colours flying. I considered him as already our own, and that the ceremony of taking possession was all that was now requisite. I could plainly perceive that his force did not exceed ten guns and thirty men; and, as any alternative was preferable to working the drags any longer, I, to the great joy of everyone on board, gave orders for attempting her with the boats, which were soon hoisted out, manned, armed, and despatched after her. The enemy, seeing so formidable a force coming against him, fired a few guns, apparently with a view of intimidating; but finding that they continued to advance, he ceased firing and hauled down his colours. The boats had now got within three quarters of a mile of him, when a fresh breeze sprung up from the eastward, with which he made all sail to the northward, hoisted his colours, fired at our gig and whale-boat as he passed, which in return gave him volleys of musketry, and before sunset he was hull down ahead of us, while we were lying the whole time perfectly becalmed. Our boats continued the chase, with the hope that it would again fall calm, and made flashes occasionally to guide me in the pursuit, which I was enabled to renew when the breeze struck me, which was not till after sundown. I came up with the boats at nine o'clock at night; we had all lost sight of the enemy, and the apprehension of losing all my boats and officers, and the greater part of my crew, induced me to heave to and take them on board."

The sequel to this episode six weeks later is related in another chapter:

"At daylight in the morning, the men at the masthead descried a strange sail to the southward. . . . I believed her to be an English whaler. I consequently directed . . . the ship to be disguised in every respect as a merchantman, and kept plying to windward for the stranger under easy sail, as he continued to lie to, drifting down on us very fast. At meridian, we were sufficiently near to ascertain that she was a whale-ship, and then employed in cutting up whales. From her general appearance, some were of opinion that it was the same ship that had given us so long a chase, and put us to so much trouble, near Abington Island. She was, however, painted very differently, and from her showing no appearance of alarm, I had my doubts on the subject. I had got possession of some of the whalemen's signals, and made one which had been agreed on between a captain William Porter and the captain of the *New Zealander*, in case they should meet. I did not know but this might be captain Porter's ship, and that the signal might be the means of shortening the chase by inducing him to come to us.

"At one o'clock we were at the distance of four miles from the chase, when she cast off from the whales she had alongside, and made all sail from us. Everything was now set to the best advantage on board the *Essex* and at four o'clock we were within gunshot, when after firing six or eight shot at her, she bore down under our lee and struck her colours. She proved to be the British letter of marque ship, *Sir Andrew Hammond*, . . . and commanded by the identical captain Porter whose signal I had hoisted. But the most agreeable circumstance of the whole was, that this was the same ship we had formerly chased; and the captain assured me, that our ship had been so strangely altered, that he supposed her to be a whale-ship until we were within three or four miles of him, and it was too late to escape. Nor did he suppose her to be a frigate until we were within gunshot, and indeed never would have suspected her to be the same ship that had chased him before, as she did not now appear above one half the size she did formerly.

"The decks of this ship were full of the blubber of the whales they had cut in, but had not time to try out. The captain informed me there was as much as would make from eighty to ninety barrels, and that it would require three days to try it out. But as I understood that it would be worth between two and three thousand dollars, I determined that it should not be lost. I therefore put on board her a crew who had been accustomed to the whaling business, and placed the ship in the charge of Mr. Adams, the chaplain, with directions to try out and stow away the oil with all possible expedition."

A large quantity of provisions was also taken from this prize:

"I hauled her alongside and took from her as much beef, pork, bread, water, wood, and other stores, as we required. But what was more acceptable to our men than all the rest, I took from her two puncheons of choice Jamaica spirits, which was greatly relished by them, as they had been without any ever since our departure from Túmbez. Whether it was the great strength of the rum, or the length of time they had been without, I cannot say; but our seamen were so affected by the first allowance served out to them, that many were taken to their hammocks perfectly drunk. . . . I did not conceive it expedient to resort to rigid measures. Considering the long time they had been deprived of it without murmuring, and the great propensity of seamen for spiritous liquors, and as no evil was likely to result from a little inebriety, provided they conducted themselves in other respects with propriety, I felt disposed to give them a little latitude, which in no instance was productive of unpleasant consequences. . . ."

It is difficult to imagine a duel as a possible occurrence in the United States Navy, but on James Island two officers fought and one died. Porter speaks of it thus:

"I have now the painful task of mentioning an occurrence which gave me the utmost pain, as it was attended by the premature death of a promising young officer, whereby the service at this time has received an irreparable injury, and by a practice which disgraces human nature. I shall, however, throw a veil over the whole previous proceedings and merely state, that without my knowledge the parties met on shore at daylight and at the third fire Mr. Cowan fell dead. His remains were buried the same day in the spot where he fell, and the following inscription was placed over his tomb:

Sacred to the memory
of LIEUT. JOHN S. COWAN,
Of the U. S. Frigate *Essex*,
Who died here anno 1813,
Aged 21 years."

One of Captain Porter's favourite tricks was that of leaving misleading messages in the hope that they would be found by the enemy and encourage him to attack. Witness the following, left in a bottle suspended to a finger-post at James Bay:

"The United States frigate *Essex* arrived here on the 21st June, 1813, her crew much afflicted with the scurvy and ship-fever, which attacked them suddenly, out of which she lost the first lieutenant, surgeon, sailing-master, two midshipmen, gunner, carpenter, and thirty-six seamen and marines.

". . . The *Essex* leaves this in a leaky state, her foremast very rotten in the partners, and her mainmast sprung. . . . Should any American vessel, or indeed a vessel of any nation, put in there, and meet with this notice, they would be doing an act of great humanity to transmit a copy of it to America, in order that our friends may know of our distressed and hopeless situation, and be prepared for worse tidings, if they should ever again hear from us. . . .

The following is a list of the names of those who died as above mentioned, to wit:"

There follows a list of forty-three names. Fortunately no ship seems to have embraced the suggestion of doing "an act of great humanity." The possibilities are endless of the Enoch Arden situations that might have arisen in consequence of such a message actually reaching the families of these supposedly defunct sailors.

An arresting feature of the Galápagos fauna is the number of descendants from once-domestic animals.. Captain Porter is at least partly responsible for the presence of one common form, and his surmise as to their probable future multiplication has proved correct. He tells of putting three female goats and one young male ashore to graze on James Island, and as they were very tame, no trouble was taken to hobble them, even at night. One morning, when the sailor who attended to them went ashore as usual to give them water, the goats had disappeared, and though they were searched for during several days, they were not seen again.

"They undoubtedly took to the mountains in the interior, where unerring instinct led them to the springs or reservoirs from whence the tortoises obtain their supply. Owing to this circumstance, future navigators may perhaps obtain here an abundant supply of goats' meat; for unmolested as they will be in the interior of this island, to which they will no doubt confine themselves on account of the water, it is probable their increase will be very rapid. Perhaps nature, whose ways are mysterious, has embraced this first opportunity of stocking this island with a race of animals, who are, from their nature, almost as well enabled to withstand the want of water as the tortoises with which it now abounds."

The crew of the *Essex* made use of oth er natural resources of the Galápagos than tortoises; our shipwrecked taxidriver's makeshift footgear had been tried a century previous, for Porter's men killed "a number of seals, the skins of which were very serviceable to us as mockasons, made after the manner of those of the North American Indians, and were a very good substitute for shoes, of which we began to stand in need." This is easy to believe, since they had spent some weeks in scrambling over these islands, where the walking is unspeakably difficult.

They also utilized the fruit of the prickly pear; "their juice, when stewed with sugar, made a delicious sirup, while their skins afforded a most excellent preserve, with which we made pies, tarts, etc."

Having captured the entire British whaling-fleet which frequented these waters, prizes which in value amounted to two and a half million dollars, Porter left the Galápagos to go to the Marquesas, where he took possession of Nukahiva, largest island of this group, in the name of his government. The United States, however, showed no interest in this acquisition and took no further steps to secure the property. Thirty years later it passed to the French, together with the rest of the archipelago.

Captain Porter, surrounded by a fleet of prizes, considered himself automatically promoted and forthwith ran up a commodore's flag. On his return voyage the *Essex* was captured by two British war-vessels, and Porter was allowed on parole to return to New York on one of his prizes. There were courteous fighters in those days. As acknowledgment of his services, unparalleled in the War of 1812, he was reprimanded by Congress for exceeding his orders in rounding the Horn at all. A few years later he was suspended from service for having forced from the Spanish commander at Porto Rico an apology for an insult, without having first reported the affront to Congress. Possibly he knew the usual fate of Congressional reports.

After this public rebuke, Porter resigned his commission and went to Mexico, where he became commander-in-chief of the Mexican Navy in the war of independence with Spain. Later he was appointed United States Minister to Constantinople, where he died. It is interesting to know that for many years he controlled a vast tract of land in Tehuantepec, hoping that this was the spot where some day an inter-oceanic canal would be constructed.

The Galápagos saw a brief recrudescence of piratical days when in 1816 Buchard and Brown, two corsairs who had for some time been partners in nefarious enterprises in Argentinian waters, came here for the sole purpose of dissolving their partnership. Differences had arisen between them which resolved them to part company, and here they divided their loot and arranged their business affairs. Brown had for his share the corvet *Halcon* and Buchard took the frigate *Consecuencia*. The former elected to continue his adventurous life in the waters of the East Indies, and the latter returned to the Rio Plata.

In 1822 Basil Hall, in H. M. S. *Conway*, made a short visit to the archipelago to experiment with an invariable pendulum on the equator. He camped for a few days on the south end of Abingdon, and suffered the usual discomforts for lack of water.

Benjamin Morrell, of Rye and Stamford, made two visits to the Galápagos. The first one was in 1823, and his description of the islands differs not at all from those of former voyagers. In 1825 he was there again and witnessed a terrific convulsion of nature, which inspired him to write the following detailed account:

"February 14th. On Monday the fourteenth, at two o'clock, A.M. while the sable mantle of night was yet spread over the mighty Pacific, shrouding the neighbouring islands from our view, and while the stillness of death reigned everywhere about us, our ears were suddenly assailed by a sound that could only be equalled by ten thousand thunders bursting upon the air at once; while, at the same instant, the whole hemisphere was lighted up with a horrid glare that might have appalled the stoutest heart! I soon ascertained that one of the volcanoes of Narborough Island, which had quietly slept for the last ten years, had suddenly broken forth with accumulated vengeance.

" The sublimity, the majesty, the terrific grandeur of this scene baffle description and set the powers of language at defiance. Had the fires of Milton's hell burst its vault of adamant, and threatened the heavens with conflagration, his description of the incident would have been appropriate to the present subject. No words that I can command will give the reader even a faint idea of the awful splendour of the great reality.

"Had it been the 'crack of doom' that aroused them, my men could not have been sooner on deck, where they stood gazing like 'sheeted spectres,' speechless and bewildered with astonishment and dismay. The heavens appeared to be one blaze of fire, intermingled with millions of falling stars and meteors; while the flames shot upward from the peak of Narborough to the height of at least two thousand feet in air. All hands soon became sensible of the cause of the startling phenomenon, and on recovering from their first panic could contemplate its progress with some degree of composure.

" But the most splendid and interesting scene of this spectacle was yet to be exhibited. At about half-past four o'clock, A.M. the boiling contents of the tremendous caldron had swollen to the brim, and poured over the edge of the crater in a cataract of liquid fire. A river of melted lava was now seen rushing down the side of the mountain, pursuing a serpentine course to the sea, a distance of about three miles from the blazing orifice of the volcano. This dazzling stream descended in a gully, one-fourth of a mile in width, presenting the appearance of a tremendous torrent of melted iron running from the furnace. Although the mountain was steep, and the gully capacious, the flaming river could not descend with sufficient rapidity to prevent its overflowing its banks in certain places, and forming new rivers, which branched out in almost every direction, each rushing downward as if eager to cool its temperament in the deep caverns of the neighbouring ocean. The demon of fire seemed rushing to the embraces of Neptune; and dreadful indeed was the uproar occasioned by their meeting. The ocean boiled and roared and bellowed, as if a civil war had broken out in the Tartarean gulf.

"At three A.M., I ascertained the temperature of the water, by Fahrenheit's thermometer, to be 61°, while that of the air was 71°. At eleven A.M., the air was 113°, and the water 100°, the eruption still continuing with unabated fury. The *Tartar's* anchorage was about ten miles to the northward of the mountain, and the heat was so great that the melted pitch was running from the vessel's seams, and the tar dropping from the rigging.

"In order to give the reader a correct idea of our situation, it will be necessary to remind him of the relative position of these two islands. Albemarle Island is the most extensive of the whole Galápagos group, being about ninety miles in length from north to south, narrow and nearly straight on its eastern shore; but on the western side it hollows in from Christopher's Point on the south, to Cape Berkley on the north; and within this space lies the island of Narborough, its eastern point approaching nearest to Albemarle. The *Tartar* lay in a cove of Banks's Bay, on the western shore of Albemarle, directly opposite the northeast point of Narborough; and this cove could be approached from the northwest through Banks's Bay, or from the southwest through Elizabeth Bay.

"Our situation was every hour becoming more critical and alarming. Not a breath of air was stirring to fill a sail, had we attempted to escape; so that we were compelled to remain idle and unwilling spectators of a pyrotechnic exhibition which evinced no indications of even a temporary suspension. All that day the fires continued to rage with unabating activity, while the mountain still continued to belch forth its melted entrails in an unceasing cataract.

"The mercury continued to rise till four P.M., when the temperature of the air had increased to 123°, and that of the water to 105°. Our respiration now became difficult, and several of the crew complained of extreme faintness. It was evident that something must be done and that promptly. 'O for a cap-full of wind!' was the prayer of each. The breath of a light zephyr from the continent, scarcely perceptible to the cheek, was at length announced as the welcome signal for the word, 'All hands, unmoor!' This was a little before eight P.M. The anchor was soon apeak, and every inch of canvass extended along the spars, where it hung in useless drapery.

"All was again suspense and anxious expectation. Again the zephyr breathed and hope revived. At length it was announced from aloft that the lighter canvass began to feel the air; and in a few minutes more the topsails began gradually to fill, when the anchor was brought to the bow, and the

Tartar began to move. At eight o'clock we were wafted by a fine little easterly breeze, for which we felt grateful to Heaven.

"Our course lay southward, through the little strait or sound that separated the burning mountain from Albemarle Island; my object being to get to windward of Narborough as soon as possible. It is true that the northwest passage from Banks's Bay, by Cape Berkley, would have been a shorter route to the main ocean; but not the safest, under existing circumstances. I therefore chose to run south, to Elizabeth Bay, though in doing so we had to pass within about four miles of those rivers of flaming lava, which were pouring into the waters of the bay. Had I adopted the other course, and passed to the leeward of Narborough, we might have got clear of the island, but it would have been impossible to prevent the sails and rigging taking fire; as the whole atmosphere on the lee side of the bay appeared to be one mass of flame. The deafening sounds accompanying the eruption still continued; indeed the terrific grandeur of the scene would have been incomplete without it.

"Heaven continued to favour us with a fine breeze, and the *Tartar* slid along through the almost boiling ocean at the rate of about seven miles an hour. On passing the currents of melted lava, I became apprehensive that I should lose some of my men, as the influence of the heat was so great that several of them were incapable of standing. At that time the mercury in the thermometer was at 147° but on immersing it in water, it instantly rose to 150°. Had the wind deserted us here, the consequences must have been horrible. But the mercy of Providence was still extended toward us—the refreshing breeeze still urged us forward towards a more temperate atmosphere; so that at eleven P.M. we were safely anchored at the south extremity of the bay, while the flaming Narborough lay fifteen miles to the leeward.

"Here the temperature of the air was 110°, and that of the water 102°; but at eight o'clock the next morning, the 16th, there being no abatement in the rage of the vomiting volcano, the heat had increased to such an alarming degree that we found it necessary again to get under way and abandon the bay entirely. At twelve meridian we passed the south point of Albemarle Bay, called Christopher's Point, at which time I found the mercury at 122° in the air and at 98° in the water. We now steered for Charles's Island, which lies about forty miles southeast of Albemarle, and came to anchor in its north-west harbour at eleven P.M. Fifty miles and more to the lee-ward, in the northwest, the crater of Narborough appeared like a colossal beacon-light, shooting its vengeful flames high into the gloomy atmosphere, with a rumbling noise like distant thunder.

"February 17th. . . . The sea was here literally covered with pumice-stone, some pieces of which were quite large, sup-posed to have been ejected from the volcano of Narborough."

In October, when he returned to the Galápagos from the Sandwich Islands, "the volcano of Narborough, which broke out in February, was still burning but very moderately."

After a day or two spent in Elizabeth Bay, at the south end of Albemarle, he went to Indefatigable Island for a stock of "Galápagos mutton," and in four days had collected one hundred and eighty-seven of these tortoises.

The next visitor, or at least the next recorded visitor to the archipelago, was brought there by a combination of the most unusual and incongruous circumstances that it is possible to imagine. In 1824 the King and Queen of the Sandwich Islands, inspired by the wonderful stories they had heard of the far-off land of England, determined to pay a visit to this country of strange inventions and of seeming magic. With their retinue they arrived in London, where they were royally entertained and where every effort was made to cement the friendly feeling between the two island kingdoms. But these visitors from the South Seas were little fitted to cope with the harsh climate and hitherto unencountered ailments, and late in August the royal pair succumbed to an attack of measles.

The seventh Lord Byron, who succeeded to the title after the death of the poet, was a Captain in the British Navy, and he was deputed to escort the bodies back to their native land. Accordingly, on September 8th, 1824, H. M. S. *Blonde*, commanded by Captain the Right Honourable Lord Byron, sailed for the Sandwich Islands, with the dead king and queen, their grieving native retinue, and a royal guard of English officers. Rounding the Horn, they stopped at the Galápagos hoping to find water, of which they stood in some need, and on March 25th, 1825, anchored in Tagus Cove, where they observed the continued activity of Narborough, whose fierce explosion Morrell had witnessed a month earlier.

Lord Byron writes of his impressions as follows:

"The place is like a new creation; the birds and beasts do not get out of our way; the pelicans and sea-lions look in our faces as if we had no right to intrude on their solitude; the small birds are so tame that they hop upon our feet; and all this amidst volcanoes which are burning round us on either hand. Altogether it is as wild and desolate a scene as imagination can picture.

"27th March. Our first care this morning was to search for the water with which we were to complete the ship, but to our mortification we found the springs, which are usually abundant, nearly dried up, and were therefore obliged to put the ship's company on an allowance. . . . Our party to Narborough Island landed among an innumerable host of sea-guanas, the ugliest living creatures we ever beheld. They are like the alligator but with a more hideous head and of a dirty sooty black color, and sat on the black lava rocks like so many imps of darkness. As far as the eye could reach we saw nothing but rough fields of lava, that seemed to have hardened while the force of the wind had been rippling its liquid surface. . . . About halfway down the steep southeast side of the Island, a volcano burns day and night; and near the beach a crater was pouring forth streams of lava, which on reaching the sea caused it to bubble in an extraordinary manner.

"March 29th. We were employed in cutting wood and

procured a sufficiency for three weeks, but as usual in hot climates, brought on board with it scorpions and centipedes.

". . . with the exception of the common balsam tree and a species of acacia, most of the vegetation is dwarfish. The land birds are few here, but the brown sea-guana, and a red-breasted lizard are to be seen in great numbers. We saw only one green snake, quite harmless, and found but few insects."

There is a foot-note referring to the "brown sea-guana," written by John Murray, publisher of the volume, in which he explains that at first he had supposed this creature to be the female of the black one, but on comparison the two appeared to be so different that he was compelled to believe them separate species. Observation of their habits, had he had an opportunity so to employ himself, would have convinced him that the brown guana had no claim to the qualifying adjective "sea."

For almost three hundred years after their discovery, the Galápagos remained unclaimed territory. No one saw in these desolate bits of land anything sufficiently desirable to claim. To be sure, the Bishop, their first recorded visitor, planted the Spanish flag on one of them, but that was merely a gesture. In 1831, after Ecuador had won her freedom from Spanish sovereignty and had seceded from the Colombian Confederation, General José Villamil, who had fought for his country's independence, conceived the idea of colonizing the islands. He formed a colonization company, and the following year Colonel Hernandez assembled the crews of several whalers that happened to be in the bay at Charles Island and in their presence took possession of the archipelago in the name of the government of Ecuador. The first colonists were eighty soldiers who had been condemned to execution for taking part in a rebellion. General Villamil intervened in their behalf and obtained their pardon on condition that they should become workers in his projected settlement. He selected a site at a thousand-foot elevation, five miles from the coast, where there was a spring sufficient to supply their needs. It was called Asilo de la Paz, but it proved to be anything but peaceful. The colony seemed to have a promising future for a while, but because of the difficulty of obtaining labour, the mistake was made of converting it into a penal settlement, which inevitably led to trouble.

Three years after the colony was founded, the memorable visit of the *Beagle* was made to the islands, with the expedition of which Charles Darwin, then a very young man, was a member. The *Beagle* remained in these waters for about five weeks, and from his observations of the peculiar fauna of the archipelago Darwin derived his first inspiration for his theory of Natural Selection. At this time there were about three hundred people resident at Asilo de la Paz.

In this same year there was found on James Island, the skull of the captain of a whaling vessel, who had been murdered by his crew.

After five years of hard and enthusiastic work, General Villamil resigned his post as Governor and a Colonel Williams was appointed in his place. This man committed such abuses in his position as pseudo-ruler over this isolated settlement that in 1842 the colonists rose against him and he fled. By this time the population was reduced to about eighty persons, and three years later the remnants of the colony were transported to Chatham Island. In 1849 only twenty-five men were living on Charles, together with the herds of cattle which had vastly increased from the first live stock imported by Villamil. These men were criminals who had been exiled for indeterminate periods, dependent on whether or not they were ever remembered by the authorities.

There have been several projects to sell the Galápagos,

the first in 1851, when it was proposed to discharge Ecuador's debt to England by the transfer of the archipelago, but Peru, France and Spain made such representations to the Government that the matter was dropped. Three years after, there were entirely unfounded reports of the existence of great guano deposits here, and an agreement was made with the United States for their exploitation. This time England joined with Peru, France and Spain in protesting, and negotiations were broken off.

In 1852 piracy raised its head once more. A man named Briones, known as the Pirate of Guayas, was deported to Charles. Here he gathered to his side seven companions, as criminally reckless as himself, and they awaited an opportunity of escaping from their exile. This opportunity came when an American whaling ship, the *George Howland*, dropped anchor in the harbour. Briones and his ruffians seized the ship, and sailed to Chatham, where an aged General, named Mena, was trying an experiment in agriculture. The Pirate of Guayas took Mena prisoner, together with five other men, and returned to the ship, where they murdered Mena.

At this time General Flores was in Peru, recruiting for an expedition which was to overthrow the existing Government in Guayaquil. As Briones sailed toward the mainland he encountered two sloops which were going to Ecuador, having aboard some sixty or seventy persons who had enlisted under the banner of Flores. Briones attacked one of these ships and ruthlessly slaughtered most of those aboard; those on the other sloop escaped by beaching their vessel and hiding on shore. They had planned a revolution, but such a holocaust was more than they had bargained for.

Briones went to Guayaquil, announcing that he had "destroyed the vanguard of Flores," and actually believed that he would be received with honours. His disappointment must have been considerable when instead of the acclamations he had expected so confidently, he and his companions in crime were hung.

In 1870 another attempt was made to colonize Charles. The man in charge of this project was Señor de Valdizian, who not only engaged in agriculture but also gathered orchilla moss, which is used in dye manufacture. While he was patrón of the island, two scientific expeditions visited the archipelago, one aboard the U. S. S. *Hassler*, and the other directed by Dr. Wolf, the geologist. Wolf made two voyages to the Galápagos, one in 1875, the second three years later.

History repeated itself in the case of this second colony. At first it flourished; then the Government decided to make it a penal settlement, and soon after Wolf's last visit, Valdizian was murdered by the convicts, who then instituted a reign of terror. The majority of the prisoners, obtaining weapons from the store-house, committed a series of murders, and pillaged and sacked in a drunken orgy. However, all the colonists were not of this calibre; in fact, they were divided into two camps, one of which was loyal either from ingrained righteousness or in self-defense. What might be described as a pitched battle was fought, and only one of the mutineers escaped with his life.

Captain Thomas Levick, master of a vessel owned by Valdizian, arrived after the combat, and did what he could to restore order and prevent the destruction of the colony, but in vain. The remaining colonists were scattered and Charles was once more abandoned. Four years later the Italian corvet Vettor Pisani stopped at Charles, where the crew went ashore. They found deserted orchards covered with fruit, vineyards, and many cattle in the best of condition, but not an inhabitant remained.

Not until 1893 was another attempt made to colonize the

ill-fated Charles Island. In that year Antonio Gil tried to establish a settlement there, but after four years he abandoned it in favour of Albemarle, where he located on the southeast point and called the place Santo Tomás, the name of the largest crater on this island, which reaches an elevation of more than 3,000 feet. In 1905 there were almost two hundred people living there, but at present there are said to be not more than sixty. They earn a scanty livelihood by shipping sulphur to Guayaquil from the deposits found in the crater, and also from the manufacture of lime, produced by burning coral on the beach. They are dependent for drinking water on the brackish liquid to be found in occasional water-holes near the coast. An eruption took place in this largest volcano about twelve years ago.

The port from which Santo Tomás is reached is called Villamil. In 1902 a small garrison was established here under a Jefe Territorial, and two years afterwards eleven soldiers deserted. They knew so little of the country that they did not trouble to provide themselves with water, but struck into the interior with perfect confidence of finding it. Search was made for them without result, and it was finally decided that they must have been taken off by a passing boat. The affair was almost forgotten when one day a man staggered into Santo Tomás in the last stages of exhaustion and was recognized as one of the deserters. His story was one of horrible sufferings from thirst, as well as from the pangs of hunger. After a day or two, the soldiers had scattered in a desperate search for water and never met again. Unprovided as they were, it is almost certain that the other ten perished; at any rate, nothing more was ever heard of them.

In 1835, Darwin, on the *Beagle*, visited the Galápagos and made them known to the scientific world. In 1891, the *Albatross* came to these islands, and Agassiz, leader of the expedition, amplified the knowledge of them which the English savant had published. In 1835 Captain Fitz-Roy was in charge of the hydrographic work on the *Beagle;* fifty-six years later C. L. Yauer did the same sort of exploration on the *Albatross.* In 1897 the schooner *Lila and Mattie* spent some time in the archipelago with a scientific expedition, led by Charles Miller Harris. It was at this time that an eruption took place on James, when a crater opened on the southeast side and with floods of molten rock ravaged anew the already barren slopes.

In 1905 the California Academy of Sciences organized an expedition to the Galápagos, under the leadership of Mr. R. H. Beck, who had been a member of the *Lila and Mattie* expedition. This party spent almost a year among the islands and gathered a tremendous quantity of specimens of all sorts. And in 1923 we, of the Williams Galápagos expedition spent an all-too-brief time in this desolate and yet fascinating desert land.

The island of Chatham has been inhabited since 1869. A Señor Manuel Cobos started the settlement there, and the few inhabitants cleared and planted land, and gathered the orchilla moss. After the assassination of Valdizian, on Charles, about a hundred peons were transferred to Chatham, bringing the population to one hundred and fifty. In 1880 Señor Cobos decided to make his home at the plantation and devote his efforts to the raising of sugarcane. He was really king of the island, since the wretched prisoners were often completely forgotten by the authorities even when their terms of punishment were served, and the far-off penal settlement was seldom visited. Cobos made and circulated his own money, elliptical copper coins of small value, and his rule was practically sovereign over the three hundred souls who laboured on his plantation. Because of the irregular communication, no one on the mainland knew in detail the lives of the inhabitants, and though there was the regularly appointed Jefe Territorial, it afterward appeared that he was completely under Cobos' domination and let this petty prince do as he pleased. Now and then a rumor drifted to Guayaquil that all was not well with the convicts on the plantation called Progreso, but no one was sufficiently interested to inquire further into the lot of the unfortunates.

Early in the year of 1904 a sloop arrived at a Colombian port. She was called *The Liberty*, and had on board seventy-seven men and eight women. They were unable to produce any ship's papers, and because of this and other suspicious circumstances, they were detained. Little by little their story was extracted from them. They were prisoners from Progreso who had resolved to put an end to their apparently endless stay in the penal settlement by committing the crimes which, they declared, they "believed it necessary to commit." They had murdered Cobos and the Jefe Territorial, seized the ship which Cobos had kept concealed at an anchorage on the further side of the island, and rather pitifully re-christening her *The Liberty*, they had made this desperate and ineffectual attempt to escape. When brought to trial they made terrible charges against the dead Cobos, in recounting their lives under his rule. He had caused six men to be flogged to death, had shot five others for small offenses, and had banished various others to some of the near-by uninhabited islands, where several had died of hunger and thirst. On the day that he condemned one man to five hundred lashes on his bare back, "all the slaves, that is to say, the majority, decided to put an end to this nauseous and tyrannical life, and organized a plot." Having killed Cobos, they buried him in a spot chosen because at that very place five men had been executed by his orders.

The Ecuadorian gunboat *Cotopaxi* was despatched to the island to investigate the situation. They found conditions even worse than they had expected, and their first care was to go to Indefatigable to search for one Camilo Casanova, who was reported to them as having been exiled there by Cobos about three years before. He had been a soldier and was sent to Progreso in punishment for some disobedience. Here he was a special target for the patrón's tendency to severity, and at last was taken to Indefatigable and abandoned with nothing but a small vessel of fresh water, a kitchen knife without a point and a machete. His fellow-convicts who were compelled to leave him on the island surreptitiously gave him some matches and several pieces of clothing.

This man deserves the name of the Robinson Crusoe of the Galápagos. For the first few days he suffered agonies of despair in contemplating the slow death that he was sure would be his fate. He lived on raw turtles and lizards, and to save his scanty store of water, he drank the blood of these creatures. He slept on the beach, where he was an object of great curiosity to the sea-lions and was frequently visited by them. After existing in this fashion in a kind of daze for some three weeks, he built himself a hut, and also found that he could allay his thirst by chewing the cactus pads. Then he tried to go inland in search of a better place to live, and for fourteen days he painfully made his thorn-beset way toward the interior, only to abandon the attempt at last.

For three years and a half he stayed on the island. He made a calendar; the days he marked on one tree, the months on another and the years on a third. Twice during those years he spoke to human beings. Two English ships came there and the crews went ashore. They neither spoke nor understood Spanish, and he was equally ignorant of English. He made them understand by signs that he begged them to take him away, but in the same manner they told him that they could not do so. This doubly-exiled convict prayed and wept, but they left him there, having deposited a little store of biscuits, cigarettes and matches on the beach beside his hut.

When the *Cotopaxi* came, he was found alive, contrary to all expectations, and the reason for the Englishmen's refusal to rescue him was made clear. On the other side of the island a board had been erected, on which was written in Spanish and English, "Do not take this man away. He is twenty times a criminal."

The expedition of which Mr. Beck was a member found on James Island the skeleton of a man, with a few little implements scattered about, and a tattered bit of canvas spread on sticks as a protection from the blazing sun. This was almost certainly the body of one Raimundo Guardado, another of the convicts whom Cobos had deported and doomed to a lingering death.

In such stories, and in fact in the atmosphere of the islands themselves, there is an element of the horrible and the bizarre which recalls some of Poe's tales. While this author never availed himself of the Galápagos as a mise-en-scène, he evidently profited by reading the accounts of Amasa Delano and Captain Porter, for in the narrative of A. Gordon Pym's strange adventures, the "Gallipago" tortoise is described in almost the identical words of those two earlier and less famous writers, with the addition of some embroidery of detail, as when Poe says that these creatures weigh from twelve to fifteen hundred pounds.

Herman Melville, of posthumous fame, wrote an excellent account of the Galápagos in a little-known book called *Piazza Tales*. The chapter dealing with the archipelago is entitled "The Encantadas," and the descriptive powers that made *Moby Dick* and *Typee* fascinating had suffered no diminution when this was written. It can hardly be classed as history, for Melville has not hesitated to draw upon his imagination to supply missing facts, or make a good story better. Indeed, in recounting the tale of Patrick Watkins, he frankly says that he has altered it to suit better his idea of the character of this strange man who seemed half beast. But though his ineradicable desire to spin a good yarn makes some of his statements decidedly doubtful, no better description of these arid lands has ever been written than the opening paragraphs of "The Encantadas":

"Take five-and-twenty heaps of cinders dumped here and there in an outside city lot; imagine some of them magnified into mountains, and the vacant lot the sea; and you will have a fit idea of the general aspect of the Encantadas, or Enchanted Isles. A group rather of extinct volcanoes than of isles; looking much as the world at large might, after a penal conflagration.

"It is to be doubted whether any spot on earth can, in desolateness, furnish a parallel to this group. Abandoned cemeteries of long ago, old cities by piecemeal tumbling to their ruin, these are melancholy enough; but like all else which has but once been associated with humanity, they still awaken in us some thoughts of sympathy, however sad. Hence, even the Dead Sea, along with whatever other emotions it may at times inspire, does not fail to touch in the pilgrim some of his less unpleasurable feelings.

"And as for solitariness; the great forests of the north, the expanses of unnavigated waters, the Greenland ice-fields, are the profoundest of solitudes to a human observer; still the magic of their changeable tides and seasons mitigates their terror; because, though unvisited by men, those forests are visited by May; the remotest seas reflect familiar stars even as Lake Erie does; and in the clear air of a fine Polar day, the irradiated azure ice shows beautifully as malachite.

"But the especial curse, as one may call it, of the Encantadas, that which exalts them in desolation above Idumea and the Pole, is, that to them change never comes; neither the change of seasons nor of sorrows. Cut by the Equator, they know not autumn and they know not spring; while already reduced to the lees of fire, ruin itself can work little more upon them. The showers refresh the deserts; but in these isles rain never falls. Like split Syrian gourds left withering in the sun, they are cracked by an everlasting drought beneath a torrid sky. 'Have mercy upon me,' the wailing spirit of the Encantadas seems to cry, 'and send Lazarus that he may dip the tip of his finger in water and cool my tongue, for I am tormented in this flame.'

"Another feature in these isles is their emphatic uninhabitableness. It is deemed a fit type of all-forsaken overthrow, that the jackal should den in the wastes of weedy Babylon; but the Encantadas refuse to harbour even the outcasts of the beasts. Man and wolf alike disown them. Little but reptile life is here found; tortoises, lizards, immense spiders, snakes, and that strangest anomaly of outlandish nature, the aguano. No voice, no low, no howl is heard; the chief sound of life here is a hiss."

CHAPTER XVII

THE GALÁPAGOS ARCHIPELAGO

IN this final general chapter I have no intention of pretending a thorough solution or even an adequate presentation of the problems offered by the Galápagos Archipelago. The fact that I spent less than one hundred hours on its shores would make such a thing impotent and valueless. I wish merely to state some of the general facts and my own inferences, and faintly to outline certain researches, especially among the vertebrates, which I shall elaborate in forthcoming numbers of *Zoologica*, the scientific publication of the New York Zoological Society. As with my theory of the origin of flight through *Tetrapteryx* and my classification of *Phasianidæ* by tail moult, so with all my points of view which in our present state of knowledge must be wholly or in part theories, I hold them in readiness to be relinquished at the first hint of better proof on the opposite side.

The Galápagos Islands belong to Ecuador and they are situated directly upon the equator, five small ones being slightly north of the line, and the others on or just south of it. The group is five hundred miles distant from the nearest point in South America, and six hundred and fifty miles from the nearest headland of Costa Rica.

The problems of greatest interest which these islands offer,

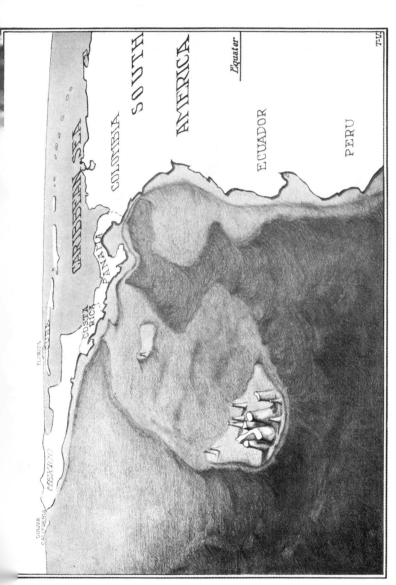

Fig. 80

DIAGRAMMATIC CHART OF THE PACIFIC OCEAN BOTTOM

Showing the 2000 metre depth about the Cocos and Galápagos Islands, and the 3000 metre line connecting these islands with the Central American Mainland

Based on charts of the Prince of Monaco

FIG. 81

GIANT RAY OR DEVIL-FISH OF AN UNKNOWN SPECIES SEEN
NEAR TOWER ISLAND

FIG. 82

THE EFFECT OF GALÁPAGOS LIFE ON SOME INSECTS

Frittilery Butterfly, *Agraulis vanillæ* Linné

White-lined Sphinx Moth, *Deilephila lineata* Fabr
The smaller specimens are from the Galápagos, the larger ones from the United States

have to do with the origin of their fauna and flora. Three points of view are possible:

I. They were geologically recently uplifted from mid-ocean as separate volcanic peaks, never in connection with one another or with the American continent.

II. The archipelago, while always oceanic, was at one time a single island, separated in comparatively recent times, by submergence, into the present number of isolated peaks.

III. The islands were at one time connected, not only with one another, but also with the American mainland.

The first theory, especially as to the interrelation of the islands themselves, I am convinced is not supported by the facts either of geology or of organic distribution. With such long continued volcanic activity as the islands indicate, there must have been many recurrences of emergence and subsidence. Distinct marine deposits have been found many feet up on the elevations of some of the islands. Darwin Bay, the submerged crater which I discovered on Tower Island, has a central depth of one hundred fathoms, and if this were re-elevated even to sea-level, all the islands of the main group, except the small northern ones, would be reunited into a single land mass. That the islands were at one time continuous dry land I consider as practically certain.

The third theory of a one-time connection with the American mainland is still a moot question.

When I went to the islands I had no firm convictions or even bias, except that when I knew that Charles Darwin held to the oceanic theory, I was ready to follow him if any interpretation of observed facts permitted. All that I saw at first hand and what I have since read and discussed, leads me to a strong belief in the theory of a former land bridge, once connecting a single large Galápagos island with the mainland of Central America. I can find no facts which irrefutably contradict the possibility of such a belief.

The chief question at issue, therefore, is whether the fauna and flora have come gradually to these islands over the five hundred miles of ocean from South America, over the six hundred and fifty miles from Central America, as well as over greater distances from intermediate and more southerly parts of the mainland. To make this oceanic theory at all tenable it is necessary to invoke wide and fundamental changes in geology and meterology, conditions, however, which are quite conceivable. A recent able supporter of the oceanic theory postulated, in a symposium on the question, the prerequisites of migration in pre-Andean days, with an opening in the South American continent through to eastern waters, and, in place of the present zone of year-long calms about the islands, a condition of very severe cyclonic storms capable of lifting and carrying seeds and animal organisms for great distances. Insistence was at the same time placed on the very recent character of the whole archipelago, thus introducing another necessary condition, of sufficiently recent eruption, elevation and cooling, to admit of the successful reception and sustention of the storm-carried organisms.

Not a few of the Galápagean forms of life differ more among themselves on the various islands than they do from their nearest relatives on the mainland and if the development of giant tortoises be reasonably inferred as from more normally sized ancestors, we are warranted in assuming a reasonably long organic isolation—long biologically if not in terms of geological epochs. The actual life of the archipelago at present however is not to me one of the important questions. For all we know, the present islands may be only a few centuries old, and have received their inhabitants from contiguous land areas which have since subsided.

The first thing is to review the more general conditions surrounding the islands. Dr. Alban Stewart, although I have scolded him roundly for not giving us any kind of a key to the Galápagos flora, yet as botanist of the expedition in 1905 of the California Academy of Sciences, has written a most excellent account of the more abstract facts relating to the flora. He has supplemented this with a few significant paragraphs dealing with the theory of migration and the present conditions of plant life.

I have thought it worth while to devote some consideration to the botanical aspect of the question, for it must be admitted that seeds are far more likely to be safely transported by currents of wind and water than are animal organisms, and a satisfactory discussion of the subject from this angle will suffice to cover many of the other questions.

Dr. Stewart is a staunch advocate of the oceanic character of the islands, although he is willing to admit sufficient elevation and subsidence to connect and separate the individual islands. But he insists that elevation stops with their uniting and will have none of a mainland connection. Now if the transportation of seeds to the islands by air and water can be proved, the similar origin of all the rest of the organic life will have at least a foundation.

He says "If it be assumed that the Galápagos Islands are of oceanic origin, there are but three means by which seeds and spores could have been brought to the islands, outside of the agency of man. These are: winds, oceanic currents, and migratory birds.

WINDS

"If winds were an important agent in bringing seeds and spores to these islands, those families of plants which have the smallest seeds and spores would be the most apt to be distributed in this way. Of all the families of vascular plants none are better adapted for wind distribution than are the ferns. Such being the case, there should be a larger number of species of ferns on the islands common to the region from which the prevailing winds blow than from any other. As the winds around these islands are almost constantly from the southeast, the fern flora should be most closely related to that of the central and southern part of South America. Such is not the case, however, for outside of fifteen species which are of wide distribution, the fern flora shows nearly as strong affinities with that of Mexico as it does with that of South America. There are on the islands fifty-four ·species common to Mexico and fifty-six to South America. Moreover, the majority of the latter belong only to the northern part of the continent.

"Devices for wind dissemination are not common on the seeds of Galápagos plants, the *Compositæ* being the only one of the larger families which has this character pronounced to any extent.

OCEANIC CURRENTS

"The northern islands of the group, viz, Abingdon, Bindloe, Culpepper, Tower and Wenman, lie in the direct path of the Panama current, and the water surrounding them is several degrees warmer than that around the southern islands, which are bathed by the Humboldt current. If oceanic currents were an important factor in the transport of seeds to the Galápagos, those islands which are washed by the Panama current should be more closely related botanically to the Mexican and Central American regions than the islands lying in the Humboldt current; and the latter islands, on the other hand, should have a flora more closely related to that of the western coast of South America. Furthermore, the several islands of each group should have a larger floral element common among themselves than with any of the islands of the other group. . . . In the majority of instances the islands of the northern group have a larger percentage of their floras common with the islands of the southern group than with each other, a condition hardly to be expected if oceanic currents were an important factor in transporting seeds to them. Robinson has already mentioned the small chance that many seeds would have of surviving even if they were washed up on the shores of the islands, a fact that cannot be too strongly emphasized. While it is entirely possible that the seeds of xerophytic plants might be able to grow if they were cast up in this way, it is hardly likely that mesophytic plants would be able to survive, because there are but two places on the islands—at the present time—where conditions at sea-level are such as to offer them a suitable habitat. One of these places is Iguana Cove, Albemarle Island, and the other is Villamil, on the same island, at neither of which places are there plants which do not have a wide distribution over the islands. While it is possible that the Humboldt current may be responsible for much of the xerophytic flora, it is hardly likely that the Panama stream could have played much of a rôle in this respect, as it flows from a region in which the flora is anything but xerophytic in character."

This proof of the total unfitness of Galápagos seeds and plants for aërial and aquatic transportation leaves only the third factor, that of introduction by migratory birds. Twenty species are listed by Dr. Stewart, of which only three are common and five fairly common; five more are merely chance visitors, and three others are so rare that they are represented by only five specimens. The seeds must either be swallowed and retained until the birds complete their long journey, or else become attached with muddy débris to their legs and feet, and on landing be washed or brushed off upon a suitable spot for germination.

The species of birds under consideration are almost all fish or shrimp or insect feeders, and only two could ever be classed as even casual berry eaters. So according to Dr. Stewart's conclusions we must acknowledge that precarious attachment to the legs of this scattering of migratory birds is "the most important cause" accounting for the presence of the six hundred odd species of plants living on the islands.

If we admit the necessity for the pre-Andean theory with its assumed alteration in meteorological and other conditions, there is no further need to study the whole problem. We can either admit that we do not know and cannot find out, or we can smother our desire in the blanket of pure abstract, unproven theory. But if we wish to probe present as well as past conditions we must, like Dr. Stewart, take the former into consideration.

His conclusions, honestly reached, show that among the plants there is an overwhelming percentage of relationship with the Mexican, West Indian and Central American flora compared with that of South America. The ferns, for example, show almost twice as many identical with mainland species from Mexico and the West Indies as with South American ones, and many of the latter are northern in extent of distribution.

This same thing can be duplicated in almost every group of organisms on the islands. In refutation of migration by means of the oceanic currents of today, while fifty-six of the Galápagos shore fishes are common to the American mainland, only four are South American.

Against avian migration by prevailing winds we have the following data: Of twenty-two genera of endemic birds, one, *Spheniscus*, the penguin, is undoubtedly of Antarctic origin, and together with the southern sea-lion must have reached the Galápagos by means of the Humboldt current. But although this current flows steadily and strongly, and in spite of the fact that closely related penguins are found on the western South American coast as far north as Peru, yet the Galápagos birds are quite distinct. This means that today, with every-thing in the favour of constant migration, via a cold current in well-stocked waters, flowing from the haunts of tens of thou-sands of penguins of many species, there has apparently been only a single stocking of the island with a pair or more of these birds.

Of twenty-one remaining endemic genera, the most closely related mainland representatives of every one are found in Central America. Seven of these range south into northern South America and thirteen extend over most of that conti-nent, but are chiefly confined to the humid part east of the Andes.

As to the geophysical factors in support of the land-bridge theory, we must depend upon the soundings which have been made in this part of the world.

Thirty years ago the *Albatross*, in charge of Alexander Agassiz, visited the Galápagos. Many soundings were made and of the general results, George McCutcheon McBride says:

"The survey of the ocean topography between Central America and the Galápagos carried out by the *Albatross* expedition in 1885–91 showed the islands to be the highest points of a submarine elevation, more or less circular in form sloping up very gradually from 1,800 fathoms below the sur-face. This elevation in turn is the summit of an immense submerged plateau that extends northward past Cocos Island almost to the Isthmus of Panama."

A careful study of the maps and charts of the route of the *Albatross* shows, however, that no soundings were made be-tween Cocos Islands and the Galápagos nor nearer than this point (360 miles), in any direction except due east. The course away from the Galápagos was set for Acapulco and un-fortunately no soundings at all were taken on this northerly route. So for a better confirmation of any hint of a submarine plateau we must consult the maps published by the Prince of Monaco.

The sea-bottom off the western coast of South America becomes deep almost at once, but if its abrupt descent to two thousand metres or six thousand feet be taken as an absolute refutation of any corresponding elevation in past time, surely we have a corresponding right to question the razing of such Andean heights as the twenty thousand feet of Chimborazo.

Although a few miles farther from shore off Panama and Colombia we find depths of four thousand or more metres, there is no need to invoke the raising of such bottoms for our beasts to travel over. The little island of Malpelo, two hundred and fifty miles off Colombia, is still on the two thousand-metre plateau, but beyond this again are great depths. Going north we find the thousand-metre coastal plateau of wide extent off Costa Rica, and following this out southwest, generously enclosing the small thousand-metre plateau of Cocos and on to the larger area of equal depth surrounding the Galápagos, we need never descend below the three-thousand metre line. On all sides, southeast, south, west and north the bottom falls away to much greater depths, but according to the carefully tabulated charts of the Prince of Monaco, the three-thousand metre sounding line outlines a direct connection, from the farthest Galápagos Island to the very shore of Costa Rica, but allows no connection with the mainland of South America.

Hundreds of additional soundings must be taken before this plateau can be more definitely delineated. Of special importance to both theories is the record of any annual southerly set of the Panama Current. For in spite of Dr. Stewart's assertion, this current does not usually reach even the most northerly of the Galápagos. At Tower and far north of Abingdon, a one or even two knot cold current sets as strongly northwest as it does along the south shore of Chatham.

The effects of insularity on the various forms of life will require long study, but every fact that I discovered was potent with interest. Of twenty genera of endemic birds, exactly half are, in general, smaller than closely related forms on the mainland, and still more, sixty-five per cent, are decidedly darker, more melanistic. Twenty per cent show an increase in stoutness of build, but not in actual size. Legs, feet and bill show an increase in length and strength, in forty, twenty-five and twenty per cent respectively, while twenty-five per cent have acquired a shorter tail. All this is easily explained on the basis of several factors; the constant rains of the interior and the omnipresent ocean and trades provide humidity, and this together with the black lava everywhere has resulted in the very general darkening of color; a general scarcity of food, especially at certain times of the year, may, as in the lepidoptera, be the cause of the reduction in size of so many forms, while the difficulty of search and securing of food in this land of a billion crevices may have helped along the increase in the size of beaks. The long legs and feet and the shortened tails are undoubtedly due to the general tendency of many of the birds to walk more and fly less, the mockingbirds being the most conspicuous, but only one of a number of examples.

Another effect, of a degenerative character, is seen in the numerous examples of albinism. Half of the birds show one or many adventitious white feathers, and still more remarkable is the wide-spread prevalence of diseased bills, legs and feet of the birds of certain islands. I found many examples of the breaking down of usual rules, such as males breeding in immature plumage, as in the martins and mockingbirds, dichromatism, and the occasional mating of birds exhibiting characters of at least subspecific diversity.

As the antithesis of this I must quote a sentence by Dr. E. W. Gifford concerning the use of a tool by the pallid tree finch on Albemarle. One of these birds was

"feeding in a leafless, dead tree. It was apparently searching for insects, for it inspected every hole carefully. Finally it found one too deep for its bill. It then flew to a neighbouring tree and broke off a small twig, about half an inch in length. Returning to the hole, the bird inserted the little stick as a probe, holding it lengthwise in its bill. It proceeded to examine other holes by the same method. Mr. Beck and Mr. King said they had noticed similar instances elsewhere."

This is a very remarkable thing—the deliberate use of a tool by a wild bird, and all the more astonishing in the finches, a group which is far from being at the summit of avian intelligence.

By far the most noticeable effect of insular isolation on the birds is also psychological—the extreme tameness of all except the migratory species. Ever since the time of the very first visitors to the archipelago this has been remarked, and most human observers have celebrated this fearlessness by knocking as many as possible on the head.

The first scientific expedition to visit the Galápagos was the voyage of the *Beagle,* with young Charles Darwin in charge of the scientific work. In 1835 he spent thirty-five days here, and has written a chapter on them, which for general grasp and for sheer interest has not since been equalled.

The last and by far the most elaborate expedition was sent out by the California Academy of Sciences eighteeen years ago, in a schooner, which remained over a year in the field, although much time was lost by drifting with contrary currents. Large collections were made and carried back, many of which have been studied and reported upon.

Although over a dozen other scientific expeditions have

visited the islands, yet the amount of work still to be done is
well attested by the results of some of the scientific papers
which I am publishing in *Zoologica*. These are based on casual
collecting, carried on during the less than one hundred hours
which we spent on shore.

Fish	of 60 species,	2 are new,	20 new to the islands.
Moths	" 52 "	26 "	" 22 " " " "
Shrimps	" 10 "	2 "	"
Lepismids	" 3 "	2 "	"
Homoptera	" 8 "	3 "	"
Mallophaga	" 6 "	2 "	"
Diptera	" 31 "	3 "	"
Spiders	" 31 "	4 "	"
Ants	" 19 "	8 "	"
Centipedes	" 4 "	2 "	"
Coccids	" 7 "	6 "	"
Coleoptera	" 21 "	7 "	"

Of greatest value now, would be thorough collections of
fish and other marine life and of terrestrial invertebrates,
combined with intensive study of the habits and inter-relationships of the birds and reptiles, of which latter groups more
than enough specimens have already been collected.

When the problems of migration, of establishment and of
adaptation to life on these islands have been solved, we will
know much more about corresponding conditions not only in
mainland jungles but in the turmoil of Broadway and Regent
Streets.

APPENDIX A

TEXT IDENTIFICATIONS

Page	Line	
13	4	Filefish, *Monacanthus hispidus* (Linné)
16	3	Sargassum Crab, *Planes minutus* (Linné)
16	14	Sargassum Shrimp, *Palæmon tenuicornis* Say
17	22	Sargassum Anemone, *Anemonia sargassensis* Hargitt
18	28	Herring Gull, *Larus argentatus* Ponto.
19	31	Atlantic Dolphin, *Delphinus* sp.
23	3	Atlantic White Gannet, *Sula bassana* (Linné)
26	15	Robin, *Planesticus m. migratorius* (Linné
26	15	Purple Martin, *Progne subis subis* (Linné)
26	19	Maryland Yellowthroat, *Geothlypis t. trichas* (Linné)
26	19	Parula Warbler, *Compsothlypis a. americana* (Linné)
26	20	Chuck-will's-widow, *Antrostomus carolinensis* (Gmelin)
27	21	Northern Phalarope, *Lobipes lobatus* (Linné)
29	33	Vesper Sparrow, *Poœcetes g. gramineus* (Gmelin)
36	14	Fish-hawk, *Pandion halietus carolinensis* (Gmelin)
36	23	Grey-breasted Kingbird, *Tyrannus melancholicus satrapa* (Cab. and H.)
36	29	Snowy Egret, *Egretta c. candidissima* (Gmelin)
37	6	Spotted Sandpiper, *Actitis macularia* (Linné)
37	7	Louisiana Heron, *Hydranassa tricolor* (Müller)
37	8	Little Blue Heron, *Florida cærulea* (Linné)
37	13	Groove-billed Ani, *Crotophaga sulcirostris* Swainson
37	27	Colon Crayfish, *Macrobrachium jamaicense* (Herbst)
38	26	Yapock, *Chironectes panamensis* Goldman
39	27	Blue Land Crab, *Cardisoma guanhumi* Latreille
39	29	Scarlet Land Crab, *Goniopsis cruentata* (Latreille)
40	25	Motmot, *Momotus subrufescens conexus* (Thayer and Bangs)

APPENDIX A

APPENDIX A

APPENDIX A

433

APPENDIX A

APPENDIX B

BIBLIOGRAPHY OF THE MORE IMPORTANT PUBLICATIONS RELATING TO THE GALÁPAGOS ARCHIPELAGO

1592   De Bry. *Grands Voyages.* Pars. III.

1593   Hawkins, Sir Richard. *Observations in a Voiage into the South Seas.*

1699   Cowley, Ambrose. *Voyage Round the World.* London.

1699   Wafer, Lionel. *A New Voyage and Description of the Isthmus of America.* London.

1718   Rogers, Captain Woodes. *A Cruising Voyage Round the World.* London.

1729   Dampier, William. *A Collection of Voyages.* London.

1798   Colnett, Captain James. *A Cruising Voyage Around the World.* London.

1810   *History of the Buccaneers of America, Containing the Narratives of Joseph Esquemeling, Basil Ringrose and the Sieur Ravenau de Lussan.* London.

1816   Burney, James. *History of the Buccaneers of America, together with A History of Discoveries in the South Seas.* London.

1817   Delano, Captain Amasa. *Narrative of Voyages and Travels in the Northern and Southern Hemispheres.* Boston.

1822   Porter, Captain David. *Journal of a Cruise made to the Pacific Ocean.* New York.

1824   Hall, Captain Basil. *Extracts from a Journal.* Edinburgh.

1826   Byron, Captain Lord. *Voyage of H. M. S. Blonde.* London.

APPENDIX B

1832   *Lives and Voyages of Early Navigators.* J. and J. Harper. New York.

1832   Morrell, Benjamin. *A Narrative of Four Voyages to the South Sea and South Pacific Ocean.* New York.

1839   Fitz-Roy, Captain Robert. *Narrative of the Surveying Voyages of H. M. Ships 'Adventure' and 'Beagle'.* Vol. II. London.

1839   Darwin, Charles. *Journal of Researches into the Geology and Natural History of the Various Countries Visited by H. M. S. 'Beagle' from 1832–1836.* London.

1841   Gould, J., and Charles Darwin. *Zoology of the Voyage of the Beagle During the Years 1832–1836.* Vol. III. Birds. London.

1841   Thomas, Captain Du Petit. *Voyage autour du Monde sur la Frégate 'Vénus', pendant les années 1836–1839.* Vol. II. Paris.

1856   Melville, Herman. *The Piazza Tales.* New York.

1876   Wallace, A. R. *Geographic Distribution of Animals.* New York.

1877   Salvin, Osbert. "Avifauna of the Galápagos Archipelago." *Trans. Zoological Society,* Vol. IX. London.

1877   Günther, A. C. L. G. *Gigantic Land Tortoises Living and Extinct in the Collection of the British Museum.*

1879   Wolf, Theodor. *Ein Besuch der Galápagos Inseln.* Heidelberg.

1881   Wallace, A. R. *Island Life,* Chap. XIII., Harper Bros. New York.

1891   Baur, G. "On the Origin of the Galápagos Islands." *American Naturalist,* pp. 217–229, 307–326.

1892   Baur, G. "Ein Besuch der Galápagos Inseln," in *Beilage zur Allgemeinen Zeitung,* Febr. 1–4, München.

1892   Agassiz, A. "The Galápagos Islands," in *Bull. Museum Comparative Zoology,* Vol. XXIII., Boston.

1892   Markham, C. R. "Discovery of the Galápagos Islands." *Proc. of the Royal Geog. Soc.,* London.

APPENDIX B

1893   Baur, G. "The Differentiation of Species on the Galápagos Islands and the Origin of the Group." In *Biological Lectures, Marine Biological Laboratory, Woods Holl,* pp. 67–78.

1895   Wolf, Theodor. "Die Galápagos Inseln." *Verhandl. der Gesell. fur Erdkunde zu Berlin,* Vol. XXII.

1897   Ridgway, Robert. "Birds of the Galápagos Archipelago." *Proc. U. S. National Museum,* Vol. XIX., pp. 459–670.

1899   Rothschild, Walter, and Ernst Hartert. "A Review of the Ornithology of the Galápagos Islands, with Notes on the Webster-Harris Expedition." *Novitates Zoologicæ,* Vol. VI.

1899   Linell, M. L. "On the Coleopterous Insects of the Galápagos Islands." *Proc. U. S. National Museum,* Vol. XXI.

1901–1905   Hopkins-Stanford Galápagos Expedition, 1898–1899. Results published in *Proc. Washington Academy of Sciences.* Vols. III–VI.

1902   Robinson, B. L. "Flora of the Galápagos Islands." *Proc. of American Academy of Arts and Sciences.* Vol. XXXVIII., pp. 77–269.

1904   *Las Islas Encantadas ó el Archipielago de Colon;* por José A. Bognoly y José Moisés Espinosa. Guayaquil.

1907–1918   Expedition of the California Academy of Sciences, 1905–1906. Results published in *Proc. California Academy of Sciences,* Vols. I–II.

1911   Walle, Paul. "Les Isles Galápagos et le Canal de Panama." *Société de Geographie Commerciale.* Paris.

1912   Scharff, R. F. *Distribution and Origin of Life in America,* pp. 295–335. New York.

1916   Kroeber, A. L. "Floral Relations among the Galápagos Islands." *University of California Publications in Botany.* Vol. VI., No. 9.

1918   McBride, G. M. "The Galápagos Islands." *Geogr. Review.* Vol. VI.

1922   Stock, Ralph. *The Cruise of the Dream Ship.* New York.

INDEX

INDEX

INDEX

INDEX

441

INDEX

Villamil, Gen. José, 407, 408
Vulture, 37

W

Wafer, Lionel, 343, 344, 354
Warbler, Parula, 26
Warbler, Yellow, 155
Wasp, 242
Wasp, Black, 327
Water-strider, Ocean, 83–86
Watkins, Patrick, 371–376
Weevils, *Pantomorus*, 101

Whalers, 365–371
Wolf, Dr. Teodor, 410

Y

Yapock, 38, 46, 48, 49, 55
Yellow Leg, Greater, 153
Yellowthroat, Maryland, 7, 26

Z

Zeus, 138

A CATALOG OF SELECTED DOVER
BOOKS IN ALL FIELDS OF INTEREST

DRAWINGS OF REMBRANDT, edited by Seymour Slive. Updated Lippmann, Hofstede de Groot edition, with definitive scholarly apparatus. All portraits, biblical sketches, landscapes, nudes. Oriental figures, classical studies, together with selection of work by followers. 550 illustrations. Total of 630pp. 9⅛ × 12¼.
21485-0, 21486-9 Pa., Two-vol. set $29.90

GHOST AND HORROR STORIES OF AMBROSE BIERCE, Ambrose Bierce. 24 tales vividly imagined, strangely prophetic, and decades ahead of their time in technical skill: "The Damned Thing," "An Inhabitant of Carcosa," "The Eyes of the Panther," "Moxon's Master," and 20 more. 199pp. 5⅜ × 8½. 20767-6 Pa. $3.95

ETHICAL WRITINGS OF MAIMONIDES, Maimonides. Most significant ethical works of great medieval sage, newly translated for utmost precision, readability. Laws Concerning Character Traits, Eight Chapters, more. 192pp. 5⅜ × 8½.
24522-5 Pa. $4.50

THE EXPLORATION OF THE COLORADO RIVER AND ITS CANYONS, J. W. Powell. Full text of Powell's 1,000-mile expedition down the fabled Colorado in 1869. Superb account of terrain, geology, vegetation, Indians, famine, mutiny, treacherous rapids, mighty canyons, during exploration of last unknown part of continental U.S. 400pp. 5⅜ × 8½. 20094-9 Pa. $7.95

HISTORY OF PHILOSOPHY, Julián Marías. Clearest one-volume history on the market. Every major philosopher and dozens of others, to Existentialism and later. 505pp. 5⅜ × 8½. 21739-6 Pa. $9.95

ALL ABOUT LIGHTNING, Martin A. Uman. Highly readable non-technical survey of nature and causes of lightning, thunderstorms, ball lightning, St. Elmo's Fire, much more. Illustrated. 192pp. 5⅜ × 8½. 25237-X Pa. $5.95

SAILING ALONE AROUND THE WORLD, Captain Joshua Slocum. First man to sail around the world, alone, in small boat. One of great feats of seamanship told in delightful manner. 67 illustrations. 294pp. 5⅜ × 8½. 20326-3 Pa. $4.95

LETTERS AND NOTES ON THE MANNERS, CUSTOMS AND CONDI- TIONS OF THE NORTH AMERICAN INDIANS, George Catlin. Classic account of life among Plains Indians: ceremonies, hunt, warfare, etc. 312 plates. 572pp. of text. 6⅛ × 9¼. 22118-0, 22119-9, Pa. Two-vol. set $17.90

ALASKA: The Harriman Expedition, 1899, John Burroughs, John Muir, et al. Informative, engrossing accounts of two-month, 9,000-mile expedition. Native peoples, wildlife, forests, geography, salmon industry, glaciers, more. Profusely illustrated. 240 black-and-white line drawings. 124 black-and-white photographs. 3 maps. Index. 576pp. 5⅜ × 8½. 25109-8 Pa. $11.95

THE BOOK OF BEASTS: Being a Translation from a Latin Bestiary of the Twelfth Century, T. H. White. Wonderful catalog real and fanciful beasts: manticore, griffin, phoenix, amphivius, jaculus, many more. White's witty erudite commentary on scientific, historical aspects. Fascinating glimpse of medieval mind. Illustrated. 296pp. 5⅜ × 8¼. (Available in U.S. only) 24609-4 Pa. $6.95

FRANK LLOYD WRIGHT: ARCHITECTURE AND NATURE With 160 Illustrations, Donald Hoffmann. Profusely illustrated study of influence of nature—especially prairie—on Wright's designs for Fallingwater, Robie House, Guggenheim Museum, other masterpieces. 96pp. 9¼ × 10¾. 25098-9 Pa. $7.95

FRANK LLOYD WRIGHT'S FALLINGWATER, Donald Hoffmann. Wright's famous waterfall house: planning and construction of organic idea. History of site, owners, Wright's personal involvement. Photographs of various stages of building. Preface by Edgar Kaufmann, Jr. 100 illustrations. 112pp. 9¼ × 10. 23671-4 Pa. $8.95

YEARS WITH FRANK LLOYD WRIGHT: Apprentice to Genius, Edgar Tafel. Insightful memoir by a former apprentice presents a revealing portrait of Wright the man, the inspired teacher, the greatest American architect. 372 black-and-white illustrations. Preface. Index. vi + 228pp. 8¼ × 11. 24801-1 Pa. $10.95

THE STORY OF KING ARTHUR AND HIS KNIGHTS, Howard Pyle. Enchanting version of King Arthur fable has delighted generations with imaginative narratives of exciting adventures and unforgettable illustrations by the author. 41 illustrations. xviii + 313pp. 6⅛ × 9¼. 21445-1 Pa. $6.95

THE GODS OF THE EGYPTIANS, E. A. Wallis Budge. Thorough coverage of numerous gods of ancient Egypt by foremost Egyptologist. Information on evolution of cults, rites and gods; the cult of Osiris; the Book of the Dead and its rites; the sacred animals and birds; Heaven and Hell; and more. 956pp. 6⅛ × 9¼. 22055-9, 22056-7 Pa., Two-vol. set $21.90

A THEOLOGICO-POLITICAL TREATISE, Benedict Spinoza. Also contains unfinished *Political Treatise*. Great classic on religious liberty, theory of government on common consent. R. Elwes translation. Total of 421pp. 5⅜ × 8½. 20249-6 Pa. $6.95

INCIDENTS OF TRAVEL IN CENTRAL AMERICA, CHIAPAS, AND YUCATAN, John L. Stephens. Almost single-handed discovery of Maya culture; exploration of ruined cities, monuments, temples; customs of Indians. 115 drawings. 892pp. 5⅜ × 8½. 22404-X, 22405-8 Pa., Two-vol. set $15.90

LOS CAPRICHOS, Francisco Goya. 80 plates of wild, grotesque monsters and caricatures. Prado manuscript included. 183pp. 6⅝ × 9⅝. 22384-1 Pa. $5.95

AUTOBIOGRAPHY: The Story of My Experiments with Truth, Mohandas K. Gandhi. Not hagiography, but Gandhi in his own words. Boyhood, legal studies, purification, the growth of the Satyagraha (nonviolent protest) movement. Critical, inspiring work of the man who freed India. 480pp. 5⅜ × 8½. (Available in U.S. only) 24593-4 Pa. $6.95

ILLUSTRATED DICTIONARY OF HISTORIC ARCHITECTURE, edited by Cyril M. Harris. Extraordinary compendium of clear, concise definitions for over 5,000 important architectural terms complemented by over 2,000 line drawings. Covers full spectrum of architecture from ancient ruins to 20th-century Modernism. Preface. 592pp. 7½ × 9⅜. 24444-X Pa. $15.95

THE NIGHT BEFORE CHRISTMAS, Clement Moore. Full text, and woodcuts from original 1848 book. Also critical, historical material. 19 illustrations. 40pp. 4⅝ × 6. 22797-9 Pa. $2.50

THE LESSON OF JAPANESE ARCHITECTURE: 165 Photographs, Jiro Harada. Memorable gallery of 165 photographs taken in the 1930's of exquisite Japanese homes of the well-to-do and historic buildings. 13 line diagrams. 192pp. 8⅜ × 11¼. 24778-3 Pa. $10.95

THE AUTOBIOGRAPHY OF CHARLES DARWIN AND SELECTED LETTERS, edited by Francis Darwin. The fascinating life of eccentric genius composed of an intimate memoir by Darwin (intended for his children); commentary by his son, Francis; hundreds of fragments from notebooks, journals, papers; and letters to and from Lyell, Hooker, Huxley, Wallace and Henslow. xi + 365pp. 5⅜ × 8. 20479-0 Pa. $6.95

WONDERS OF THE SKY: Observing Rainbows, Comets, Eclipses, the Stars and Other Phenomena, Fred Schaaf. Charming, easy-to-read poetic guide to all manner of celestial events visible to the naked eye. Mock suns, glories, Belt of Venus, more. Illustrated. 299pp. 5¼ × 8¼. 24402-4 Pa. $7.95

BURNHAM'S CELESTIAL HANDBOOK, Robert Burnham, Jr. Thorough guide to the stars beyond our solar system. Exhaustive treatment. Alphabetical by constellation: Andromeda to Cetus in Vol. 1; Chamaeleon to Orion in Vol. 2; and Pavo to Vulpecula in Vol. 3. Hundreds of illustrations. Index in Vol. 3. 2,000pp. 6⅛ × 9¼. 23567-X, 23568-8, 23673-0 Pa., Three-vol. set $41.85

STAR NAMES: Their Lore and Meaning, Richard Hinckley Allen. Fascinating history of names various cultures have given to constellations and literary and folkloristic uses that have been made of stars. Indexes to subjects. Arabic and Greek names. Biblical references. Bibliography. 563pp. 5⅜ × 8½. 21079-0 Pa. $8.95

THIRTY YEARS THAT SHOOK PHYSICS: The Story of Quantum Theory, George Gamow. Lucid, accessible introduction to influential theory of energy and matter. Careful explanations of Dirac's anti-particles, Bohr's model of the atom, much more. 12 plates. Numerous drawings. 240pp. 5⅜ × 8½. 24895-X Pa. $5.95

CHINESE DOMESTIC FURNITURE IN PHOTOGRAPHS AND MEASURED DRAWINGS, Gustav Ecke. A rare volume, now affordably priced for antique collectors, furniture buffs and art historians. Detailed review of styles ranging from early Shang to late Ming. Unabridged republication. 161 black-and-white drawings, photos. Total of 224pp. 8⅜ × 11¼. (Available in U.S. only) 25171-3 Pa. $13.95

VINCENT VAN GOGH: A Biography, Julius Meier-Graefe. Dynamic, penetrating study of artist's life, relationship with brother, Theo, painting techniques, travels, more. Readable, engrossing. 160pp. 5⅜ × 8½. (Available in U.S. only) 25253-1 Pa. $4.95

HOW TO WRITE, Gertrude Stein. Gertrude Stein claimed anyone could understand her unconventional writing—here are clues to help. Fascinating improvisations, language experiments, explanations illuminate Stein's craft and the art of writing. Total of 414pp. 4⅝ × 6⅜. 23144-5 Pa. $6.95

ADVENTURES AT SEA IN THE GREAT AGE OF SAIL: Five Firsthand Narratives, edited by Elliot Snow. Rare true accounts of exploration, whaling, shipwreck, fierce natives, trade, shipboard life, more. 33 illustrations. Introduction. 353pp. 5⅜ × 8½. 25177-2 Pa. $8.95

THE HERBAL OR GENERAL HISTORY OF PLANTS, John Gerard. Classic descriptions of about 2,850 plants—with over 2,700 illustrations—includes Latin and English names, physical descriptions, varieties, time and place of growth, more. 2,706 illustrations. xlv + 1,678pp. 8½ × 12¼. 23147-X Cloth. $75.00

DOROTHY AND THE WIZARD IN OZ, L. Frank Baum. Dorothy and the Wizard visit the center of the Earth, where people are vegetables, glass houses grow and Oz characters reappear. Classic sequel to *Wizard of Oz*. 256pp. 5⅜ × 8. 24714-7 Pa. $5.95

SONGS OF EXPERIENCE: Facsimile Reproduction with 26 Plates in Full Color, William Blake. This facsimile of Blake's original "Illuminated Book" reproduces 26 full-color plates from a rare 1826 edition. Includes "The Tyger," "London," "Holy Thursday," and other immortal poems. 26 color plates. Printed text of poems. 48pp. 5¼ × 7. 24636-1 Pa. $3.50

SONGS OF INNOCENCE, William Blake. The first and most popular of Blake's famous "Illuminated Books," in a facsimile edition reproducing all 31 brightly colored plates. Additional printed text of each poem. 64pp. 5¼ × 7. 22764-2 Pa. $3.50

PRECIOUS STONES, Max Bauer. Classic, thorough study of diamonds, rubies, emeralds, garnets, etc.: physical character, occurrence, properties, use, similar topics. 20 plates, 8 in color. 94 figures. 659pp. 6⅛ × 9¼. 21910-0, 21911-9 Pa., Two-vol. set $15.90

ENCYCLOPEDIA OF VICTORIAN NEEDLEWORK, S. F. A. Caulfeild and Blanche Saward. Full, precise descriptions of stitches, techniques for dozens of needlecrafts—most exhaustive reference of its kind. Over 800 figures. Total of 679pp. 8⅛ × 11. Two volumes. Vol. 1 22800-2 Pa. $11.95
Vol. 2 22801-0 Pa. $11.95

THE MARVELOUS LAND OF OZ, L. Frank Baum. Second Oz book, the Scarecrow and Tin Woodman are back with hero named Tip, Oz magic. 136 illustrations. 287pp. 5⅜ × 8½. 20692-0 Pa. $5.95

WILD FOWL DECOYS, Joel Barber. Basic book on the subject, by foremost authority and collector. Reveals history of decoy making and rigging, place in American culture, different kinds of decoys, how to make them, and how to use them. 140 plates. 156pp. 7⅞ × 10¾. 20011-6 Pa. $8.95

HISTORY OF LACE, Mrs. Bury Palliser. Definitive, profusely illustrated chronicle of lace from earliest times to late 19th century. Laces of Italy, Greece, England, France, Belgium, etc. Landmark of needlework scholarship. 266 illustrations. 672pp. 6⅛ × 9¼. 24742-2 Pa. $14.95

ILLUSTRATED GUIDE TO SHAKER FURNITURE, Robert Meader. All furniture and appurtenances, with much on unknown local styles. 235 photos. 146pp. 9 × 12. 22819-3 Pa. $8.95

WHALE SHIPS AND WHALING: A Pictorial Survey, George Francis Dow. Over 200 vintage engravings, drawings, photographs of barks, brigs, cutters, other vessels. Also harpoons, lances, whaling guns, many other artifacts. Comprehensive text by foremost authority. 207 black-and-white illustrations. 288pp. 6 × 9. 24808-9 Pa. $8.95

THE BERTRAMS, Anthony Trollope. Powerful portrayal of blind self-will and thwarted ambition includes one of Trollope's most heartrending love stories. 497pp. 5⅜ × 8½. 25119-5 Pa. $9.95

ADVENTURES WITH A HAND LENS, Richard Headstrom. Clearly written guide to observing and studying flowers and grasses, fish scales, moth and insect wings, egg cases, buds, feathers, seeds, leaf scars, moss, molds, ferns, common crystals, etc.—all with an ordinary, inexpensive magnifying glass. 209 exact line drawings aid in your discoveries. 220pp. 5⅜ × 8½. 23330-8 Pa. $4.95

RODIN ON ART AND ARTISTS, Auguste Rodin. Great sculptor's candid, wide-ranging comments on meaning of art; great artists; relation of sculpture to poetry, painting, music; philosophy of life, more. 76 superb black-and-white illustrations of Rodin's sculpture, drawings and prints. 119pp. 8⅜ × 11¼. 24487-3 Pa. $7.95

FIFTY CLASSIC FRENCH FILMS, 1912–1982: A Pictorial Record, Anthony Slide. Memorable stills from Grand Illusion, Beauty and the Beast, Hiroshima, Mon Amour, many more. Credits, plot synopses, reviews, etc. 160pp. 8¼ × 11. 25256-6 Pa. $11.95

THE PRINCIPLES OF PSYCHOLOGY, William James. Famous long course complete, unabridged. Stream of thought, time perception, memory, experimental methods; great work decades ahead of its time. 94 figures. 1,391pp. 5⅜ × 8½. 20381-6, 20382-4 Pa., Two-vol. set $23.90

BODIES IN A BOOKSHOP, R. T. Campbell. Challenging mystery of blackmail and murder with ingenious plot and superbly drawn characters. In the best tradition of British suspense fiction. 192pp. 5⅜ × 8½. 24720-1 Pa. $3.95

CALLAS: PORTRAIT OF A PRIMA DONNA, George Jellinek. Renowned commentator on the musical scene chronicles incredible career and life of the most controversial, fascinating, influential operatic personality of our time. 64 black-and-white photographs. 416pp. 5⅜ × 8¼. 25047-4 Pa. $8.95

GEOMETRY, RELATIVITY AND THE FOURTH DIMENSION, Rudolph Rucker. Exposition of fourth dimension, concepts of relativity as Flatland characters continue adventures. Popular, easily followed yet accurate, profound. 141 illustrations. 133pp. 5⅜ × 8½. 23400-2 Pa. $4.95

HOUSEHOLD STORIES BY THE BROTHERS GRIMM, with pictures by Walter Crane. 53 classic stories—Rumpelstiltskin, Rapunzel, Hansel and Gretel, the Fisherman and his Wife, Snow White, Tom Thumb, Sleeping Beauty, Cinderella, and so much more—lavishly illustrated with original 19th century drawings. 114 illustrations. x + 269pp. 5⅜ × 8½. 21080-4 Pa. $4.95

SUNDIALS, Albert Waugh. Far and away the best, most thorough coverage of ideas, mathematics concerned, types, construction, adjusting anywhere. Over 100 illustrations. 230pp. 5⅜ × 8½. 22947-5 Pa. $4.95

PICTURE HISTORY OF THE NORMANDIE: With 190 Illustrations, Frank O. Braynard. Full story of legendary French ocean liner: Art Deco interiors, design innovations, furnishings, celebrities, maiden voyage, tragic fire, much more. Extensive text. 144pp. 8⅞ × 11¾. 25257-4 Pa. $10.95

THE FIRST AMERICAN COOKBOOK: A Facsimile of "American Cookery," 1796, Amelia Simmons. Facsimile of the first American-written cookbook published in the United States contains authentic recipes for colonial favorites—pumpkin pudding, winter squash pudding, spruce beer, Indian slapjacks, and more. Introductory Essay and Glossary of colonial cooking terms. 80pp. 5⅜ × 8½. 24710-4 Pa. $3.50

101 PUZZLES IN THOUGHT AND LOGIC, C. R. Wylie, Jr. Solve murders and robberies, find out which fishermen are liars, how a blind man could possibly identify a color—purely by your own reasoning! 107pp. 5⅜ × 8½. 20367-0 Pa. $2.50

THE BOOK OF WORLD-FAMOUS MUSIC—CLASSICAL, POPULAR AND FOLK, James J. Fuld. Revised and enlarged republication of landmark work in musico-bibliography. Full information about nearly 1,000 songs and compositions including first lines of music and lyrics. New supplement. Index. 800pp. 5⅜ × 8¼. 24857-7 Pa. $15.95

ANTHROPOLOGY AND MODERN LIFE, Franz Boas. Great anthropologist's classic treatise on race and culture. Introduction by Ruth Bunzel. Only inexpensive paperback edition. 255pp. 5⅜ × 8½. 25245-0 Pa. $6.95

THE TALE OF PETER RABBIT, Beatrix Potter. The inimitable Peter's terrifying adventure in Mr. McGregor's garden, with all 27 wonderful, full-color Potter illustrations. 55pp. 4¼ × 5½. (Available in U.S. only) 22827-4 Pa. $1.75

THREE PROPHETIC SCIENCE FICTION NOVELS, H. G. Wells. *When the Sleeper Wakes, A Story of the Days to Come* and *The Time Machine* (full version). 335pp. 5⅜ × 8½. (Available in U.S. only) 20605-X Pa. $6.95

APICIUS COOKERY AND DINING IN IMPERIAL ROME, edited and translated by Joseph Dommers Vehling. Oldest known cookbook in existence offers readers a clear picture of what foods Romans ate, how they prepared them, etc. 49 illustrations. 301pp. 6⅛ × 9¼. 23563-7 Pa. $7.95

SHAKESPEARE LEXICON AND QUOTATION DICTIONARY, Alexander Schmidt. Full definitions, locations, shades of meaning of every word in plays and poems. More than 50,000 exact quotations. 1,485pp. 6½ × 9¼. 22726-X, 22727-8 Pa., Two-vol. set $29.90

THE WORLD'S GREAT SPEECHES, edited by Lewis Copeland and Lawrence W. Lamm. Vast collection of 278 speeches from Greeks to 1970. Powerful and effective models; unique look at history. 842pp. 5⅜ × 8½. 20468-5 Pa. $11.95

THE BLUE FAIRY BOOK, Andrew Lang. The first, most famous collection, with many familiar tales: Little Red Riding Hood, Aladdin and the Wonderful Lamp, Puss in Boots, Sleeping Beauty, Hansel and Gretel, Rumpelstiltskin; 37 in all. 138 illustrations. 390pp. 5⅜ × 8½. 21437-0 Pa. $6.95

THE STORY OF THE CHAMPIONS OF THE ROUND TABLE, Howard Pyle. Sir Launcelot, Sir Tristram and Sir Percival in spirited adventures of love and triumph retold in Pyle's inimitable style. 50 drawings, 31 full-page. xviii + 329pp. 6½ × 9¼. 21883-X Pa. $7.95

AUDUBON AND HIS JOURNALS, Maria Audubon. Unmatched two-volume portrait of the great artist, naturalist and author contains his journals, an excellent biography by his granddaughter, expert annotations by the noted ornithologist, Dr. Elliott Coues, and 37 superb illustrations. Total of 1,200pp. 5⅜ × 8.
Vol. I 25143-8 Pa. $8.95
Vol. II 25144-6 Pa. $8.95

GREAT DINOSAUR HUNTERS AND THEIR DISCOVERIES, Edwin H. Colbert. Fascinating, lavishly illustrated chronicle of dinosaur research, 1820's to 1960. Achievements of Cope, Marsh, Brown, Buckland, Mantell, Huxley, many others. 384pp. 5¼ × 8¼. 24701-5 Pa. $7.95

THE TASTEMAKERS, Russell Lynes. Informal, illustrated social history of American taste 1850's–1950's. First popularized categories Highbrow, Lowbrow, Middlebrow. 129 illustrations. New (1979) afterword. 384pp. 6 × 9.
23993-4 Pa. $8.95

DOUBLE CROSS PURPOSES, Ronald A. Knox. A treasure hunt in the Scottish Highlands, an old map, unidentified corpse, surprise discoveries keep reader guessing in this cleverly intricate tale of financial skullduggery. 2 black-and-white maps. 320pp. 5⅜ × 8½. (Available in U.S. only) 25032-6 Pa. $6.95

AUTHENTIC VICTORIAN DECORATION AND ORNAMENTATION IN FULL COLOR: 46 Plates from "Studies in Design," Christopher Dresser. Superb full-color lithographs reproduced from rare original portfolio of a major Victorian designer. 48pp. 9¼ × 12¼. 25083-0 Pa. $7.95

PRIMITIVE ART, Franz Boas. Remains the best text ever prepared on subject, thoroughly discussing Indian, African, Asian, Australian, and, especially, Northern American primitive art. Over 950 illustrations show ceramics, masks, totem poles, weapons, textiles, paintings, much more. 376pp. 5⅜ × 8. 20025-6 Pa. $7.95

SIDELIGHTS ON RELATIVITY, Albert Einstein. Unabridged republication of two lectures delivered by the great physicist in 1920–21. *Ether and Relativity* and *Geometry and Experience*. Elegant ideas in non-mathematical form, accessible to intelligent layman. vi + 56pp. 5⅜ × 8½. 24511-X Pa. $2.95

THE WIT AND HUMOR OF OSCAR WILDE, edited by Alvin Redman. More than 1,000 ripostes, paradoxes, wisecracks: Work is the curse of the drinking classes, I can resist everything except temptation, etc. 258pp. 5⅜ × 8½. 20602-5 Pa. $4.95

ADVENTURES WITH A MICROSCOPE, Richard Headstrom. 59 adventures with clothing fibers, protozoa, ferns and lichens, roots and leaves, much more. 142 illustrations. 232pp. 5⅜ × 8½. 23471-1 Pa. $3.95

PLANTS OF THE BIBLE, Harold N. Moldenke and Alma L. Moldenke. Standard reference to all 230 plants mentioned in Scriptures. Latin name, biblical reference, uses, modern identity, much more. Unsurpassed encyclopedic resource for scholars, botanists, nature lovers, students of Bible. Bibliography. Indexes. 123 black-and-white illustrations. 384pp. 6 × 9. 25069-5 Pa. $8.95

FAMOUS AMERICAN WOMEN: A Biographical Dictionary from Colonial Times to the Present, Robert McHenry, ed. From Pocahontas to Rosa Parks, 1,035 distinguished American women documented in separate biographical entries. Accurate, up-to-date data, numerous categories, spans 400 years. Indices. 493pp. 6½ × 9¼. 24523-3 Pa. $10.95

THE FABULOUS INTERIORS OF THE GREAT OCEAN LINERS IN HISTORIC PHOTOGRAPHS, William H. Miller, Jr. Some 200 superb photographs capture exquisite interiors of world's great "floating palaces"—1890's to 1980's: *Titanic, Ile de France, Queen Elizabeth, United States, Europa,* more. Approx. 200 black-and-white photographs. Captions. Text. Introduction. 160pp. 8⅞ × 11¾. 24756-2 Pa. $9.95

THE GREAT LUXURY LINERS, 1927–1954: A Photographic Record, William H. Miller, Jr. Nostalgic tribute to heyday of ocean liners. 186 photos of Ile de France, Normandie, Leviathan, Queen Elizabeth, United States, many others. Interior and exterior views. Introduction. Captions. 160pp. 9 × 12. 24056-8 Pa. $10.95

A NATURAL HISTORY OF THE DUCKS, John Charles Phillips. Great landmark of ornithology offers complete detailed coverage of nearly 200 species and subspecies of ducks: gadwall, sheldrake, merganser, pintail, many more. 74 full-color plates, 102 black-and-white. Bibliography. Total of 1,920pp. 8⅝ × 11¼. 25141-1, 25142-X Cloth. Two-vol. set $100.00

THE SEAWEED HANDBOOK: An Illustrated Guide to Seaweeds from North Carolina to Canada, Thomas F. Lee. Concise reference covers 78 species. Scientific and common names, habitat, distribution, more. Finding keys for easy identification. 224pp. 5⅜ × 8½. 25215-9 Pa. $6.95

THE TEN BOOKS OF ARCHITECTURE: The 1755 Leoni Edition, Leon Battista Alberti. Rare classic helped introduce the glories of ancient architecture to the Renaissance. 68 black-and-white plates. 336pp. 8⅜ × 11¼. 25239-6 Pa. $14.95

MISS MACKENZIE, Anthony Trollope. Minor masterpieces by Victorian master unmasks many truths about life in 19th-century England. First inexpensive edition in years. 392pp. 5⅜ × 8½. 25201-9 Pa. $8.95

THE RIME OF THE ANCIENT MARINER, Gustave Doré, Samuel Taylor Coleridge. Dramatic engravings considered by many to be his greatest work. The terrifying space of the open sea, the storms and whirlpools of an unknown ocean, the ice of Antarctica, more—all rendered in a powerful, chilling manner. Full text. 38 plates. 77pp. 9¼ × 12. 22305-1 Pa. $4.95

THE EXPEDITIONS OF ZEBULON MONTGOMERY PIKE, Zebulon Montgomery Pike. Fascinating first-hand accounts (1805–6) of exploration of Mississippi River, Indian wars, capture by Spanish dragoons, much more. 1,088pp. 5⅜ × 8½. 25254-X, 25255-8 Pa. Two-vol. set $25.90

A CONCISE HISTORY OF PHOTOGRAPHY: Third Revised Edition, Helmut Gernsheim. Best one-volume history—camera obscura, photochemistry, daguerreotypes, evolution of cameras, film, more. Also artistic aspects—landscape, portraits, fine art, etc. 281 black-and-white photographs. 26 in color. 176pp. 8⅜ × 11¼. 25128-4 Pa. $13.95

THE DORÉ BIBLE ILLUSTRATIONS, Gustave Doré. 241 detailed plates from the Bible: the Creation scenes, Adam and Eve, Flood, Babylon, battle sequences, life of Jesus, etc. Each plate is accompanied by the verses from the King James version of the Bible. 241pp. 9 × 12. 23004-X Pa. $9.95

HUGGER-MUGGER IN THE LOUVRE, Elliot Paul. Second Homer Evans mystery-comedy. Theft at the Louvre involves sleuth in hilarious, madcap caper. "A knockout."—Books. 336pp. 5⅜ × 8½. 25185-3 Pa. $5.95

FLATLAND, E. A. Abbott. Intriguing and enormously popular science-fiction classic explores the complexities of trying to survive as a two-dimensional being in a three-dimensional world. Amusingly illustrated by the author. 16 illustrations. 103pp. 5⅜ × 8½. 20001-9 Pa. $2.50

THE HISTORY OF THE LEWIS AND CLARK EXPEDITION, Meriwether Lewis and William Clark, edited by Elliott Coues. Classic edition of Lewis and Clark's day-by-day journals that later became the basis for U.S. claims to Oregon and the West. Accurate and invaluable geographical, botanical, biological, meteorological and anthropological material. Total of 1,508pp. 5⅜ × 8½. 21268-8, 21269-6, 21270-X Pa. Three-vol. set $26.85

LANGUAGE, TRUTH AND LOGIC, Alfred J. Ayer. Famous, clear introduction to Vienna, Cambridge schools of Logical Positivism. Role of philosophy, elimination of metaphysics, nature of analysis, etc. 160pp. 5⅜ × 8½. (Available in U.S. and Canada only) 20010-8 Pa. $3.95

MATHEMATICS FOR THE NONMATHEMATICIAN, Morris Kline. Detailed, college-level treatment of mathematics in cultural and historical context, with numerous exercises. For liberal arts students. Preface. Recommended Reading Lists. Tables. Index. Numerous black-and-white figures. xvi + 641pp. 5⅜ × 8½. 24823-2 Pa. $11.95

HANDBOOK OF PICTORIAL SYMBOLS, Rudolph Modley. 3,250 signs and symbols, many systems in full; official or heavy commercial use. Arranged by subject. Most in Pictorial Archive series. 143pp. 8¼ × 11. 23357-X Pa. $6.95

INCIDENTS OF TRAVEL IN YUCATAN, John L. Stephens. Classic (1843) exploration of jungles of Yucatan, looking for evidences of Maya civilization. Travel adventures, Mexican and Indian culture, etc. Total of 669pp. 5⅜ × 8½. 20926-1, 20927-X Pa., Two-vol. set $11.90

DEGAS: An Intimate Portrait, Ambroise Vollard. Charming, anecdotal memoir by famous art dealer of one of the greatest 19th-century French painters. 14 black-and-white illustrations. Introduction by Harold L. Van Doren. 96pp. 5⅜ × 8½.
25131-4 Pa. $4.95

PERSONAL NARRATIVE OF A PILGRIMAGE TO ALMANDINAH AND MECCAH, Richard Burton. Great travel classic by remarkably colorful personality. Burton, disguised as a Moroccan, visited sacred shrines of Islam, narrowly escaping death. 47 illustrations. 959pp. 5⅜ × 8½.   21217-3, 21218-1 Pa., Two-vol. set $19.90

PHRASE AND WORD ORIGINS, A. H. Holt. Entertaining, reliable, modern study of more than 1,200 colorful words, phrases, origins and histories. Much unexpected information. 254pp. 5⅜ × 8½.   20758-7 Pa. $5.95

THE RED THUMB MARK, R. Austin Freeman. In this first Dr. Thorndyke case, the great scientific detective draws fascinating conclusions from the nature of a single fingerprint. Exciting story, authentic science. 320pp. 5⅜ × 8½. (Available in U.S. only)   25210-8 Pa. $6.95

AN EGYPTIAN HIEROGLYPHIC DICTIONARY, E. A. Wallis Budge. Monumental work containing about 25,000 words or terms that occur in texts ranging from 3000 B.C. to 600 A.D. Each entry consists of a transliteration of the word, the word in hieroglyphs, and the meaning in English. 1,314pp. 6⅜ × 10.
23615-3, 23616-1 Pa., Two-vol. set $31.90

THE COMPLEAT STRATEGYST: Being a Primer on the Theory of Games of Strategy, J. D. Williams. Highly entertaining classic describes, with many illustrated examples, how to select best strategies in conflict situations. Prefaces. Appendices. xvi + 268pp. 5⅜ × 8½.   25101-2 Pa. $5.95

THE ROAD TO OZ, L. Frank Baum. Dorothy meets the Shaggy Man, little Button-Bright and the Rainbow's beautiful daughter in this delightful trip to the magical Land of Oz. 272pp. 5⅜ × 8.   25208-6 Pa. $5.95

POINT AND LINE TO PLANE, Wassily Kandinsky. Seminal exposition of role of point, line, other elements in non-objective painting. Essential to understanding 20th-century art. 127 illustrations. 192pp. 6½ × 9¼.   23808-3 Pa. $5.95

LADY ANNA, Anthony Trollope. Moving chronicle of Countess Lovel's bitter struggle to win for herself and daughter Anna their rightful rank and fortune— perhaps at cost of sanity itself. 384pp. 5⅜ × 8½.   24669-8 Pa. $8.95

EGYPTIAN MAGIC, E. A. Wallis Budge. Sums up all that is known about magic in Ancient Egypt: the role of magic in controlling the gods, powerful amulets that warded off evil spirits, scarabs of immortality, use of wax images, formulas and spells, the secret name, much more. 253pp. 5⅜ × 8½.   22681-6 Pa. $4.50

THE DANCE OF SIVA, Ananda Coomaraswamy. Preeminent authority unfolds the vast metaphysic of India: the revelation of her art, conception of the universe, social organization, etc. 27 reproductions of art masterpieces. 192pp. 5⅜ × 8½.
24817-8 Pa. $5.95

CHRISTMAS CUSTOMS AND TRADITIONS, Clement A. Miles. Origin, evolution, significance of religious, secular practices. Caroling, gifts, yule logs, much more. Full, scholarly yet fascinating; non-sectarian. 400pp. 5⅜ × 8½.
23354-5 Pa. $6.95

THE HUMAN FIGURE IN MOTION, Eadweard Muybridge. More than 4,500 stopped-action photos, in action series, showing undraped men, women, children jumping, lying down, throwing, sitting, wrestling, carrying, etc. 390pp. 7⅞ × 10⅝.
20204-6 Cloth. $21.95

THE MAN WHO WAS THURSDAY, Gilbert Keith Chesterton. Witty, fast-paced novel about a club of anarchists in turn-of-the-century London. Brilliant social, religious, philosophical speculations. 128pp. 5⅜ × 8½.
25121-7 Pa. $3.95

A CEZANNE SKETCHBOOK: Figures, Portraits, Landscapes and Still Lifes, Paul Cezanne. Great artist experiments with tonal effects, light, mass, other qualities in over 100 drawings. A revealing view of developing master painter, precursor of Cubism. 102 black-and-white illustrations. 144pp. 8¾ × 6⅝.
24790-2 Pa. $5.95

AN ENCYCLOPEDIA OF BATTLES: Accounts of Over 1,560 Battles from 1479 B.C. to the Present, David Eggenberger. Presents essential details of every major battle in recorded history, from the first battle of Megiddo in 1479 B.C. to Grenada in 1984. List of Battle Maps. New Appendix covering the years 1967–1984. Index. 99 illustrations. 544pp. 6½ × 9¼.
24913-1 Pa. $14.95

AN ETYMOLOGICAL DICTIONARY OF MODERN ENGLISH, Ernest Weekley. Richest, fullest work, by foremost British lexicographer. Detailed word histories. Inexhaustible. Total of 856pp. 6½ × 9¼.
21873-2, 21874-0 Pa., Two-vol. set $17.00

WEBSTER'S AMERICAN MILITARY BIOGRAPHIES, edited by Robert McHenry. Over 1,000 figures who shaped 3 centuries of American military history. Detailed biographies of Nathan Hale, Douglas MacArthur, Mary Hallaren, others. Chronologies of engagements, more. Introduction. Addenda. 1,033 entries in alphabetical order. xi + 548pp. 6½ × 9¼. (Available in U.S. only)
24758-9 Pa. $13.95

LIFE IN ANCIENT EGYPT, Adolf Erman. Detailed older account, with much not in more recent books: domestic life, religion, magic, medicine, commerce, and whatever else needed for complete picture. Many illustrations. 597pp. 5⅜ × 8½.
22632-8 Pa. $8.95

HISTORIC COSTUME IN PICTURES, Braun & Schneider. Over 1,450 costumed figures shown, covering a wide variety of peoples: kings, emperors, nobles, priests, servants, soldiers, scholars, townsfolk, peasants, merchants, courtiers, cavaliers, and more. 256pp. 8⅜ × 11¼.
23150-X Pa. $9.95

THE NOTEBOOKS OF LEONARDO DA VINCI, edited by J. P. Richter. Extracts from manuscripts reveal great genius; on painting, sculpture, anatomy, sciences, geography, etc. Both Italian and English. 186 ms. pages reproduced, plus 500 additional drawings, including studies for *Last Supper, Sforza* monument, etc. 860pp. 7⅞ × 10¾. (Available in U.S. only) 22572-0, 22573-9 Pa., Two-vol. set $31.90

THE ART NOUVEAU STYLE BOOK OF ALPHONSE MUCHA: All 72 Plates from "Documents Decoratifs" in Original Color, Alphonse Mucha. Rare copy-right-free design portfolio by high priest of Art Nouveau. Jewelry, wallpaper, stained glass, furniture, figure studies, plant and animal motifs, etc. Only complete one-volume edition. 80pp. 9⅜ × 12¼. 24044-4 Pa. $9.95

ANIMALS: 1,419 COPYRIGHT-FREE ILLUSTRATIONS OF MAMMALS, BIRDS, FISH, INSECTS, ETC., edited by Jim Harter. Clear wood engravings present, in extremely lifelike poses, over 1,000 species of animals. One of the most extensive pictorial sourcebooks of its kind. Captions. Index. 284pp. 9 × 12. 23766-4 Pa. $9.95

OBELISTS FLY HIGH, C. Daly King. Masterpiece of American detective fiction, long out of print, involves murder on a 1935 transcontinental flight—"a very thrilling story"—NY Times. Unabridged and unaltered republication of the edition published by William Collins Sons & Co. Ltd., London, 1935. 288pp. 5⅜ × 8½. (Available in U.S. only) 25036-9 Pa. $5.95

VICTORIAN AND EDWARDIAN FASHION: A Photographic Survey, Alison Gernsheim. First fashion history completely illustrated by contemporary photo-graphs. Full text plus 235 photos, 1840–1914, in which many celebrities appear. 240pp. 6½ × 9¼. 24205-6 Pa. $6.95

THE ART OF THE FRENCH ILLUSTRATED BOOK, 1700–1914, Gordon N. Ray. Over 630 superb book illustrations by Fragonard, Delacroix, Daumier, Doré, Grandville, Manet, Mucha, Steinlen, Toulouse-Lautrec and many others. Preface. Introduction. 633 halftones. Indices of artists, authors & titles, binders and provenances. Appendices. Bibliography. 608pp. 8⅜ × 11¼. 25086-5 Pa. $24.95

THE WONDERFUL WIZARD OF OZ, L. Frank Baum. Facsimile in full color of America's finest children's classic. 143 illustrations by W. W. Denslow. 267pp. 5⅜ × 8½. 20691-2 Pa. $7.95

FRONTIERS OF MODERN PHYSICS: New Perspectives on Cosmology, Rela-tivity, Black Holes and Extraterrestrial Intelligence, Tony Rothman, et al. For the intelligent layman. Subjects include: cosmological models of the universe; black holes; the neutrino; the search for extraterrestrial intelligence. Introduction. 46 black-and-white illustrations. 192pp. 5⅜ × 8½. 24587-X Pa. $7.95

THE FRIENDLY STARS, Martha Evans Martin & Donald Howard Menzel. Classic text marshalls the stars together in an engaging, non-technical survey, presenting them as sources of beauty in night sky. 23 illustrations. Foreword. 2 star charts. Index. 147pp. 5⅜ × 8½. 21099-5 Pa. $3.95

FADS AND FALLACIES IN THE NAME OF SCIENCE, Martin Gardner. Fair, witty appraisal of cranks, quacks, and quackeries of science and pseudoscience: hollow earth, Velikovsky, orgone energy, Dianetics, flying saucers, Bridey Murphy, food and medical fads, etc. Revised, expanded In the Name of Science. "A very able and even-tempered presentation."—The New Yorker. 363pp. 5⅜ × 8. 20394-8 Pa. $6.95

ANCIENT EGYPT: ITS CULTURE AND HISTORY, J. E Manchip White. From pre-dynastics through Ptolemies: society, history, political structure, religion, daily life, literature, cultural heritage. 48 plates. 217pp. 5⅜ × 8½. 22548-8 Pa. $5.95

SIR HARRY HOTSPUR OF HUMBLETHWAITE, Anthony Trollope. Incisive, unconventional psychological study of a conflict between a wealthy baronet, his idealistic daughter, and their scapegrace cousin. The 1870 novel in its first inexpensive edition in years. 250pp. 5⅜ × 8½. 24953-0 Pa. $5.95

LASERS AND HOLOGRAPHY, Winston E. Kock. Sound introduction to burgeoning field, expanded (1981) for second edition. Wave patterns, coherence, lasers, diffraction, zone plates, properties of holograms, recent advances. 84 illustrations. 160pp. 5⅜ × 8¼. (Except in United Kingdom) 24041-X Pa. $3.95

INTRODUCTION TO ARTIFICIAL INTELLIGENCE: SECOND, ENLARGED EDITION, Philip C. Jackson, Jr. Comprehensive survey of artificial intelligence—the study of how machines (computers) can be made to act intelligently. Includes introductory and advanced material. Extensive notes updating the main text. 132 black-and-white illustrations. 512pp. 5⅜ × 8½. 24864-X Pa. $8.95

HISTORY OF INDIAN AND INDONESIAN ART, Ananda K. Coomaraswamy. Over 400 illustrations illuminate classic study of Indian art from earliest Harappa finds to early 20th century. Provides philosophical, religious and social insights. 304pp. 6⅝ × 9⅝. 25005-9 Pa. $9.95

THE GOLEM, Gustav Meyrink. Most famous supernatural novel in modern European literature, set in Ghetto of Old Prague around 1890. Compelling story of mystical experiences, strange transformations, profound terror. 13 black-and-white illustrations. 224pp. 5⅜ × 8½. (Available in U.S. only) 25025-3 Pa. $6.95

PICTORIAL ENCYCLOPEDIA OF HISTORIC ARCHITECTURAL PLANS, DETAILS AND ELEMENTS: With 1,880 Line Drawings of Arches, Domes, Doorways, Facades, Gables, Windows, etc., John Theodore Haneman. Sourcebook of inspiration for architects, designers, others. Bibliography. Captions. 141pp. 9 × 12. 24605-1 Pa. $7.95

BENCHLEY LOST AND FOUND, Robert Benchley. Finest humor from early 30's, about pet peeves, child psychologists, post office and others. Mostly unavailable elsewhere. 73 illustrations by Peter Arno and others. 183pp. 5⅜ × 8½. 22410-4 Pa. $4.95

ERTÉ GRAPHICS, Erté. Collection of striking color graphics: *Seasons, Alphabet, Numerals, Aces* and *Precious Stones*. 50 plates, including 4 on covers. 48pp. 9⅜ × 12¼. 23580-7 Pa. $7.95

THE JOURNAL OF HENRY D. THOREAU, edited by Bradford Torrey, F. H. Allen. Complete reprinting of 14 volumes, 1837–61, over two million words; the sourcebooks for *Walden*, etc. Definitive. All original sketches, plus 75 photographs. 1,804pp. 8½ × 12¼. 20312-3, 20313-1 Cloth., Two-vol. set $120.00

CASTLES: THEIR CONSTRUCTION AND HISTORY, Sidney Toy. Traces castle development from ancient roots. Nearly 200 photographs and drawings illustrate moats, keeps, baileys, many other features. Caernarvon, Dover Castles, Hadrian's Wall, Tower of London, dozens more. 256pp. 5⅜ × 8¼. 24898-4 Pa. $6.95

AMERICAN CLIPPER SHIPS: 1833–1858, Octavius T. Howe & Frederick C. Matthews. Fully-illustrated, encyclopedic review of 352 clipper ships from the period of America's greatest maritime supremacy. Introduction. 109 halftones. 5 black-and-white line illustrations. Index. Total of 928pp. 5⅜ × 8½.
25115-2, 25116-0 Pa., Two-vol. set $17.90

TOWARDS A NEW ARCHITECTURE, Le Corbusier. Pioneering manifesto by great architect, near legendary founder of "International School." Technical and aesthetic theories, views on industry, economics, relation of form to function, "mass-production spirit," much more. Profusely illustrated. Unabridged translation of 13th French edition. Introduction by Frederick Etchells. 320pp. 6⅛ × 9¼. (Available in U.S. only) 25023-7 Pa. $8.95

THE BOOK OF KELLS, edited by Blanche Cirker. Inexpensive collection of 32 full-color, full-page plates from the greatest illuminated manuscript of the Middle Ages, painstakingly reproduced from rare facsimile edition. Publisher's Note. Captions. 32pp. 9⅜ × 12¼. 24345-1 Pa. $4.95

BEST SCIENCE FICTION STORIES OF H. G. WELLS, H. G. Wells. Full novel *The Invisible Man,* plus 17 short stories: "The Crystal Egg," "Aepyornis Island," "The Strange Orchid," etc. 303pp. 5⅜ × 8½. (Available in U.S. only)
21531-8 Pa. $6.95

AMERICAN SAILING SHIPS: Their Plans and History, Charles G. Davis. Photos, construction details of schooners, frigates, clippers, other sailcraft of 18th to early 20th centuries—plus entertaining discourse on design, rigging, nautical lore, much more. 137 black-and-white illustrations. 240pp. 6⅛ × 9¼.
24658-2 Pa. $6.95

ENTERTAINING MATHEMATICAL PUZZLES, Martin Gardner. Selection of author's favorite conundrums involving arithmetic, money, speed, etc., with lively commentary. Complete solutions. 112pp. 5⅜ × 8½. 25211-6 Pa. $2.95

THE WILL TO BELIEVE, HUMAN IMMORTALITY, William James. Two books bound together. Effect of irrational on logical, and arguments for human immortality. 402pp. 5⅜ × 8½. 20291-7 Pa. $7.95

THE HAUNTED MONASTERY and THE CHINESE MAZE MURDERS, Robert Van Gulik. 2 full novels by Van Gulik continue adventures of Judge Dee and his companions. An evil Taoist monastery, seemingly supernatural events; overgrown topiary maze that hides strange crimes. Set in 7th-century China. 27 illustrations. 328pp. 5⅜ × 8½. 23502-5 Pa. $6.95

CELEBRATED CASES OF JUDGE DEE (DEE GOONG AN), translated by Robert Van Gulik. Authentic 18th-century Chinese detective novel; Dee and associates solve three interlocked cases. Led to Van Gulik's own stories with same characters. Extensive introduction. 9 illustrations. 237pp. 5⅜ × 8½.
23337-5 Pa. $4.95

Prices subject to change without notice.
Available at your book dealer or write for free catalog to Dept. GI, Dover Publications, Inc., 31 East 2nd St., Mineola, N.Y. 11501. Dover publishes more than 175 books each year on science, elementary and advanced mathematics, biology, music, art, literary history, social sciences and other areas.